The Crystal as a Supramolecular Entity

Editorial Board

The Crystal as a Supramolecular Entity

Perspectives in Supramolecular Chemistry Volume 2

EDITED BY GAUTAM R. DESIRAJU

University of Hyderabad, Hyderabad, India

JOHN WILEY & SONS

Chichester · New York · Brisbane · Toronto · Singapore

Other Wiley Editorial Offices

John Wiley & Sons, Inc., 605 Third Avenue,
New York, NY 10158-0012, USA

Jacaranda Wiley Ltd, 33 Park Road, Milton,
Queensland 4064, Australia

John Wiley & Sons (Canada) Ltd, 22 Worcester Road,
Rexdale, Ontario M9W 1LI, Canada

John Wiley & Sons (Asia) Pte Ltd, 2 Clementi Loop #02-01,
Jin Xing Distripark, Singapore 0512

Library of Congress Cataloging-in-Publication Data

The crystal as a supramolecular entity / edited by Gautam R. Desiraju.
        p.   cm.—(Perspectives in supramolecular chemistry : v. 2)
    Includes bibliographical references and index.
    ISBN 0-471-95015-7 (alk. paper)
    1. Molecular crystals.   I. Desiraju, G. R. (Gautam R.)
    II. Series.
    QD921.C755   1995   95-19756
    548—dc20   CIP

British Library Cataloguing in Publication Data

A catalogue record for this book is available from the British Library

ISBN 0 471 950157

Typeset in 10/12pt Times by Dobbie Typesetting Limited, Tavistock, Devon
Printed and bound in Great Britain by Biddles Ltd, Guildford, Surrey

This book is printed on acid-free paper responsibly manufactured from sustainable forestation,
for which at least two trees are planted for each one used for paper production.

Contents

Contributors

Ning-Leh Chang, Department of Chemistry and Biochemistry, The University of Texas at Austin, Austin, TX 78712, USA

Ian Dance, School of Chemistry, University of New South Wales, Sydney 2052, Australia

Raymond E. Davis, Department of Chemistry and Biochemistry, The University of Texas at Austin, Austin, TX 78712, USA

Gautam R. Desiraju, School of Chemistry, University of Hyderabad, Hyderabad 500046, India

Jack D. Dunitz, Organic Chemistry Laboratory, Swiss Federal Institute of Technology, ETH Zentrum, CH-8092 Zurich, Switzerland

Paul J. Fagan, E. I. du Pont de Nemours and Company, Central Research and Development Department, Experimental Station, PO Box 80328, Wilmington, DE 19880-0328, USA

Jenny P. Glusker, The Institute for Cancer Research, The Fox Chase Cancer Center, 7701 Burholme Avenue, Philadelphia, PA 19111, USA

C. V. Krishnamohan Sharma, School of Chemistry, University of Hyderabad, Hyderabad 500046, India

Michael D. Ward, Department of Chemical Engineering and Materials Science, University of Minnesota, Amundson Hall, 421 Washington Avenue SE, Minneapolis, MN 55455, USA

James K. Whitesell, Department of Chemistry and Biochemistry, The University of Texas at Austin, Austin, TX 78712, USA

Man-Shing Wong, Department of Chemistry and Biochemistry, The University of Texas at Austin, Austin, TX 78712, USA

Preface

The term 'supramolecular' signifies that which is beyond the molecule, and supramolecular concepts have consequently had the greatest influence in organic chemistry where the molecule is paramount. Supramolecular chemistry has gained in importance as it represents a point of departure from classical organic chemistry which, since the time of Wöhler, has continuously emphasized that all physical and chemical properties of substances are derivable from and are delimited by their molecular structures. Supramolecular ideas have much relevance to modern organic chemistry, and it is difficult to conceive of the recent advances in bio-organic chemistry, catalysis and transport phenomena, chemistry at interfaces and organic materials chemistry relying solely on molecule-based thought.

This kaleidoscope of contemporary research interests reveals that another distinctive feature of supramolecular chemistry is its ability to unite areas with seemingly widely differing perceptions. In keeping with such a feature, structural chemists and crystallographers have had little difficulty in recognizing a molecular crystal as the ultimate example of a supermolecule. Consequently, supramolecular chemistry today encompasses the study of molecular crystals with all the applications and ramifications that such study implies in the fields of solid-state chemistry, crystal engineering and materials science. This then is the theme of this volume. Crystals constitute one end of the supramolecular continuum and may be viewed as 'hard' supermolecules in contrast to the 'softer' supramolecular aggregates which exist in solution.

The historical molecular bias of organic chemistry necessitates the reiteration that supermolecules are not just collections of molecules and that their structure and characteristic properties are distinct from the aggregate properties of their molecular constituents. Such a statement is, however, hardly necessary in inorganic crystal chemistry where structures have traditionally been viewed in terms of networks and connectivities and where the very definition of molecularity is excitingly different. And yet, these 'organic' and 'inorganic' viewpoints are only parts of the whole, and another manifestation of the synthesizing aspect of supramolecular chemistry is its ability to bridge organic and inorganic structural chemistry with the result that it will soon probably be difficult to distinguish between organic, inorganic and organometallic viewpoints in solid-state supramolecular chemistry.

The initial motivation behind supramolecular chemistry was to design chemical systems which would mimic biological processes, drawing inspiration, as it were, from nature itelf, for the living cell is a wonderful example of a highly ordered supramolecular species embodying a close relationship between structure, information and function. The crystallography of biological macromolecules has been explored in great detail since the 1930s, and if an organic (small-molecule) crystal is the ultimate supermolecule, a biomolecular crystal is the complete supermolecule, for in it the relationship between structure and function is so much more clearly apparent. Whether it be supersecondary structures in a protein or supramolecular synthons in a small-molecule crystal, however, the motivation behind such identification and classification is the same—to improve our relatively poor present understanding of supramolecular algorithms, i.e. the protocols which connect molecular and supramolecular structure, the operational aspects, as it were, of molecular recognition.

Many of the prerequisites for such an improved understanding have been discussed in this volume. Central to the issue is the nature of weak intermolecular forces, the supramolecular glue, as it were. Surprisingly, this is incompletely known even for the ubiquitous forces; hydrogen bonding, herringbone, $\pi \cdots \pi$ and ionic interactions. Another key element in supramolecular engineering is the ability to dissect and insulate different interaction types, or alternatively the ability to exploit the interference between different interaction types in the design strategy. Again, is it possible to distinguish clearly between a chemical bond and an intermolecular interaction? Even as supramolecular chemistry has sought to demarcate between what is within and without the molecule, it has demonstrated as deficient the classification of forces as 'bonded' and 'nonbonded'. The weakest covalent bonds are indeed feebler than the strongest intermolecular interactions such as those between some metal atoms in organometallic crystals, interactions which confound attempts at distinguishing between molecules and supermolecules. The efficiency of computational methods in the development of supramolecular algorithms is another open question. So, whether it be crystal engineering of an organic zeolite or a frequency doubler for materials science applications or the prediction of the tertiary fold of a protein, the emphasis is on the collective properties of molecules mediated by intermolecular interactions.

This book, which is intended to clarify our perception of a crystal as a supramolecular entity, consists of six chapters which illustrate the diversity and scope of structural supramolecular chemistry. Neither the selection of topics nor the treatment within the individual chapters is exhaustive, and this is entirely intended. In a fast-moving subject such as this, it was felt that it would be more important that the reader obtain an accurate and critical appraisal of important developments in the field rather than a comprehensive coverage of the literature. Such an ethos, it was felt, would also more accurately justify the appearance of this volume in a series entitled *Perspectives in Supramolecular Chemistry*. The chapters convey, in this sense, the respective authors' points of view, and it is hoped that such a presentation will stimulate further discussion, debate and, of course, new work.

I would like to thank the authors for their cooperation, the other series editors for their helpful suggestions and Professor J.-M. Lehn for his encouragement. I am most grateful to the staff at John Wiley & Sons Ltd for their assistance and to Dr C. B. Aakeröy, Queen's University, Belfast, who prepared the cover illustration.

Gautam R. Desiraju
Hyderabad
March 1995

Chapter 1

Thoughts on Crystals as Supermolecules

JACK D. DUNITZ

Swiss Federal Institute of Technology, Zurich, Switzerland

1. CRYSTALS VIEWED AS SUPERMOLECULES

The crystal is, in a sense, the supermolecule *par excellence*: a lump of matter, of macroscopic dimensions, millions of molecules long, held together in a periodic arrangement by just the same kind of noncovalent bonding interactions as are responsible for molecular recognition and complexation at all levels. Indeed, the crystallization process itself is an impressive display of supramolecular self-assembly, involving specific molecular recognition at an amazing level of precision. Long-range periodicity is a product of directionally specific short-range interactions, nothing more. Crystals are ordered supramolecular systems.

1.1 Polymorphism

Polymorphism, the existence of a given compound in more than one crystal form, is widespread. If a crystal is a supermolecule, then polymorphic modifications are superisomers and polymorphism is a kind of superisomerism; thus, diamond, graphite and the fullerenes are an extreme example of a family of superisomers. Even with less extreme examples, polymorphs may differ markedly in colour, hardness, solubility, density and other physical properties. As a general rule, the polymorph stable at $0\,K$, i.e. the one with the lowest potential energy, has the smallest volume. It may be displaced by another polymorph with a higher potential energy if the latter has a larger heat capacity and hence a larger entropy increase as the temperature is raised. Free energy

The Crystal as a Supramolecular Entity. Edited by G. R. Desiraju
©1996 John Wiley & Sons Ltd

differences between polymorphs are usually quite small, a matter of a few hundred calories per mole, and have different temperature dependences, so that over quite a small range of temperature, and particularly between room temperature and the melting point of the crystals, one polymorph or another can be the thermodynamically stable form. A metastable form can persist for years (e.g. diamond!) or it can undergo spontaneous transformation to the stable form — a solid–solid phase transformation or, in supramolecular terms, an isomerization reaction.

1.2   Cooperative and Noncooperative Processes

Even if crystals are supermolecules, there are obviously important differences between phase transformations in solids and isomerization reactions in solution or the gas phase. One difference is that in a reversible chemical reaction there is at any temperature an equilibrium mixture of reactants and products whereas in a crystal the transformation, once it is triggered on either side of the transition temperature, usually goes practically to completion. In other words, either there is no reaction or one isomer is transformed completely into the other. Secondly, there is the difference in the temperature dependence of the reaction rate. The rate constant of a 'normal' chemical reaction increases smoothly with temperature. With first-order phase transformations (those involving a discontinuous change in free enthalpy or entropy or unit cell parameters), on warming through the thermodynamic transition temperature (where the free energies of the two polymorphs are equal) nothing much happens on the low-temperature side, but once this temperature is passed the transformation rate suddenly increases and the reaction goes rapidly to completion. On cooling from above to below the transition temperature, the reverse transformation may occur or it may not, depending on various factors that influence the kinetics of the transformation. High-temperature crystal modifications can often be kept indefinitely at temperatures far below the thermodynamic transition point. Second-order phase transformations (those involving a more gradual change in thermodynamic properties and unit cell parameters) are usually associated with a change from a more ordered low-temperature phase to a more disordered high-temperature phase. Here the process takes place within a temperature range of a few degrees around the transition temperature, reversibly.

These differences between the 'normal' chemical reaction and the crystal transformation result from the importance of cooperativity in the crystal but not in the liquid or gaseous state. In a 'normal' chemical reaction, molecules react more or less independently of one another; what happens to one molecule has little effect on what happens to its neighbours. In a phase transformation, cooperativity is the essence. Within a crystal, every displacement of a molecule from its equilibrium conformation, position, and orientation is immediately communicated to its immediate neighbours and thence to more distant neighbours and so on, so that molecular motions are coupled in a set of lattice vibrations that extend through the entire crystal. In a liquid, on the other hand, although motions among neighbouring molecules are coupled, there is no long-range correlation between molecular positions or orientations; there are only local effects. It is this difference that is responsible for the different kinds of temperature dependence between normal reactions and the highly cooperative ones that are typical of phase transitions. In general, polymorphic transitions are associated with changes in molecular packing arrangements but there are often radical changes in molecular conformation and hydrogen-bonding patterns as well.

From physics books one might get the impression that as far as phase transitions are concerned there are no problems left. This is very far from the truth. It may be true that no *fundamental* problems are left, but it is also true that practically nothing is understood about the actual mechanisms of solid–solid phase transitions in molecular crystals. It is likely that defects in the ordered crystal structure play a vital role. Indeed, in the absence of defects, the solid–solid transition may be completely inhibited; crystals may be held at temperatures well above their thermodynamic range of stability almost indefinitely. On warming, instead of transforming to the thermodynamically stable form they simply undergo melting at their fusion temperature [1]. For phase transformations in molecular solids there is essentially no theory worth speaking of at the molecular level, only a few broad thermodynamic generalizations and *ad hoc* explanations; we understand hardly anything.

1.3 Self-Recognition of Molecules

One other apparent difference between molecular interactions in crystals and in other chemical and supramolecular systems turns out on closer scrutiny to be nugatory. In chemical recognition processes, including supramolecular and biomolecular ones, we are concerned mainly with interactions among chemically different molecules, and not so much with interactions among identical (or enantiomeric) molecules. With crystals it is usually the other way round. Although some crystals—co-crystals or 'molecular compounds'—are built from more than one kind of molecule, these are the exceptions; most crystals are built from identical (or enantiomeric) copies of the same molecule*.

---

*From the point of view of the phase rule, a crystalline racemate, built from enantiomeric molecules, should be regarded as a co-crystal if the interconversion rate of the enantiomers is slow compared with the time-scale of the crystallization, melting or solution processes. The liquid phase in equilibrium with the crystalline one then has to be described as a two-component (melt) or three-component (solution) system, whereas the pure enantiomer yields a one-component (melt) or two-component (solution) system.

This problem, or rather a closely related one, was discussed by Pauling and Delbrück [2] in their critique of Jordan's views on 'special' forces between identical or nearly identical molecules or parts of molecules. In discussing molecular interactions, they wrote:

> . . . in order to achieve the maximum stability, the two molecules must have complementary surfaces, like die and coin, and also a complementary distribution of active groups. The case might occur in which the two complementary structures happen to be identical; however, in this case also the *stability of the complex between two molecules would be due to their complementariness rather than their identity*[†]

Thus, even when all the molecules are identical (or enantiomeric), an A-part of one molecule can interact with a D-part of a second, and the A-part of the second can interact in exactly the same manner with the D-part of a third, and so on. Indeed, several ways of describing such mutually complementary interactions are commonly encountered (e.g. locks in keys, bumps in hollows, interactions between opposite charges or favourably oriented dipoles, donor–acceptor interactions, acid–base types of interaction involved in hydrogen bonding, etc). The important thing is that they are local effects with strongly directional properties.

If the geometrical aspects of these interactions are preserved from molecule to molecule, one comes naturally to periodicity — to translational symmetry relationships among the molecular units. There are only a limited number of ways in which these relationships can be achieved. In general, identical parts of different molecules avoid one another, which is why space groups containing rotation axes and mirror planes are infrequent compared with those that contain screw axes or glide planes. The contrast is striking. Among the 130 000 or so known crystal structures of organic compounds, only a few % occur in space groups that contain rotation axes or mirror planes, and the overwhelming majority of these crystal structures involve molecules that themselves contain the symmetry elements in question [3]. It is very uncommon

---

From this point of view, in the slow-conversion limit the two solid enantiomers and the crystalline racemate would have to be classed as three different compounds (stereoisomers in this case). In the fast-conversion limit, they would be three polymorphic modifications of the same compound.

[†]Emil Fischer's 'lock-and-key' analogy is just 100 years old [E. Fischer, *Ber. Dtsch. Chem. Ges.*, **27**, 2985 (1894)], but the basic idea of complementariness as a factor in the formation of stable associations goes back to much earlier times. It was expressed more than 2000 years ago by Lucretius in his *De Rerum Natura*:

> Things whose textures have such a mutual correspondence, that cavities fit solids, the cavities of the first the solids of the second, the cavities of the second the solids of the first, form the closest union.

(translated by H. A. J. Munro, in *The Stoic and Epicurean Philosophers* (ed. W. J. Oates), Random House, New York, 1940).

for molecules in a crystal structure to be related by rotation axes or mirror planes.

Once the importance of molecular recognition in self-assembly processes such as crystallization is established, one comes to the possibility of influencing these processes by the introduction of mistakes in molecular recognition. This can be achieved by the addition of small amounts of 'tailor-made auxiliaries' to the solution in which the crystallization process is taking place. 'Tailor-made auxiliaries' consist of molecules that simulate the genuine crystal substrate molecules but differ from them in some crucial respect; thus, for example, aspartic acid for aparagine, ω-aminocaproic acid for lysine, D-serine for L-serine, etc. The two molecules are so similar that the guests may be incorporated on a growing crystal face of the host but then reveal themselves as impostors when the next layer of molecules is added, thus inhibiting growth normal to the face and changing the normal crystal habit. The method has been exploited to great effect in studies of crystal nucleation and growth, in the determination of the absolute configuration of chiral additive molecules and of chiral polar crystals, in the lowering of crystal symmetry, leading to alteration of physical properties, and in other applications*.

2. CRYSTAL STRUCTURES

Apart from minor details, the static aspects of crystal structures are relatively well understood. That is to say, there are no fundamental problems, only problems of complexity. Even though there is no rigorous way to do this, the potential energy of a crystal can be factorized into component parts and attributed to various kinds of interaction: electrostatic interactions, hydrogen bonds, donor–acceptor interactions, steric repulsions, van der Waals attractions, and so on. Several sets of more or less elaborate atom–atom interactions are in use (see Section 3), and computer programs are available to estimate their net contributions to the potential energy of the crystal structure — the packing energy. Most potential functions in general use have been parametrized to reproduce room-temperature crystal structures at least and sublimation enthalpies at best, but the calculated packing energies are not accurate to better than a few kilojoules per mole or so. Moreover, they do not

---

*The work summarized in this paragraph has been described in many papers and review articles by a team of crystallographers and chemists from the structural chemistry group at the Weizmann Institute of Science from 1985 onwards. See especially L. Addadi, Z. Berkovitch-Yellin, I. Weissbuch, M. Lahav and L. Leiserowitz, in *Progress in Stereochemistry*, Vol. 16 (eds E. L. Eliel, S. H. Wilen and N. L. Allinger), Wiley-Interscience, New York, 1986, p. 1 and L. Addadi, Z. Berkovitch-Yellin, I. Weissbuch, J. van Mil, L. J. W. Shimon, M. Lahav and L. Leiserowitz, *Angew. Chem., Int. Ed. Engl.*, **24**, 466 (1985).

include any allowance for the entropic contribution to the free energy and hence they cannot be used for predicting the ranges of relative stability of polymorphs.

We can usually reassure ourselves that an observed structure corresponds, or at least is quite close, to a local minimum in the packing energy as calculated with the help of atom–atom potential functions. Thus, once we know the crystal structure or structures adopted by a given molecule, we can try to analyse the factors that contribute to the stability of these observed molecular arrangements. Once the unit cell dimensions are given, i.e. once the space available to the molecules is fixed, the most important factor in determining the details of the molecular packing is usually minimization of the intermolecular repulsions (the attractive forces have already played their part in bringing the molecules from infinite separation into mutual contact). Nevertheless, given one crystal structure we are generally unable to make reliable predictions about the structures of possible polymorphs or about the packing arrangements that structurally related molecules might adopt, except perhaps for trivial variations.

Given only the molecular structure of a compound, the problem of predicting its crystal structure (or crystal structures) is still essentially unsolved despite considerable contemporary interest and effort in the problem. (See Perlstein [4] for a progress report.) Indeed, for molecules with several degrees of conformational freedom the problem may well turn out to be insoluble, at least with present-day methods. At any rate, it is a challenge that must be, and is being, taken up.

2.1 Simple Models of Intermolecular Interactions

In spite of the formidable and unsolved problems involved in predicting crystal structures, even those of quite simple compounds, it seems hard to resist the temptation to try to *explain* known crystal structures in terms of simple models, i.e. models that ignore much of the complexity and concentrate on only one or two of the factors that are involved. As an extreme example, it might be asserted that the cubic face-centred rocksalt structure is a result of the electrostatic attraction between the Na cations and the Cl anions, both having spherical symmetry. It is true that a spherical cation at the centre of a regular octahedron of anions is attracted equally to all six neighbours, but it is also equally repelled by all six. In fact, this arrangement corresponds not to a minimum in the electrostatic energy but to a local maximum; any slight displacement of the cation from its central position lowers the electrostatic energy but it raises the repulsion energy by a greater amount. The important point is that the equilibrium structure corresponds to a *balance* between attractive and repulsive forces; to concentrate on the attractive forces alone, or on only one of them, can be very misleading. This applies not only to crystals but to supramolecular complexes in general.

While on the topic of repulsive forces, it may be worth remarking here that the recent use of 'clathrate compound' as a more or less synonymous expression for 'inclusion compound' [5] does not do justice to the distinction that Powell wished to make when he introduced the phrase in the first place in 1948. Powell had studied the crystal structures of inclusion compounds formed by various small molecules with quinol and had established that these molecules were enclosed in cavities with the β-quinol framework. He recognized that while the formation of most molecular compounds was due to mutual attraction of the components, the β-quinol inclusion compounds were different; they were based mainly on repulsion. Let Powell speak for himself [6]:

> A molecule to be enclosed must be in the right place when the cage is closing. . . . For an enclosed molecule to leave its enclosure it must overcome the attraction between itself and the cage, but even if it were possible for there to be no attraction, escape will be prevented by another process for some types of cage and enclosed molecule. When this molecule approaches a possible hole of exit its outward passage will be opposed by the repulsive forces that arise when any two atoms approach closely. . . . If the molecules forming the cage are subject to strong mutual attraction they will not be pushed apart readily, and if the holes of exit are sufficiently small the enclosed molecules will therefore be repelled inwards. . . .
>
> There may thus arise a structural combination of two substances which remain associated not through strong attraction between them but because strong mutual binding of the molecules of one sort only makes possible the enclosure of the other. It is suggested that the general character of this type of combination should be indicated by the description 'clathrate' compound — 'clathratus', enclosed or protected by cross bars of a grating.

This seems as clear as it could be. Perhaps one should avoid referring to clathrate inclusion compounds as 'host–guest' compounds. After all, guests are usually free to leave, they are not prisoners.

2.2 The Benzene Crystal Structure; A Simple Example

One of the favourite crystal structures that seem to call for a simple explanation is that of benzene, a simple, highly symmetrical molecule which crystallizes at normal pressure in the orthorhombic space group *Pbca* with four molecules in the unit cell (benzene I). Results of several crystal structure analyses have been described in the literature, and we shall refer here mainly to those of the latest and most accurate study, that of perdeuterobenzene by low-temperature neutron diffraction [7]. The unit cell dimensions are $a = 7.360$ (7.398) Å, $b = 9.375$ (9.435) Å and $c = 6.703$ (6.778) Å at 15 K (and 123 K, the latter values in parentheses), and the four molecules are located at the crystallographic centres of symmetry at the origin of the cell and at the

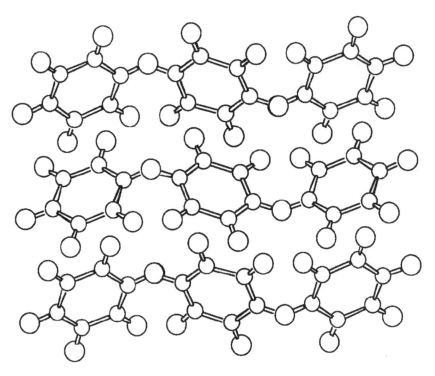

Figure 1   Crystal structure of benzene (benzene I) viewed down the *c*-axis

midpoints of the faces, i.e. at 0,0,0, 1/2,1/2,0, 1/2,0,1/2 and 0,1/2,1/2. Each molecule is surrounded by 12 others, the intercentre distances being 4.98 Å (1/2,0,1/2), 5.76 Å (0,1/2,1/2) and 5.96 Å (1/2,1/2,0). A picture of the structure is shown in Figure 1.

Benzene vapour contains a few % of dimer, and electric deflection measurements have shown that this dimer has an electric dipole moment [8, 9]. The most likely model was taken as one in which the two planes are mutually perpendicular, a so-called T-shaped arrangement 'as observed in crystalline benzene nearest-neighbor pairs' [8]. It was stressed [8] that the observed polarity of the dimer merely identifies its structure as that of an asymmetric top and that virtually all asymmetric top structures behave similarly enough to qualify for compatibility with the experimental result. In other words, the description of the dimer structure as a T-shaped structure was intended merely as a rather coarse characterization. In particular, it was clearly not intended to imply that the experimental data called for a C_{2v}-symmetric dimer with a C–H bond of one benzene molecule pointing towards the centre of the other, involving 'a kind of hydrogen bond to the π-face of a neighbouring aromatic ring', as has been invoked as a dominating factor in the packing of aromatic rings in condensed systems [10]. In any case, this particular orientation of neighbouring benzene rings is not to be found in the crystal structure of benzene itself. As we shall see, the dominating feature of the benzene crystal structure is not so much the attraction of C–H bonds for π-clouds but rather the coulombic attraction between hydrogen atoms and carbon atoms of neighouring molecules. But even if the arrangement with a C–H bond pointing to a ring centre may not correspond to optimal attraction, it could well be the one of local minimal repulsion. This could account for its occurrence in host–guest complexes where a benzene ring is tightly enclosed in the more or less rigid cavity of a host molecule.

In recent analyses of the benzene structure, Williams and Xiao [11] and Klebe and Diederich [12] have considered the pairwise interactions between a central molecule and one at each of the three face centres (Figure 2). (Note that the diagrams in this figure do not represent benzene dimers but nearest-neighbour interactions in infinite chains of molecules. Each molecular centre corresponds to a crystallographic centre of symmetry.) The shortest centre-to-centre distance (4.98 Å to the molecule at 1/2,0,1/2)* involves a nearly perpendicular pair of molecules (interplanar angle 87.6°) such that 'two adjacent CH-bonds of the first molecule point towards the core of the neighbouring benzene molecule, and a shift of the centre of the first from the normal centred on the second molecule is observed so that one hydrogen of the first is located above the centre of the second molecule' [12] (but note that the corresponding C–H bond direction does not point towards the ring centre). The other two pairwise interactions have longer centre-to-centre distances, i.e. 5.76 Å (0,1/2,1/2) and 5.96 Å (1/2,1/2,0); the longest one again involves a nearly perpendicular pair of molecules (interplanar angle 85.1°) but with a larger offset (Figure 2, centre), and the intermediate one involves rings inclined to one another by 29.4°, again with 'two CH-bonds approximately oriented toward the core of the neighbouring benzene ring'. None of these pairs has a geometry that corresponds either to a C_{2v}-symmetric perpendicular plane model dimer or to the optimized dimer, as calculated by Williams and Xiao [11], in which the orientation of the two benzene rings is intermediate between parallel and perpendicular. As these authors point out [11]:

> The structure and energy of a molecular dimer abstracted from its crystal is not expected to be optimum for the isolated gas dimer. The presence of several nearest-neighbor molecules and of long-range intermolecular energy leads one to

---

*The distances and angles given in these analyses [11, 12] differ slightly from those given here because they are based on an earlier structure analysis of C_6H_6 [G. E. Bacon, N. A. Curry and S. A. Wilson, *Proc. R. Soc. London, Ser. A*, **279**, 98 (1964)], not the later one of C_6D_6 described elsewhere [7].

10

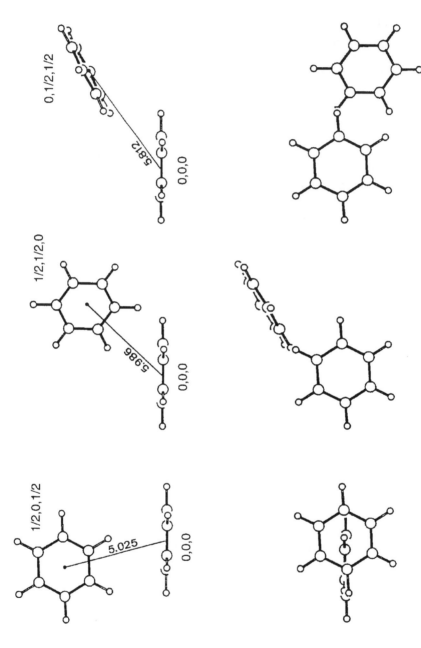

Figure 2 Two orthogonal views (upper and lower rows) of the basic nearest-neighbour motifs (distances in ångstroms) in crystalline benzene (reproduced with permission from Klebe and Diederich [12])

anticipate that these dimers will be different from optimum gas dimers, perhaps even radically.

Thus, while such dimers or other supramolecular assemblies extracted out of observed crystal structures may well be useful in host–guest chemistry in suggesting models for the design of hosts for specific aromatic guest molecules, they should not necessarily be regarded as reliable indicators of the optimal geometry of such assemblies.

Another simple model that has recently been invoked in explaining the benzene crystal structure is based on the interaction between electric quadrupole moments [13]. The benzene molecule has a quadrupole moment of $-29 \times 10^{-40} \, C \, m^2$ [14], and the negative value can be interpreted in terms of excess negative charge above and below the plane of the ring, and excess positive charge in the plane. The pattern shown in Figure 3, the periodic repetition of the perpendicular T-shaped arrangement mentioned above, has been proposed [13] as the energetically most favourable arrangement of an assemblage of identical quadrupoles.

The quadrupole interaction model for crystalline benzene can be criticized on several grounds and should not be taken too seriously. In the first place, although the charge distribution described above may be compatible with the observed quadrupole moment, it is in no way required by it. For example, the quadrupole moment is just as compatible with a charge distribution in the plane of the ring, with about $-0.15e$ on each of the six carbon atoms and $+0.15e$ on each of the hydrogens. Secondly, the pattern shown in Figure 3 is only two-dimensional and cannot be applied without reservations to a three-dimensional periodic pattern. For example, the crystal structure of carbon dioxide, which can be regarded as the prototype of a quadrupolar molecule, does not follow the pattern shown in Figure 3. The linear CO_2 molecule crystallizes in the cubic space group $Pa\overline{3}$, the four molecules in the unit cell being oriented along the body diagonals (threefold axes) [15]. Thus, each negatively charged oxygen atom does not interact with a single positively charged carbon centre (as would be suggested by Figure 3), but is equidistant from three such centres (Figure 4).

Finally, if the benzene crystal structure were indeed determined by interactions among quadrupoles, one might expect to find essentially the same crystal structure for hexafluorobenzene, a molecule that has the same hexagonal shape, nearly the same size and nearly the same quadrupole moment as benzene but with the opposite sign — the reversal of sign makes no difference to the quadrupole–quadrupole interactions. However, the C_6F_6 structure ($P2_1/n, Z=6$)[16] is different from that of benzene. Whereas all molecules in the benzene structure are crystallographically equivalent, the C_6F_6 structure contains two types of molecules in crystallographically different environments. One set of molecules (say A) occupy crystallographic centres of symmetry, while the other set (B) occupy general positions (Figure 5). It is interesting that

Figure 3 Schematic representation of a regular periodic arrangement of like quadrupolar molecules

two opposite fluorine atoms of the A-molecules point nearly at the centres of the B-molecules (the six C$\cdots$F distances are roughly equal, ranging from 3.35 to 3.50 Å). This is similar to the T-shaped arrangement that was invoked earlier for benzene–benzene interactions, only here the interplanar angle is about 60° and it is the negatively charged end of the bond dipole that interacts with the π-cloud! Most likely, this arrangement corresponds to a local repulsion minimum — the fluorine atom sits in the 'hole' in the π-cloud. In any case, from the simple quadrupole interaction model, there seems no reason why two such geometrically similar molecules as C_6H_6 and C_6F_6 should adopt such different crystal structures when they have the same quadrupole moment.

A more fundamental objection to the quadrupole interaction model for benzene derives from elementary physics. The electric field or potential at a distance r from a source can be expressed as a multipole expansion, but the

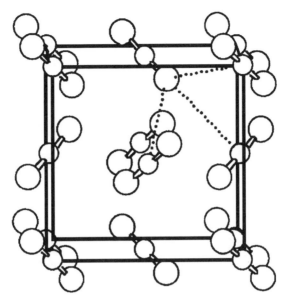

Figure 4 Crystal structure of carbon dioxide (example of a cylindrical quadrupolar molecule in space group $Pa\overline{3}$). The molecular centres form a face-centred cubic array, with the oxygen atoms on the body diagonals of the cube. Each oxygen atom is therefore equidistant (3.14 Å) from the carbon atoms of three other molecules (dotted lines)

expansion is only valid when r is large compared with the size of the source itself. This requirement is satisfied to some extent for the carbon dioxide structure, where the shortest intermolecular contact distances are about 3.1 Å, in any case larger than the longest intramolecular distance of about 2.3 Å. But it is not satisfied for the benzene crystal structure, where the shortest contact distances between molecules in the tightly juxtaposed array are smaller than the dimensions of the benzene molecules themselves.

The same objection can be raised against the notion that molecular dipole moments are important in determining molecular packing in crystals; in particular, that large molecular dipole moments are an important factor leading to centrosymmetry in crystals. Experimental evidence against this idea, if it were needed, has been provided by a recent study [17], which shows that there is no correlation between the mutual orientation of molecular pairs in crystals and the magnitudes of the molecular dipole moments. For very small molecules, e.g. diatomic ones, it is another matter; here the dipoles can be regarded in good approximation as point dipoles, and similarly for interactions between local bond dipoles.

To calculate the electric field or potential due to a molecule at distances that are not large compared with the molecular dimensions, it is necessary to use the

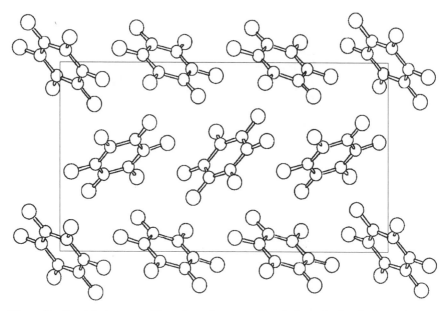

Figure 5 Crystal structure of hexafluorobenzene viewed along the *b*-axis. Note the two types of molecules: the ones at the corners and the centre of the pattern (at crystallographic centres of symmetry) and the others (in general positions)

molecular charge distribution itself, or, for practical purposes, at least a reasonable approximation to this based on local point charges. Indeed, although the C–H bonds of benzene are not usually believed to be highly polarized, the most careful available studies, such as those of Williams and Xiao [11] and Shi and Bartell [18], indicate that the observed crystal structures of benzene and other aromatic hydrocarbons cannot be reproduced in a consistent fashion without introducing local charge–charge interactions. Placing equal and opposite charges on the carbon and hydrogen atoms of benzene makes the C$\cdots$H coulombic terms attractive, all other atom–atom interactions repulsive. The benzene crystal structure is heavily determined by this pattern of atom–atom interactions.

The orthorhombic benzene I crystal structure can be altered with not too much deformation into a hypothetical cubic close-packed structure in the same space group as that described above for carbon dioxide. The cubic benzene structure with the same volume as that of benzene I and with all three axes equal (space group $Pa\bar{3}$) would have $a = 7.73$ Å, intercentre distance 5.47 Å. In this structure the molecules would occupy sites with $\bar{3}$ (S_6) crystallographic symmetry and the benzene rings would be perfect hexagons (although not necessarily planar). In the observed orthorhombic structure the rings show merely $\bar{1}$ (C_i) crystallographic symmetry, although the deviations from hexagonal symmetry are minimal*. Close packing has long been invoked as an important factor in determining the crystal structures of organic molecules [19], so one may well ask: why does benzene choose a structure that is slightly deformed from this ideally close-packed one? The answer is that there are specific attractions among bits of neighbouring molecules in the crystal that cannot be optimized in the cubic structure, which has only two degrees of freedom: the unit cell dimension and the rotation angle of the benzene molecule in its own plane. In the cubic structure, this plane is required to be perpendicular to one of the body diagonals of the cube, and hence the angle between the ring planes of different molecules is fixed by symmetry at $70° 32'$ (the supplement of the tetrahedral angle). By deviating from cubic symmetry, the structure gains degrees of freedom that allow it to lower its potential energy. However, the energy difference between the observed structure and the cubic one is probably quite small, of the order of a few kilojoules per mole only[†]. For the carbon dioxide crystal structure in $Pa\overline{3}$, with only one degree of freedom (the unit cell dimension), the potential energy is apparently not lowered by deviating from cubic symmetry, at least not at normal pressure. However, it has recently been shown [20] that at very high pressure, CO_2 undergoes a phase transformation to an orthorhombic structure in which the mutual orientation of the molecules is no longer restricted by crystal symmetry.

Benzene also undergoes a pressure-induced phase transformation to a structure of lower symmetry. In addition to the orthorhombic structure discussed so far (benzene I), there is also a monoclinic high-pressure form, benzene II, with unit cell dimensions (294 K, 25 kbar) $a = 5.417 \text{ Å}$, $b = 5.376 \text{ Å}$, $c = 7.532 \text{ Å}$ and $\beta = 110°$; the space group is $P2_1/c$ with two centrosymmetric molecules in the unit cell [21], as shown in Figure 6. The molecular volume is about 7% less than that of benzene I at the same temperature (at around 0.7 kbar pressure) and about 10% less than that of benzene I at 0 K and normal pressure (extrapolated value). If the monoclinic a-axis and c-axis are interchanged to alter the space group to $P2_1/a$, the space group in which the crystal structures of naphthalene and anthracene are usually described, comparison of unit cell dimensions suggests that the three crystal structures must be closely related (Table 1).

The naphthalene and anthracene values are taken from low-temperature X-ray analyses [22] to compensate as far as possible for the contraction in volume of benzene II due to the high pressure required for its formation. As the molecular size increases, there is only a relatively small increase in the a-axis

---

*In fact, according to the results of the accurate low-temperature neutron diffraction analyses [7], they are less than 0.001 Å for C–C distances and about 0.002 Å for C–D distances; the deuterium atoms deviate from the mean plane of the carbon atoms by about 0.01 Å to give a slight deformation in the direction of a chair conformation.

†According to a rough force field calculation kindly made by Dr Gerhard Klebe in response to a query from the author.

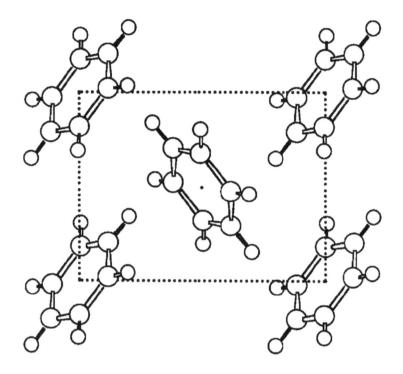

Figure 6 Crystal structure of the high-pressure modification of benzene (benzene II) viewed in projection down the *a*-axis (in the $P2_1/c$ orientation)

Table 1

	a (Å)	*b* (Å)	*c* (Å)	*β* (°)
Benzene II (294 K, 25 kbar)	7.532	5.376	5.417	110
Naphthalene (92 K, 1 bar)	8.108	5.940	8.647	124.4
Anthracene (94 K, 1 bar)	8.414	5.990	11.095	125.3

and *b*-axis, the main extension being in the length of the *c*-axis, by about 3.2–3.4 Å for each additional ring. Indeed, the three crystal structures are very similar. No polymorphs of naphthalene or anthracene are known so far, but possible structures for orthorhombic modifications have been proposed [23].

   In contrast to the orthorhombic benzene I structure in which the molecules adopt an approximately cubic close-packed arrangement, the molecular arrangement in benzene II is more like hexagonal close packing: in other words, this crystal contains definite layers of molecules (in the projection plane of Figure 6, perpendicular to the crystal c-axis in the $P2_1/c$ orientation). One can discern the same tendency in the naphthalene and anthracene structures, although the distance between the layers is now larger because of the increase in molecular size. When the benzene I structure is compressed and the interatomic distances between molecules are contracted, the repulsion energy increases faster than the attraction energy decreases, i.e. the net energy increases (to balance the work done in compressing the crystal). Hall and Williams [24] have shown that in order to model this change by atom–atom potential calculations, opposite charges of at least $0.09e$ have to be assigned to the carbon and hydrogen atoms. (To fit the experimental value of the quadrupole moment of benzene, the atomic charges should be about $0.15e$, with the inner carbons negative, the hydrogens positive.)

At the same time, it has to be remembered that molecules in condensed phases should not be regarded as systems of rigid, fixed charges. Molecules are polarizable, which means that the local charge distribution is influenced by other charges in the vicinity, i.e. by the presence of neighbouring molecules. There is therefore no fundamental, rigorous way of partitioning the total interaction energy between molecules into separate contributions (dispersion terms, van der Waals energy, coulombic terms, etc.), as has already been mentioned in Section 2. In particular, although the molecular charge density distribution and its associated potential are well-defined quantities that are available from calculation and from experiment, at least in principle, atomic charges are not observables and can be defined in different ways, such as Mulliken atom charges [25], potential-derived charges [26], Hirshfeld charges according to his stockholder recipe [27] and Bader charges [28].

There is no unique or generally accepted answer because of the lack of agreement about what constitutes an atom in a molecule or in a condensed phase. When the outer electron orbitals overlap, how should the boundaries enclosing individual atoms be defined? Nevertheless, whatever the detailed physical interpretation, it seems to be necessary to introduce coulombic interactions into the force field that expresses the short-range attractions of the carbon and hydrogen atoms. As the long-range coulombic effects are relatively unimportant because of extensive cancellations and dielectric polarization, it has been suggested that in atom–atom potential calculations little is lost and much is gained by replacing the r^{-1} dependence of the electrostatic potential energy by an r^{-2} dependence adjusted to yield the correct coulombic forces at molecular contact distances and with the term truncated at $5\,\text{Å}$ [18].

More generally, as far as the planar condensed aromatic hydrocarbons are concerned, their crystal structures seem to be determined by a competition between two main types of interaction [29]: a core–edge interaction, involving $C\cdots H$ attractions and producing a herringbone (HB) pattern; and a core–core interaction, leading to interplanar stacking. Their relative importance depends on the ratio of the size of the molecular perimeter (HB promoting) to the molecular area (stack promoting). To model the core–edge interactions in terms of atom–atom potentials, opposite charges on the carbons and hydrogens must be introduced. Models that do not possess this feature fail to reproduce this characteristic feature of the packing of aromatic hydrocarbons. For example, a model introduced by Hunter and Sanders [30] puts a charge of $+1e$ at each carbon nucleus and two compensating charges of $-0.5e$ at $0.47\,\text{\AA}$ above and below the plane of the π-system. This charge distribution is compatible with the quadrupole moment of benzene and is claimed to explain the strong geometrical requirements for interactions between aromatic molecules, but it does not reproduce the crystal structures of benzene or of other aromatic hydrocarbons.

3. FORCE FIELDS; ATOM–ATOM POTENTIALS

Visit a chemistry or biochemistry laboratory nowadays and you are likely to find a good number of researchers peering at the screen of a computer monitor instead of busying themselves with experiments with actual materials. Ask what they are doing and you are likely to be informed that they are trying to predict by calculation the outcome of possible experiments or at least obtain some guidance as to which experiments are likely to be worthwhile and which a waste of time. Among the most popular types of such calculation are those that fall into the general class of molecular mechanics calculations. Given an atomic arrangement compatible with a molecular (or supramolecular) structure, this type of calculation will adjust the relative atomic positions in directions corresponding to a lowering of the potential energy until a local energy minimum is reached. More elaborate versions may be able to sample regions of the parameter space and find not just a single energy minimum but several. Almost all such types of calculation are based on a summation of interaction energies between pairs of atoms — atom–atom potentials.

The use of atom–atom potentials in molecular mechanics calculations on strained molecules goes back to the late 1940s and early 1950s [31–34]. The idea is that the geometrical structure of a molecule is described by some set of ideal bond distances, bond angles and torsion angles. Deviations from these ideal values may occur, but only at the cost of an increase in potential energy, as estimated by some assumed force field, usually a fairly simple one, including mainly harmonic terms and neglecting cross-terms among the variables. In addition, it was found necessary to introduce non-bonded interactions as perturbations on the hypothetical, unstrained molecular structure. The sum of the deformation energies and of the non-bonded interactions is the strain energy, which is a function of the relative atomic positions. These are varied until an energy minimum is found. The ideal values of the structural parameters and the functions describing the energy cost of deviations from these ideal values were initially based mainly on empirical information about the structures of simple molecules then available, but they are now based on a miscellany of structural, thermochemical and spectroscopic information. Many programs itemize the separate contributions to the total strain energy, i.e. from bond-stretching energy, angle-bending energy, torsional strain, non-bonded repulsion, etc., but the itemized values should not be taken too seriously. The individual terms in the energy expansion are strongly coupled, which means that as one bond angle, say, is changed from its assumed equilibrium value it may become more or less easy to deform another bond distance or angle, but the cross-terms that express such couplings are often inadequate or even absent in the construction of the force field. As computers become cheaper, more powerful and more generally available, molecular mechanics calculations are being increasingly backed by quantum mechanical calculations, at least for simple molecules, and by molecular dynamics calculations for more complex systems.

The connection between molecular mechanics and crystal structures came about in the attempt to quantify the non-bonded interactions. These were first taken over from intermolecular interaction potentials of rare-gas-type molecules. They start from the premise, contained in the van der Waals equation of state for real gases, that atoms are not localized at points, i.e. not at their respective nuclei. They occupy a volume of space and can be assigned, at least as a first step, more or less definite radii, by custom called van der Waals radii, which were initially estimated for many types of atom mainly from packing radii in crystals*. Mutual approach of non-bonded atoms to distances less than the sum of these radii leads to strong repulsive forces. The empirical atom–atom potentials that were introduced to describe the balance between atom–atom attractions and repulsions were assumed to be characteristic of the atom types and independent of the molecules they are embedded in. They were assumed to hold equally for interactions between non-bonded atoms in

---

*These van der Waals radii are only distantly related to the constant b in the van der Waals equation of state $(p + a/v^2)\,(V - b) = RT$, and the great Dutch chemist would undoubtedly have been surprised to know that his name was to be associated with them. Indeed, it is difficult to trace how the term established itself in the everyday language of structural chemistry. One likely source is Pauling's influential text *The Nature of the Chemical Bond*. In the section entitled 'Van der Waals and Nonbonded Radii of Atoms', Pauling wrote:

In a crystal of the substance (chlorine) the molecules are attracted together by their van der Waals interactions and assume equilibrium positions at which the attractive forces are balanced by the characteristic repulsive forces between atoms, resulting from interpenetration of their electron shells. Let us call one-half of the equilibrium internuclear distance between two chlorine atoms in such van der Waals contact, corresponding to the relative positions of two molecules, the van der Waals radius of chlorine.

Pauling's table of van der Waals radii of atoms still stands as a set of convenient rule-of-thumb values.

the same molecule and for those between atoms in different molecules. The very simplicity of the idea is appealing; it would mean, for example, that interaction energies for hydrocarbon crystals and strain energies for cycloalkanes could be evaluated from characteristic interaction energies for carbon and hydrogen atoms. Thus there has been a historic link between molecular mechanics calculations designed to model the structures and energies of molecules on the one hand and atom–atom potential calculations designed to model the structures and energies of molecular crystals on the other. To a good approximation, the same functions can be used for both models.

The basic assumptions underlying the use of most atom–atom potential calculations are that only central forces operate between pairs of atoms and that the total interaction energy is the sum of the interactions between all pairs of atoms — the additivity assumption. The individual atom–atom interaction energies include a repulsive term with a steep rise in the energy at small interatomic distances, an attractive term designed to allow for London-type dispersion attractions and, sometimes, an additional coulombic interaction as well. With an exponential function as the repulsive term, the interaction energy between a pair of atoms can be written as

$$E_{ij} = A \exp(-Br_{ij}) - Cr_{ij}^{-6} + q_i q_j r_{ij}^{-1}$$

where A, B and C are empirical parameters and r_{ij} is the interatomic distance between atoms i and j; q_i and q_j are the assumed point charges for the coulombic contribution. At the equilibrium distance r_0 the positive and negative terms exactly balance. Different authors prefer different functional forms for the repulsive and attractive terms, and there is also no consensus about the coulombic terms. Some authors prefer to omit them altogether, while others include them but with no agreement about how the point charges are to be evaluated nor about how the coulombic interaction should be attenuated at long distances by something resembling an effective dielectric constant. As mentioned in the previous section, one suggestion is to replace the r^{-1} coulombic potential by an r^{-2} dependence adjusted to yield the correct coulombic forces at molecular contact distances and with the term truncated at 5 Å [18].

In calculations involving binding energies of host–guest complexes the interaction energy has to be summed over all non-bonded pairs of atoms, and similarly for the cohesive energy of a cluster or a crystal. Fortunately, the summations converge fairly quickly (except for the coulombic terms). For benzene, for example, the packing energies calculated with Kitaigorodskii's atom–atom potentials at 7 Å and 15 Å cut-offs provide respectively about 85% and 99% of the value at full convergence [35].

A few words about the scaling of energy units may be in order. For simplicity, consider a cluster of N monatomic molecules. The energy summation has to be made over $N(N-1)/2$ interatomic distances. If the resultant cohesive energy is to be expressed in kilocalories per mole of monomeric unit, then the sum has to be divided by N, the number of atoms in the cluster. For the extended periodic crystal where all the atoms can be considered as equivalent, the same result is obtained by choosing an arbitrary atom as a reference atom and summing the interaction energy over the $N-1$ distances to the remaining atoms and then dividing by 2. This will give the calculated cohesive energy of the crystal per mole, often referred to also as the lattice energy or the packing energy (PE), in any case the quantity relevant for comparison with the experimental sublimation energy. It is the energy released per mole when the molecules are brought from infinity into contact in the crystal.

When estimating quantities such as activation barriers to the rotation of molecules in crystals, one is interested in the change in the interaction energy of a single given molecule with respect to its neighbours. For such problems it would be inappropriate to divide the energy sum by 2. In certain contexts it is therefore useful to distinguish between the packing energy (PE) and what has been termed the packing potential energy (PPE), which is exactly double the packing energy. The PPE is the energy released on bringing one molecule from infinity to its location in the crystal and is obtained by summing the interaction terms from all atoms of a reference molecule to all atoms of other molecules within the cut-off radius [36].

Besides the question of units there are also questions about what is actually being calculated in atom–atom calculations. It is clear that what Kitaigorodskii originally had in mind [35] was to calculate the potential energy of particle interaction, equal to the heat of sublimation of the crystal at absolute zero (apart from a correction for the zero-point vibrational energy, which is generally expected to be small). The crystal structure at 0 K should then correspond to the coordinates of the minimum of the potential energy well. Unfortunately, hardly any crystal structures of reasonably complex molecules have been determined at sufficiently low temperatures to allow reliable extrapolation to 0 K (benzene is an exception), and heats of sublimation are not measured at 0 K but at considerably higher temperatures. Most crystal structures are measured far from 0 K. The structure at temperature T does not correspond to a minimum in the potential energy but to a minimum in the free energy $G = \text{PE} + E(\text{vib}) - TS$, where $E(\text{vib})$ is the vibrational energy, usually only a small fraction of the packing energy, and S is the entropy.

What should be done about the entropy contribution? To ignore it in atom–atom calculations of crystals would be equivalent to assuming that the 'energy' calculated for any given structure is independent of temperature — that the structure of lowest energy remains the structure of lowest energy throughout the entire temperature range. But we know that this is not true. In some molecular mechanics programs (such as MM3 [37], for example), the molecular thermodynamic quantities $H - H_0$, S and $G - H_0$ are estimated from the calculated vibrational levels of the molecule using standard formulae [38]. Thus, by including also the entropy contributions for molecular translation and rotation, the thermodynamic quantities for the hypothetical ideal gas are obtainable. We shall not discuss here the problems of parametrization. Liquids and solutions present more serious problems, but for crystals, in principle, the vibrational entropy can be estimated in a similar way from vibration frequencies, as calculated by lattice dynamics. In fact, lattice dynamics treats the crystal as a giant molecule, as a supermolecule in effect, and applies exactly the same formalism as used in molecular vibrational analysis. In many applications of atom–atom potentials, however, no attention is paid to the thermodynamic functions, and the interpretation of results obtained with these potentials is therefore often left obscure. This may be fairly innocuous for calculations involving isolated molecules, but in estimating the strength of molecular associations, for example, in host–guest or enzyme–substrate complexes, it could be very serious.

For crystals built of fairly rigid molecules, such as the aromatic hydrocarbons, lattice dynamical calculations reproduce the frequencies of the molecular translational vibrations and librations [39] reasonably well. For naphthalene and anthracene, the mean square amplitudes of rigid-body vibrations and librations and their temperature dependences [22] are also well produced by calculations based on Williams's set IV atom–atom potentials [40] (in which the long-range electrostatic interactions are not included explicitly but are absorbed into the other parameters, which are adapted for a 5.5 Å cut-off; increasing the cut-off limit in a lattice dynamical calculation for naphthalene worsens the agreement [22]).

The good agreement for aromatic hydrocarbons is encouraging. But most molecules contain other atoms besides carbon and hydrogen, and for most of these other atoms the quality of the available atom–atom potentials is much inferior. Indeed, the empirical parameters in many atom–atom potential functions have tended to be fitted to a miscellany of information from experimental crystal structures done mostly around room temperature. The equilibrium distances r_0 corresponding to the minima of the curves have mostly been fitted to some selection of appropriate contact distances in crystals, but the depths of the energy wells are often less securely founded and their curvatures even less so.

Moreover, most molecules are not rigid but have many low-frequency internal motions. For such non-rigid molecules it is generally difficult to place much reliance in the outcome of crystal calculations based on atom–atom potentials. Even the results of molecular mechanics calculations for non-rigid molecules should be viewed with caution and, if possible, checked by repeating the calculation with a different program using different potential functions. One should try to ascertain whether the energy functions have been calibrated against room-temperature data or whether they are supposed to refer to potential energy or whatever.

One would like to ignore the entropy contribution S, but the existence of polymorphs and of polymorphlic transitions is an effective demonstration that this cannot be done. As the temperature is increased past some transition point, the polymorph stable at 0 K may become metastable and then transform to a different polymorph. The entropy always increases with temperature faster in the polymorphic form that becomes stable at higher temperature.

Another difficulty is that sublimation energies are known only for about 1000 organic crystals [41], for most of which the crystal structure has never been determined. Even when the crystal structure and sublimation energy of an organic compound are both known, there is the question of whether the compound is polymorphic, and, if so, whether the same polymorph was used for the two kinds of experiment. Another source of uncertainty in the comparison is that for non-rigid molecules the molecular conformation in the crystal may be different from that in the vapour where, indeed, a mixture of conformations can well occur. One authority has come to the somewhat disappointing but realistic conclusion that the discrepancy between observed and calculated quantities cannot be expected to be less than a few kilocalories per mole, i.e. around 10% of the experimental sublimation energy [36].

4.   CAN ATOM–ATOM POTENTIALS BE IMPROVED?

Is it necessary to develop better atom–atom potentials? At one level, we can be satisfied with a potential such as Kitaigorodskii's 'universal potential'

$$f(r) = 3.5[8600 \exp(-13r/r_0 - 0.04(r/r_0)^6]$$

which contains just one parameter, the equilibrium distance r_0, and has been shown to yield quite good results in several contexts [42]. What is the advantage of adding additional parameters? Should one use different functions to describe intramolecular non-bonded interactions and intermolecular ones? Do we need different functions for the same atom in different chemical environments, e.g. carbonyl oxygen and ether oxygen? What are the criteria by which we should compare the various sets of atom–atom potentials that are available? It is easier to ask these questions than to answer them. After all, atom–atom potentials are used in many different contexts to estimate many quite different kinds of quantity — cell dimensions, sublimation energies and thermodynamic properties of crystals, molecular conformations and strain energies, vibration frequencies of molecules and crystals, and no doubt others — and the demands in accuracy and reliability can vary from one application to another. Let us be idealistic and ask: what can we ask from the best atom–atom potentials and how can we improve the available ones?

The confrontation between the calculated PE and the experimental heat of sublimation of a crystal would seem to be one of the most direct tests of atom–atom potentials. While the ability of a set of model potentials to reproduce the experimental structure and PE is no guarantee that the potentials are correct, their inability to do so shows that something is wrong. In view of this possibility of testing potential functions, it seems a pity that so little effort is spent nowadays in systematic measurements of the thermodynamic properties of selected crystals showing key structural features. Unfortunately, careful thermodynamic measurements require much time and patience and are, let us admit it, somewhat old-fashioned and unexciting. In any case, it seems likely (at least to this author) that a research project along these lines would encounter formidable difficulties in securing even minimal financial support from funding agencies.

A programme setting out to derive atom–atom potentials from *ab initio* quantum mechanical calculations would seem much more in keeping with contemporary attitudes and more likely to be funded. Yet the difficulties in obtaining reliable theoretical estimates of such potential functions are severe. These difficulties arise mainly from two opposing requirements: the need for a large set of basis functions to model the diffuse electron density in regions far from the atomic nuclei, and the need to include the effect of electron correlation. At this level of theory the attractive force between closed-shell atoms originates entirely from correlation effects; at the Hartree–Fock level the interaction is repulsive at all distances. As an example, one can point to the difficulties encountered in the attempt to estimate the C–H$\cdots$O interaction energy in some simple systems by high level *ab initio* calculations [43].

One obvious imperfection in past and present atom–atom potentials is the assumption of central forces, i.e. the assumption that the force field round an atom is spherically symmetrical. As justification for this assumption, it has been argued that in accurate X-ray analysis the total electron density in a crystal is very close to the superposition of densities of spherically symmetrical atoms; furthermore, that the atomic multipole moments resulting from redistribution of the electron density would lead to electrostatic interaction energies that would be negligible compared with the sublimation energy (or with the other errors in the calculation). However, in contradiction to these arguments, we know that even if the deviations from spherically symmetrical atomic charge distributions in molecules may be small, they are responsible for the binding energy of the molecule, which is very large indeed compared with the non-bonded interactions we are interested in here.

In fact, although the assumption of spherical symmetry of the atomic force field has the advantage of simplicity, it is plainly wrong. There is an abundance of evidence to show that atom–atom non-bonded interactions are highly directional. Hydrogen bonding is an extreme example; it has been known for many years that in intermolecular hydrogen bonds the A–H···B angle tends to be close to 180° and that the approach of the proton to the basic atom B is preferentially along a lone pair direction. Empirical potential functions have been constructed that introduce energy penalties for deviations from the preferred patterns [44].

But highly directional interactions are known for many other types of donor–acceptor systems as well. One of the most instructive examples is provided by crystalline iodine (Figure 7). This structure is interesting in the first place because the interatomic distances do not quite fall into the usual categories. The intramolecular distance of 2.72 Å (at 110 K) [45] is slightly larger than the corresponding distance of 2.66 Å in the free iodine molecule. On the other hand, the two shortest intermolecular distances, 3.50 Å and 3.97 Å, are markedly different and considerably shorter than 4.3 Å, twice the accepted van der Waals radius of iodine, indicative of strong bonding interactions between the I_2 molecules. Essentially the same type of packing is found in crystalline bromine and chlorine, except that the intermolecular bonding is weaker in Br_2 and weaker still in Cl_2 [46]. The clue to understanding these structures comes from the crystalline molecular complexes formed by halogenated molecules with various electron donors, studied first by Hassel and his school [47]*. Among these complexes are many examples of linear (or almost linear) X–Hal···D arrangements in which X is any atom, Hal is chlorine, bromine or iodine, and D is oxygen, nitrogen, sulfur or selenium. Typically, the X–Hal distance is slightly longer than the normal covalent distance and the Hal···D distance is shorter, and sometimes very much shorter, than the sum of the van der Waals radii. Moreover, the Hal···D direction is always roughly along the expected lone pair direction of D. The interpretation of these interactions as electron donor–acceptor interactions is convincing, and the structures illustrate the rule: electron acceptors accept preferentially along the extension of the existing covalent bond, or, in other words, the nucleophile approaches along the extension of the existing bond [in modern parlance, so that the incoming electron pair overlaps optimally with the X–Hal σ*-orbital, the lowest unoccupied molecular orbital (LUMO)]. Another striking example is found in the crystal structure of diiodoacetylene [48], where each molecule points at both ends towards the middles of two adjacent molecules (Figure 8; incidentally, one of the best examples I know of a quadrupolar molecule crystallizing in the pattern shown in Figure 3). Here, it is the carbon–carbon triple bond that is the electron donor. If the iodine molecule is an electron acceptor along the bond direction, Figure 7 suggests it is an electron donor perpendicular to this direction. No set of atom–atom potentials with central

---

*See particularly also the prescient review of donor–acceptor interactions by H. Bent, *Chem. Rev.*, **68**, 587 (1968).

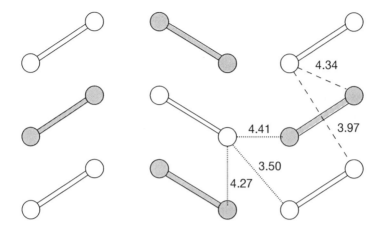

Figure 7 Crystal structure of iodine viewed in projection down the *a*-axis with the short intermolecular distances (in ångstroms) indicated. The shaded molecules are displaced half way along the *a*-axis

forces, i.e. spherical atoms, can possibly reproduce these simple structures of the elements.

The regularity mentioned in the previous paragraph seems to apply quite generally for interactions of electron donors with σ-bonds. Apart from the well-known chemical example of S_N2 displacement at carbon leading to inversion of configuration (Walden inversion), analyses of packing patterns in many crystal structures show that the shortest intermolecular approach distance between an electron-donating centre D (Lewis base) and an electron-accepting centre A (Lewis acid) tends to be approximately collinear with one of the existing σ-bonds formed by A. For example, this is true for the approach of a nucleophilic centre towards divalent sulfur [49] or sulfonium sulfur [50]. Similarly, when one or two electron-donating centres approach a tetrahedral SnX_4 molecule, the direction of approach of each such centre is always towards the centre of a face, i.e. the tin atom acts as an electron acceptor along the direction opposite to an Sn–X bond. When an electron-donating centre approaches a trigonal pyramidal SnC_2X_3 ensemble, the approach is towards the middle of an equatorial edge, again opposite to an Sn–X bond [51]. On the other hand, approaches of *electrophilic* centres to divalent sulfur [49] and to ether oxygen [52] show a quite different kind of directional dependence.

One way of coping with these complications in atom–atom potentials is to shift the attraction–repulsion centre in atom–atom potentials away from the atomic nucleus in the bond direction [53]. Another is to say that the van der Waals radii of atoms are direction dependent. Thus, in his review of van der Waals radii, Bondi stated that many atoms are 'pear-shaped' [54], and Nyburg

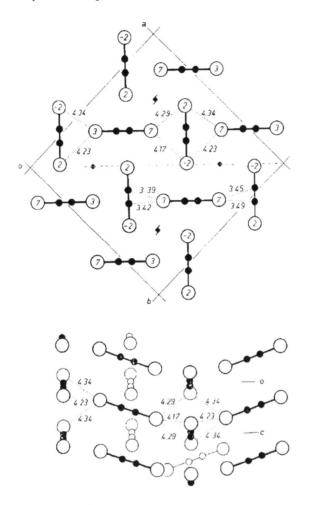

Figure 8 Two projections of the crystal structure of diiodoacetylene with interatomic distances in Ångstroms (from Dunitz *et al.* [48])

noted that in non-bonded interactions where the two atoms are each covalently bonded to only one other atom, some atoms behave as if 'flattened at the poles' [55]. This applies, for example, to $C=S\cdots S$, $C-Cl\cdots Cl$, $C-Br\cdots Br$ and $C-I\cdots I$ contacts, where the van der Waals surfaces are markedly direction dependent, with their smallest extension along the $C-X$ bond vector. On the other hand, the corresponding surfaces for nitrile nitrogen and carbonyl oxygen are virtually spherical [56]. In a similar study, it was found that $C-I\cdots I$ (and $C-I\cdots O$) contacts tend to be short for linear arrangements and longer for perpendicular approaches [57]. Thus the directional dependence appears to increase as we move towards heavier (more polarizable) atoms. Quite generally, the variable van der Waals radius can be expected to depend not only on the nature of the atom concerned, but also on the atom(s) to which it is covalently bonded, the non-bonded atoms with which it interacts and on their mutual orientations. However, in a sense it may be misleading to talk about a direction-dependent van der Waals radius. One may conjecture that any directional preference in intermolecular approaches in crystals corresponds not so much to changes in repulsive forces between the atoms concerned but to an increase in the attractive forces in this direction.

The occurrence of short intermolecular approach distances in crystals has been interpreted as corresponding to weak 'secondary' bonding between the molecules and hence to incipient stages of chemical reactions [58]. The directional preferences shown in such interactions are thus guides to the preferred (low-energy) directions of approach leading to transition states. More quantitative expressions of the directional dependence of atom–atom interactions are obviously required.

The main source of information about the directional dependence of atom–atom interactions is likely to come from the careful study of the results of crystal structure analyses. Some of this will come from structures of host–guest complexes, but much more by examining the details of the packing in small-molecule structures. It is worth emphasizing here that crystal structures contain information that can be of importance in quite different contexts and for quite different purposes from those that prompted the initial structure determination. Analyses of packing patterns in groups of related molecules can sometimes lead to rather general conclusions that could not have been drawn from the individual analyses. Yet published accounts of crystal structure analyses today rarely comprise more than a schematic picture of the molecule and a condensed list of intramolecular geometric parameters, i.e. bond distances and angles. The packing is usually passed over. It may be recovered if the unit cell dimensions, space group and atomic parameters are listed in the publication or deposited with the Cambridge Crystallographic Data Centre*, otherwise the information will usually be irrecoverable. Careful analysis and interpretation of systematic features of packing patterns in crystals is likely to be a rewarding area of supramolecular chemistry for some time to come.

---

*The Cambridge Crystallographic Data Centre, 12 Union Road, Cambridge CB2 1EZ, UK, produces and distributes the Cambridge Structural Database (CSD), which contains, in computer-readable form, unit cell dimensions, space groups, atomic coordinates and bibliographic data for organic and organometallic crystals of known structure. The CSD contains more than 10^5 entries and is expanding by about 10^4 entries annually.

REFERENCES

1. M. F. Richardson, Q.-C. Yang, E. Novotny-Bregger and J. D. Dunitz, *Acta Crystallogr.*, *Part B*, **46**, 653 (1990).
2. L. Pauling and M. Delbrück, *Science*, **92**, 77 (1940).
3. C. P. Brock and J. D. Dunitz, *Chem. Mater.*, **6**, 1118 (1994).
4. J. Perlstein, *J. Am. Chem. Soc.*, **116**, 455 (1994).
5. E. Weber, *Topics in Current Chemistry*, **140**, 1 (1987).
6. H. M. Powell, *J. Chem. Soc.*, **61** (1948).
7. G. J. Jeffrey, J. R. Ruble, R. K. McMullan and J. A. Pople, *Proc. R. Soc. London, Ser. A*, **414**, 47 (1987).
8. K. C. Janda, J. C. Hemminger, J. S. Winn, S. E. Novick, S. J. Harris and W. Klemperer, *J. Chem. Phys.*, **63**, 1419 (1975).
9. J. M. Steed, T. A. Dixon and W. Klemperer, *J. Chem. Phys.*, **70**, 4940 (1979).
10. J. F. Stoddart, in *Host–Guest Molecular Interactions: From Chemistry to Biology* (eds D. J. Chadwick and K. Widdows), Ciba Foundation, London, 1991, p. 19.
11. D. E. Williams and Y. Xiao, *Acta Crystallogr.*, *Part A*, **49**, 1 (1993).
12. G. Klebe and F. Diederich, *Philos. Trans. R. Soc. London, Ser. A*, **345**, 37 (1993).
13. J. H. Williams, *Acc. Chem. Res.*, **26**, 593 (1993).
14. M. R. Battaglia, A. D. Buckingham and J. H. Williams, *Chem. Phys. Lett.*, **78**, 421 (1981).
15. R. W. G. Wyckoff, *Crystal Structures*, 2nd edn, Vol. 1, Wiley-Interscience, New York, 1963, p. 368.
16. N. Boden, P. P. Davis, C. H. Stam and G. A. Wesselink, *Mol. Phys.*, **25**, 81 (1973).
17. J. K. Whitesell, R. E. Davis, L. L. Saunders, R. J. Wilson and J. P. Feagins, *J. Am. Chem. Soc.*, **113**, 3267 (1991).
18. X. Shi and L. S. Bartell, *J. Phys. Chem.*, **92**, 5667 (1988).
19. A. I. Kitaigorodskii, *Organic Chemical Crystallography*, Consultants Bureau, New York, 1961, Chapters 3 and 4; A. I. Kitaigorodskii, *Molecular Crystals and Molecules*, Academic Press, New York, 1973, Chapter 1.
20. K. Aoki, H. Yamawaki, M. Sakashita, Y. Gotoh and K. Takemura, *Science*, **263**, 356 (1994).
21. G. J. Piermarini, A. D. Mighell, C. E. Weir and S. Block, *Science*, **165**, 1250 (1969).
22. C. P. Brock and J. D. Dunitz, *Acta Crystallogr.*, *Part B*, **38**, 2218 (1982); **46**, 795 (1990).
23. J. Bernstein, J. A. R. P. Sarma and A. Gavezzotti, *Chem. Phys. Lett.*, **174**, 361 (1990).
24. D. Hall and D. E. Williams, *Acta Crystallogr.*, *Part A*, **31**, 56 (1975).
25. R. S. Mulliken, *J. Chem. Phys.*, **23**, 1833 (1955) and the paper immediately following.
26. S. R. Cox and D. E. Williams, *J. Comput. Chem.*, **2**, 304 (1981).
27. F. L. Hirshfeld, *Theor. Chim. Acta*, **44**, 129 (1977).
28. R. F. W. Bader, T.-H. Tang, Y. Tal and F. W. Biegler-König, *J. Am. Chem. Soc.*, **104**, 946 (1982).
29. G. R. Desiraju and A. Gavezzotti, *Acta Crystallogr.*, *Part B*, **45**, 473 (1989).
30. C. A. Hunter and J. K. M. Sanders, *J. Am. Chem. Soc.*, **112**, 5525 (1990).
31. T. L. Hill, *J. Chem. Phys.*, **14**, 465 (1946).
32. F. H. Westheimer and J. E. Mayer, *J. Chem. Phys.*, **14**, 733 (1946).
33. A. I. Kitaigorodskii, *Izv. Akad. Nauk SSSR, Ser. Fiz.*, **15**, 157 (1951).
34. J. D. Dunitz and V. Schomaker, *J. Chem. Phys.*, **20**, 1703 (1952).

35. A. I. Kitaigorodskii, *Molecular Crystals and Molecules*, Academic Press, New York, 1973, p. 167.
36. A. Gavezzotti, in *Structure Correlation*, Vol. 2 (eds H.-B. Bürgi and J. D. Dunitz), VCH, Weinheim, 1994, p. 509.
37. N. L. Allinger, Y. H. Yuh and J.-H. Lii, *J. Am. Chem. Soc.*, **111**, 8551 (1989) and the two papers immediately following.
38. J. H. Knox, *Molecular Thermodynamics*, Wiley-Interscience, London, 1971.
39. F. Vovelle, M.-P. Chedin and G. G. Dumas, *Mol. Cryst. Liq. Cryst.*, **48**, 261 (1978).
40. D. E. Williams, *J. Chem. Phys.*, **47**, 4680 (1967).
41. J. S. Chickos, in *Molecular Structure and Energetics*, Vol. 2 (eds J. F. Liebman and A. Greenberg), VCH, New York, 1987, p. 67.
42. A. I. Kitaigorodskii, *Tetrahedron*, **14**, 230 (1961).
43. P. Seiler, G. R. Weisman, E. O. Glendening, F. Weinhold, V. B. Johnson and J. D. Dunitz, *Angew. Chem.*, **99**, 1216 (1988); *Angew. Chem., Int. Ed. Engl.*, **26**, 1175 (1988).
44. A. Vedani and J. D. Dunitz, *J. Am. Chem. Soc.*, **107**, 7653 (1985).
45. F. van Bolhuis, P. B. Koster and T. Mighelsen, *Acta Crystallogr.*, **23**, 90 (1967).
46. J. Donohue, *The Structures of the Elements*, Wiley-Interscience, New York, 1974.
47. O. Hassel and C. Rømming, *Q. Rev. Chem. Soc.*, **16**, 1 (1962).
48. J. D. Dunitz, H. Gehrer and D. Britton, *Acta Crystallogr., Part B*, **28**, 1989 (1972).
49. R. E. Rosenfield, R. Parthasarathy and J. D. Dunitz, *J. Am. Chem. Soc.*, **99**, 4860 (1977).
50. D. Britton and J. D. Dunitz, *Helv. Chim. Acta*, **63**, 1068 (1980).
51. D. Britton and J. D. Dunitz, *J. Am. Chem. Soc.*, **103**, 2971 (1981).
52. P. Chakrabarti and J. D. Dunitz, *Helv. Chim. Acta*, **65**, 1482 (1982).
53. D. E. Williams, *J. Chem. Phys.*, **43**, 4424 (1965).
54. A. Bondi, *J. Phys. Chem.*, **68**, 441 (1964).
55. S. C. Nyburg, *Acta Crystallogr., Part A*, **35**, 641 (1979).
56. S. C. Nyburg and C. H. Faerman, *Acta Crystallogr., Part B*, **41**, 274 (1985).
57. P. Murray-Rust and W. D. S. Motherwell, *J. Am. Chem. Soc.*, **101**, 4374 (1979).
58. H.-B. Bürgi and J. D. Dunitz, *Acc. Chem. Res.*, **16**, 153 (1983); H.-B. Bürgi and J. D. Dunitz, in *Structure Correlation*, Vol. 1 (eds H.-B. Bürgi and J. D. Dunitz), VCH, Weinheim, p. 163.

Chapter 2

Crystal Engineering and Molecular Recognition— Twin Facets of Supramolecular Chemistry

GAUTAM R. DESIRAJU AND C. V. KRISHNAMOHAN SHARMA

University of Hyderabad, India

1. INTRODUCTION

Supramolecular chemistry, i.e. chemistry beyond the molecule, has grown around Lehn's analogy that 'supermolecules are to molecules and the intermolecular bond what molecules are to atoms and the covalent bond' [1]. This analogy represents a departure from classical chemical thought in that it discards the notion that the molecular structure of an organic substance embodies all of its chemical and physical properties. The initial inspiration underlying the study of supermolecules was to mimic biological structure and function, but supramolecular chemistry today encompasses many other novel species and processes. The subject has therefore become a highly interdisciplinary field with a wide range of applications in the physical, chemical and biological sciences [2]. This field is developing rapidly because it has been realized that many interesting properties of materials (conduction, magnetism, non linear optics) and biomolecules (receptor–protein binding, enzyme functions) are derived from a common origin—the nature and properties of intermolecular interations.

Supermolecules are not just collections of molecules and their structures and characteristic properties are distinct from the aggregate properties of their

The Crystal as a Supramolecular Entity. Edited by G. R. Desiraju
©1996 John Wiley & Sons Ltd

molecular constituents. Now, the constituents of a supermolecule are held together by intermolecular interactions, and so supramolecular chemistry may also be viewed as the chemistry of the intermolecular bond, just as molecular chemistry is that of the covalent bond. The construction of a supermolecule from molecular components implies a detailed and precise knowledge of the chemical and geometrical properties of intermolecular interactions. Central to the theme of supramolecular assembly is the concept of molecular recognition which is 'the strategy by which a molecule bears supramolecular functions' [3]. Molecules recognize one another geometrically (shape, size) and chemically (hydrogen bonding, stacking interactions, electrostatic forces), though it must be recognized that in the limit it is impossible to demarcate strictly geometrical and chemical effects [4]. Molecular recognition studies are typically carried out in solution and the effects of intermolecular interactions are probed spectroscopically even as one attempts to correlate supramolecular structure with biological function.

Supramolecular chemistry is, however, not restricted to solution, and structural chemists and crystallographers have been quick to realize that an organic crystal is the perfect supermolecule: according to Dunitz, 'a supermolecule *par excellence*', an assembly of literally millions of molecules self-crafted by mutual recognition at an 'amazing level of precision' [5]. This indeed is the theme of this volume. If molecules are built by connecting atoms with covalent bonds, solid-state supermolecules (crystals) are built by connecting molecules with intermolecular interactions. This traditional view of a molecular crystal follows from the work of Kitaigorodskii [6], and the study of organic crystal structures by X-ray crystallography provides precise and unambiguous data on intermolecular interactions. Such information, which is essential for the development of systematic supramolecular chemistry, is conveniently retrieved from the Cambridge Structural Database (CSD) [7]. Analysis of these geometrical data by statistical procedures results in extremely reliable conclusions pertaining to the nature of intermolecular interactions [8]. A statistical approach such as is possible with the CSD is actually mandatory for the weaker interactions, i.e. those feebler than the O–H$\cdots$O and N–H$\cdots$O hydrogen bonds, because mutual interfering effects often deform interaction geometries considerably. One such weakly attractive force is the Cl$\cdots$Cl interaction [9–11]; and indeed, the systematic study of this particular interaction in crystal structures dates back to the late 1960s when it was proposed that it could be used to control crystal packing to design solid-state reactions [12]. Such design aims defined a new subject known as crystal engineering. Today, the need for rational approaches towards solid-state structures of fundamental and practical importance has led to a renewed interest in crystal engineering which seeks to understand intermolecular interactions and recognition phenomena in the context of crystal packing [13].

It is clear from the above discussion that crystal engineering and molecular recognition are closely related. The process of crystallization is one of the most precise and impressive examples of molecular recognition. Alternatively, crystal engineering aims towards 'recognition-directed spontaneous assembly of a supramolecular strand' [1]. Crystal engineering and molecular recognition are twin facets of supramolecular chemistry that depend on multiple matching of functionalities among molecular components so as to optimize a number of intermolecular interactions which may have varying strengths, directionalities and distance-dependent properties. This chapter attempts to provide a brief description of the well-known intermolecular interactions, and to reveal the complementary approaches of crystal engineering and molecular recognition towards supramolecular architecture. An appreciation of the relationship between these twin facets of the subject may well accelerate the pace of research in supramolecular chemistry.

2. INTERMOLECULAR INTERACTIONS

The design of supermolecules requires a correct assessment of the energetic and spatial attributes of the noncovalent, intermolecular forces which will result from the assembly of a given molecular structure or structures. The understanding of the nature and strength of intermolecular interactions is hence of fundamental importance in supramolecular chemistry. Intermolecular interactions in organic compounds can be classified as isotropic, medium-range forces, which define molecular shape, size and close packing; and anisotropic, long-range forces, which are electrostatic and involve heteroatom interactions [13]. In general, isotropic forces include $C \cdots C$, $C \cdots H$ and $H \cdots H$ interactions and anisotropic interactions include ionic forces, strongly directional hydrogen bonds ($O-H \cdots O$, $N-H \cdots O$), weakly directional hydrogen bonds ($C-H \cdots O$, $C-H \cdots N$, $C-H \cdots X$, where X is a halogen, and $O-H \cdots \pi$) and other weak forces such as halogen$\cdots$halogen, nitrogen$\cdots$halogen, sulfur$\cdots$halogen and so on. In a crystal, various strong and weak intermolecular interactions coexist (sometimes uneasily, as is demonstrated by the phenomenon of polymorphism) and determine the three-dimensional scaffolding of the molecules. One may conveniently, though also somewhat subjectively, designate interactions in a crystal as being strong, moderate and weak, and thereby define respectively primary, secondary and tertiary structural motifs. When used in this text, these terms must be understood as appropriate. In the biological context, hydrogen bonds, aryl$\cdots$aryl forces and van der Waals interactions are of significance.

2.1 Isotropic Interactions

By isotropic or van der Waals interactions is usually meant dispersive and repulsive forces. The dispersive forces are attractive in nature and are caused by the interaction between fluctuating multipoles in adjacent molecules. The magnitude of these forces is proportional to the inverse sixth power of the interatomic separation (r^{-6}) and approximately proportional to the size of the molecule. The repulsive forces balance the dispersive forces and have an approximate twelfth power distance dependence. The trade-off between these two sets of interactions, with some contribution from electrostatic forces, is the cornerstone of the atom–atom potential method. The close-packing principle of Kitaigorodskii follows from such an approach and assumes that molecules in a crystal tend to assume equilibrium positions so that the potential energy of the system is minimized [6]. In other words, molecules in a crystal pack such that the projections of one molecule dovetail into the hollows of its neighbours so that a maximum number of intermolecular contacts is achieved.

The close-packing principle generally holds well for hydrocarbons. The binding features in a series of condensed aromatic hydrocarbons have been explained by Desiraju and Gavezzotti using this principle [14,15]. The packing arrangements of these molecules are governed by tendencies towards 'graphitic' stacking in the carbon-rich molecules and towards inclined, herringbone geometries in the hydrogen-rich molecules. The carbon-to-hydrogen stoichiometric ratio is thus a key structural parameter. It is also likely that electrostatic factors have a bearing on the manifestation of van der Waals interactions in aromatic hydrocarbons. This can be explained with reference to Figure 1. Planar aromatic molecules are known to interact with one another in three possible geometrical arrangements: stacking (face–face overlap in duroquinone); offset stacking (laterally shifted overlap in [18]annulene, which is isostructural with the condensed aromatic hydrocarbon coronene); and herringbone (T-shaped or edge–face interactions in benzene). It is clear that of these three arrangements, the face–face overlap results in the maximum number of C$\cdots$C intermolecular contacts and consequently a greater van der Waals interaction energy. This suggests that if van der Waals forces were solely to determine the packing of flat aromatic molecules, the offset stack and herringbone arrangements should not be commonly observed. Of course, this is not the case, and Hunter and Sanders have suggested that the offset stack arrangement, for instance, is a favoured packing mode for planar aromatics because of the electrostatic barrier to face–face stacking due to $\pi\cdots\pi$ repulsions [16]. Again, the preponderance of the herringbone interaction in a very large majority of aromatic hydrocarbons adduces evidence for the C($\delta-$)$\cdots$H($\delta+$) nature of this interaction [17]. This slightly electrostatic character of the herringbone interaction might well predispose molecules during crystallization towards inclined geometries. Whether the herringbone interaction is an incipient C–H$\cdots\pi$ hydrogen bond of the type discussed more recently by Steiner [18] is still an open question.

| Stack | Offset stack | Herringbone |

Figure 1 Stacking and herringbone arrangements in π-systems

The packing of aliphatic side chains in molecular crystals may be governed to a much greater extent by 'pure' van der Waals forces. When the side chain is longer than around five carbon atoms, H⋯H interactions predominate leading to the so-called hydrophobic effect, which leads to preferential close packing of the aliphatic side chains. Such effects have been employed in several studies in crystal engineering and molecular recognition [19, 20]. It should be noted that these interactions are also important in establishing the tertiary and quaternary structures of biological macromolecules.

2.2 Anisotropic Interactions

Molecular recognition implies strongly directional preferences in the mutual approach of molecules in solution or during crystallization, and from an engineering viewpoint this is most conveniently realized via anisotropic interactions. Generally, most interactions involving the heteroatoms nitrogen, oxygen, chlorine, bromine, iodine, phosphorus, sulfur and selenium with one another or with carbon and hydrogen are anisotropic in character [13]. An interaction such as O⋯O in crystalline hexanitrobenzene [6] is probably of the dispersive–repulsive type, but other 'symmetrical' interactions like Cl⋯Cl, Br⋯Br and S⋯S [11, 21] are clearly anisotropic and perhaps even weakly bonding in character.

(a) Hydrogen bonding

Hydrogen bonding, the master key for molecular recognition, is the most reliable directional interaction in supramolecular construction and its significance in crystal engineering and molecular recognition can be scarcely underestimated. A vast amount of work has been carried out on the various aspects of hydrogen bonding, and it would be futile to attempt even a brief review of these efforts here. Hydrogen bonds of the type O–H$\cdots$O and N–H$\cdots$O may be termed 'strong' or 'conventional' (energies 20–40 kJ mol^{-1}) to distinguish them from the weaker C–H$\cdots$O and O–H$\cdots\pi$ interactions (energies 2–20 kJ mol^{-1}) which share many of their characteristics and which are discussed in greater detail in Section 2.2(c).

Despite the tremendous relevance and utility of hydrogen bonding as the most effective supramolecular cement, the definition of a hydrogen bond is still undergoing improvements [22]! According to Pauling, a hydrogen bond is largely ionic in character and is formed only between the most electronegative atoms [23]. Atkins has stated that a hydrogen bond is a link formed by a proton lying between two electronegative atoms [24]. If one were to accept these definitions strictly, only interactions such as O–H$\cdots$O, N–H$\cdots$O, N–H$\cdots$N and F–H$\cdots$F could be considered as being hydrogen bonds. A more general definition that does not restrict the nature of the donor and acceptor atoms in a hydrogen bond was provided by Pimentel and McClellan, who stated that a hydrogen bond is said to exist when there is evidence of a bond and evidence that this bond sterically involves a hydrogen atom already bonded to another atom [25]. With such a definition, weaker interactions like C–H$\cdots$O, C–H$\cdots$N, O–H$\cdots\pi$ and C–H$\cdots\pi$ may also be considered as being hydrogen bonds along with the strong hydrogen bonds mentioned above. Steiner and Saenger have provided a general and much simpler definition for the hydrogen bond as any cohesive interaction X–H$\cdots$A where H carries a positive charge and A a negative (partial or full) charge and where the charge on X is more negative than that on H [26]. So, the hydrogen bond has to be viewed with respect to the overall electronic effects in the molecule of which the XH group forms a part.

(b)   Strong hydrogen bonds

In strong hydrogen bonds X–H$\cdots$A–Y, X is highly electronegative (F, O, N) and removes electron density from the hydrogen atoms leaving them with a significant positive charge and leading to a coulombic bonding interaction with the electronegative acceptor atom A. Strong hydrogen bonds such as O–H$\cdots$O or N–H$\cdots$O are widely prevalent in biological molecules. The distinctive geometrical attributes of the hydrogen bond X–H$\cdots$A–Y are the lengths X$\cdots$A and H$\cdots$A, the hydrogen bond angle X–H$\cdots$A or θ, the H$\cdots$A–Y angle or ϕ and the planarity of the XHAY system. All these features have been discussed extensively [27]. Typical values for the X$\cdots$A distance lie between 1.80 and 2.00 Å for N–H$\cdots$O bonds and between 1.60 and 1.80 Å for O–H$\cdots$O bonds. More relaxed definitions of these distance criteria permit a consideration of the slightly longer three-centred and four-centred bonds which seem to be very common in several hydrogen-bonded structures. Ranges for θ and ϕ occur around 150–160° and around 120–130°, respectively. This marked directionality of hydrogen bonds is well established and is the basis for efficient and reliable supramolecular construction using groups such as OH, CO_2H, CONHR and $CONH_2$.

A more interesting question concerns the distance cut-off criteria for $X \cdots A$ and $H \cdots A$ separations while assessing a possible $X-H \cdots A$ contact as a hydrogen bond. In fact, van der Waals radii are sometimes inconsistent with theory and sometimes give rise to incorrect conclusions (for example, the carbon atom has been assigned van der Waals radii of 1.53 Å, 1.70 Å and 1.75 Å by Pauling, Bondi and Allinger, respectively). For some hydrogen bonds, the potential energy minima of electrostatic attractive and repulsive forces occur at intermolecular distances greater than the sum of the respective van der Waals radii. For this reason, it is necessary to consider more relaxed values of the interatomic distances $H \cdots A$ and $X \cdots A$ while assessing hydrogen bond geometries. So, distance ranges of 1.40–2.20 Å and 2.50–3.00 Å for $H \cdots A$ and $X \cdots A$, respectively, are reasonable and take into account the electrostatic character of the hydrogen bond.

The hydrogen bond is a predominantly linear interaction ($\theta = 180°$), and yet the configuration of hydrogen bonds usually observed in crystal structures is rarely linear, because the number of possible spatial configurations with the $X-H \cdots A$ angle in the range θ to ($\theta + d\theta$) is proportional to $\sin \theta$ [28]. In practice, the median value for θ is around 165°. A further reason for this bent approach is that the hydrogen atoms are often approached by a second acceptor in an attractive fashion. This type of interaction may be termed a bifurcated-donor hydrogen bond (three-centred interaction). Similarly, three acceptor atoms may also form interactions with a hydrogen atom (trifurcated donor). These possibilities are shown in Figure 2(a). These sorts of bifurcated-donor arrangements are common in organic crystal structures because most organic molecules (barring simple amines and polyamines) are acceptor rich. We have found recently that many patterns of hydrogen bonds in organic crystals are retained largely intact in organometallic crystals [29], and while donor-rich systems are uncommon in simple organics, they may be found among organometallic compounds especially when $C-H \cdots O$ bonds are considered [30]. In such cases, the bifurcated-acceptor arrangement as shown in Figure 2(b) may become important. In this context, it should be noted that it may not be possible to detect the presence of bifurcated interactions if one does not consider the long-range electrostatic nature of hydrogen bonds [20].

Hydrogen bonding is usually associated with the property of cooperativity, and this property is more pronounced in organic when compared with inorganic crystal structures [31]. In effect, polarizability and the charge transfer nature of hydrogen bonds in molecular aggregates make the total binding energy of all hydrogen bonds in the aggregate greater than the energy sum of the individual bonds. So, hydrogen bonds will be more stable when they are

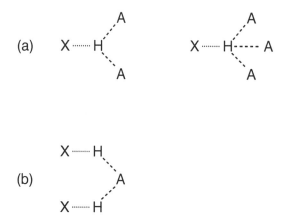

Figure 2 Multicentred hydrogen bond arrangements: (a) bifurcated and trifurcated donors; and (b) bifurcated acceptor (see Jeffrey and Saenger [27])

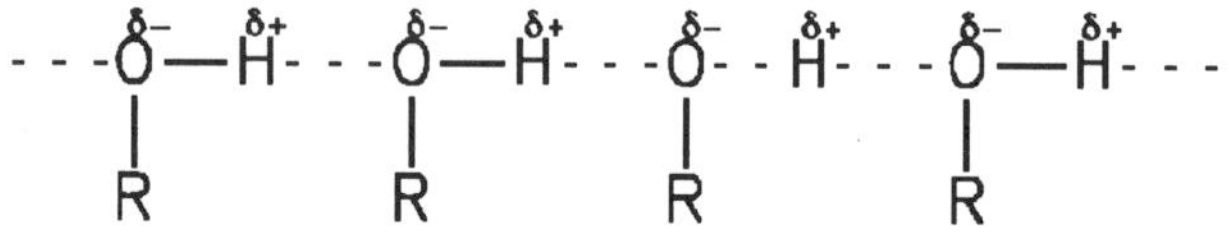

$$- - \overset{\delta-}{\underset{R}{\overset{|}{O}}} - \overset{\delta+}{H} - - - \overset{\delta-}{\underset{R}{\overset{|}{O}}} - \overset{\delta+}{H} - - - \overset{\delta-}{\underset{R}{\overset{|}{O}}} - \cdot \overset{\delta+}{H} - - - \overset{\delta-}{\underset{R}{\overset{|}{O}}} - \overset{\delta+}{H} - - -$$

Figure 3 Cooperative hydrogen bonding in an alcohol (ROH). Notice that the formation of the first hydrogen bond strengthens the second along the chain, and so on

involved in dimers, trimers, chains and infinite two- or three-dimensional structures (Figure 3).

(c)   Weak hydrogen bonds

In contrast to strong hydrogen bonds, weak hydrogen bonds are compressed or expanded and bent or straightened by the other crystal-packing forces, and the characteristic equilibrium bond geometry may be obtained only after consideration of a statistical sampling of crystal structures such as is possible by using the CSD [32]. The weaker the hydrogen bond, the less discernible are the associated directional characteristics. Here we discuss the more common C–H⋯O and C–H⋯N hydrogen bonds, while the related (N, O)–H⋯ π and C–H⋯ π hydrogen bonds are mentioned briefly.

C–H⋯O hydrogen bonds are commonly encountered in organic crystal structures owing to the frequent occurrence of the oxygen atom in simple organic molecules (72% of the entries in the 1994 version of the CSD contain an oxygen atom). These interactions play a significant role in determining crystal packing [32]. C–H···O interactions are largely electrostatic in nature, with a long-range distance fall-off character. The length of the C···O bond is more sensitive to the nature of the hydrogen atom than to the nature of the basic acceptor [33], though oxygen atom basicity also has some influence on hydrogen bond length [34]. The more acidic CH groups form shorter C···O bonds and their lengths extend well beyond the conventional van der Waals limit. Typical C···O and H···O distances lie in the ranges 3.00–4.00 Å and 2.20–3.00 Å [32]. As mentioned earlier, the so-called van der Waals separations are influenced by chemical factors associated with a particular group. Hence, chemically meaningful results can often only be obtained by observing systematic trends in intermolecular interactions within particular groups of compounds rather than by the application of 'universal' van der Waals cut-off distances. In practice, many longer C–H···O contacts (C···O 3.50–4.00 Å) have angular characteristics and effects on crystal structures which resemble the shorter contacts (3.10–3.50 Å). It should be emphasized that the C–H···O bond is not really a van der Waals contact but is primarily electrostatic, falling off much more slowly with distance and hence viable at distances which are equal to or longer than the van der Waals limit. Therefore, even long C···O separations (~4.00 Å) have to be considered seriously and these weak hydrogen bonds are expected to have orienting effects on molecules prior to nucleation and crystallization. This underlines the important role of these bonds in molecular recognition and crystal engineering [35–37].

Typical C–H···O hydrogen bond angles θ occur in the range 100–180°, but cluster around 150–160° for reasons stated earlier. Interestingly, these angles are insensitive to C···O distances unlike in strong hydrogen bonds where the shortest O···O and N···O distances are associated with linear approaches ($\theta \approx 180°$). Perhaps repulsions between negatively charged acceptor atoms in the strong hydrogen bonds cause a straightening of the O–H···O and N–H···O angles, while the longer C···O separations and smaller charges on the relevant atoms in C–H···O bonds are insufficient to exert a significant effect on angle geometry. The bending angle, ϕ at the acceptor atom is dependent on the nature of the oxygen atom (ketone, ether, epoxide), though perhaps less so than is the case for strong hydrogen bonds. Even so, the fact that such preferences are even observed for these weak interactions is noteworthy [30].

While C–H···O hydrogen bonds are widespread, the complementary O–H···π interactions are rare because carbon is not as electronegative as oxygen and also because carbon atoms are not often situated in sterically unhindered positions, unlike carbonyl and ethereal oxygen atoms which permit easy access by CH groups to form C–H···O bonds. In spite of these limitations, a sufficiently electron-rich carbon atom (alkyne, alkene, aromatic, cyclopropane) has a propensity to form a hydrogen-bond-like interaction with OH groups, thus contributing to crystal stability [38]. We have studied these interactions systematically with the CSD and around 50 unambiguous cases of intermolecular O–H⋯π and N–H⋯π hydrogen bonding were found in the 1993 update [39].

C–H⋯π interactions are even weaker than O–H⋯π and C–H⋯O hydrogen bonds because carbon is less electronegative than oxygen [18]. Indeed, there is still no well-accepted convention which considers C–H⋯π interactions as hydrogen bonds. No distinction has been generally made between carbon atoms belonging to π-systems (alkyne, alkene, arene) and to aliphatic compounds, and C–H⋯C interactions in both situations are referred to as van der Waals interactions. However, there is now an increasing body of evidence which suggests that C–H⋯C interactions in π-systems are coulombic in nature [17, 40].

(d)   Other interactions

It has long been known that the halogens chlorine, bromine and iodine form short nonbonded contacts in crystals [12]. The nature of these interactions is still somewhat obscure. Nyburg [41] and, more recently, Price [10] have held that these short contacts are the result of elliptically shaped atoms (anisotropy), while others, including ourselves, have maintained that there are specific attractive forces between halogen atoms in crystals [9, 11, 42]. Certainly, atomic polarization is an important factor here because there are clear distinctions between Cl⋯Cl, Br⋯Br and I⋯I interactions and between these symmetrical interactions on the one hand and unsymmetrical interactions such as I⋯Cl, Br⋯F and I⋯F on the other [11]. Directional interactions formed by the halogens with oxygen and nitrogen atoms are also polarization induced and have been used by us and others in systematic crystal engineering [43–45].

Like the halogens, polarization effects are important for sulfur, which is known to form short directional contacts of the type S⋯N, S⋯S and S⋯Cl. These contacts have been well studied in the crystal chemistry of the tetrathiafulvalene-7,7,8,8-tetracyanoquinodimethane (TTF–TCNQ) family of compounds [46]. Intermolecular interactions which are specific to organometallic compounds have only been the subject of very recent study [47, 48]. We have carried out a systematic survey of Au⋯Au interactions and have shown that they affect molecular conformations and crystal packing [49]. Hydrogen-bonding patterns in organometallic crystals are, in general, similar to those in simple organics, but the CO ligand is a base unique to organometallic compounds and it is perhaps more relevant to C–H⋯O than to O–H⋯O hydrogen bonding [29, 30]. This leads to the question of hardness and softness of intermolecular interactions, and it is possible that the hard and soft acids and bases (HSAB) principle may be profitably extended to supramolecular chemistry. For instance, N–H⋯O and O–H⋯O could be considered hard hydrogen bonds and C–H⋯O and C–H⋯π as soft hydrogen bonds, with perhaps the N–H⋯N hydrogen bond being of intermediate character.

Accordingly, the soft base CO prefers to hydrogen bond with the soft acid CH rather than to the hard acid OH.

3. CRYSTAL ENGINEERING

The term 'crystal engineering' was coined by Schmidt in the context of topochemical reactions of organic solids, especially the photodimerization of cinnamic acids [12]. The original objective of crystal engineering was to design organic molecules which would adopt particular crystal structures within which topochemical reactions could take place, leading to regioselective or stereoselective products. Many solid-state reactions were thus designed [50]. While solid-state organic synthesis is not yet a serious and general alternative to conventional solution synthesis, the control and prediction of organic crystal structures have emerged as important areas of materials development research because the crystal structures of solids are closely linked with their physical and chemical properties. There is an ever-growing demand for designer organic solids which may find use as catalysts, separating agents, conductors, ferromagnets and frequency doublers. Despite their mechanical weakness, organic solids are superior to inorganic compounds with respect to several of their optical, electronic and magnetic properties and can be more precisely and systematically designed. More importantly, crystal engineering principles provide insights into the nature of intermolecular interactions and it is possible to understand the structures of many complex biomolecules using these principles. In the light of its broadened scope today, it is possible to define more generally crystal engineering as 'the understanding of intermolecular interactions in the context of crystal packing and in the utilization of such understanding in the design of new solids with desired physical and chemical properties' [13].

3.1 Approaches to Crystal Engineering: Supramolecular Synthons and Pattern Recognition

In principle, there are two possible approaches to systematic crystal engineering, i.e. the prediction of organic crystal structures and of substructural packing motifs. One might examine a large number of known crystal structures and attempt to eliminate certain possibilities on the basis of established structural principles such as the need to achieve close packing or maximal hydrogen bonding. Implicit in this approach is the knowledge that there is some correlation between molecular and crystal structure, in other words an awareness that with a particular molecular structure is associated a 'molecular' valence with which the supramolecular structure is constructed. In such valence, the reader will recognize the operational aspect of molecular recognition. Accurate atom potentials are needed if quantitative results are desired. The early work of Kitaigorodskii [6] and subsequently Williams [51] has been followed by more extensive compilations of potentials by Filippini and Gavezzotti [52]. If this is a 'static' approach, static in the sense that energies are obtained for fixed structural configurations, the second 'dynamic' approach uses Monte Carlo or molecular dynamics calculations. Starting with a collection of molecules in random orientations, the final equilibrium state or stable crystal structure is obtained computationally [53]. At present, this latter approach is being used in the 'Polymorph' module of the Cerius^2 software package, commercially available from BIOSYM/Molecular Simulations [54]. Yet crystal engineering is essentially an experimental subject and both the 'static' and 'dynamic' approaches outlined above rely on comparison with experiment and on chemical intuition. Pattern recognition is therefore important in both approaches because it enables an elimination of unlikely structural possibilities. Information concerning intermolecular interactions is extracted from the CSD and deconvoluted to design specific solid-state supermolecular structures, and many present studies are focused around this most practical strategy.

The forces in an organic crystal are a complex blend of isotropic and anisotropic, short-range and long-range, and polar and non-polar interactions. However, it has been repeatedly observed that certain building blocks or 'supramolecular synthons' which consist of groups of characteristic interactions display a clear pattern preference and tend to crystallize in specific energetically favourable arrangements that can coexist with efficient close packing [55]. By identifying and classifying these synthons and by rationalizing the formation of synthon-mediated networks, it may be possible to utilize them as effective design tools in the crystal engineering of novel materials with specific structural features. In this context, the CSD provides an excellent opportunity to work backwards or retrosynthetically from observed crystal structures to formulate empirical rules about the recognition patterns of various geometrical and functional groups.

Crystal engineering may therefore be considered as the supramolecular equivalent of organic synthesis. In the same manner that molecular chemists employ a wide variety of synthetic methods to combine atoms into molecules, supramolecular chemists and crystal engineers have recognized the need for corresponding methodologies to combine molecules into supermolecules (crystals). Supramolecular synthons are substructural motifs which incorporate the chemical and geometrical characteristics of intermolecular interactions and are of the greatest importance in systematic crystal engineering. They are spatial combinations of intermolecular interactions which can be recognized clearly as design elements for solid-state architecture. Typical synthons include the carboxy dimer and catemer in Figure 4 and also the herringbone 'phenyl dimer' in Figure 1. Other synthons are illustrated in Figure 8. This analogy between crystal engineering and traditional organic synthesis is quite accurate and has been detailed elsewhere [55].

A crystal consisting of a single molecular species is the perfect example of self-recognition. Concepts of molecular recognition are intuitively applied to crystals which contain two or more molecular species, i.e. donor–acceptor and host–guest complexes, hydrates and clathrates, solid solutions and salts [56]. While there is no fundamental difference between single-component and polycomponent crystals in the crystal-engineering context, the design of a polycomponent crystal is generally considered more simple. This bias is more a matter of tradition and exists, in part, because it is often easier to predict with sufficient reliability if two components will cocrystallize in a particular manner than to predict the packing pattern of a single-component crystal. In the study of polycomponent crystals lies the meeting point of crystal engineering and molecular recognition.

The prerequisite for the formation of a molecular complex AB is that the heteromolecular species should contain stronger intermolecular interactions (A···B) than either of the homomolecular substances (A···A, B···B). This condition will be satisfied in general when A and B have complementary functionalities and in particular if (1) there are differences in the strengths of the hydrogen-bond-donating and hydrogen-bond-accepting capabilities of A and B [57]; (2) donor–acceptor interactions are possible between A and B [58]; (3) either A or B is unable to close pack on its own [56]; and (4) A and B are so similar that solid solutions are formed [59]. Here, we mention a few examples of hydrogen-bond-directed pattern recognition in single-component and polycomponent systems.

Carboxylic acids are currently the most commonly used functional groups in crystal engineering strategies. In general, carboxylic acids RCO_2H form two types of hydrogen-bonded patterns, i.e. dimers and catemers, based on the size of the R group (Figure 4). Acids containing small substitutent groups (formic acid, acetic acid) form the catemer synthon, while most others, especially aromatic carboxylic acids, form dimers (benzoic acid) [60]. The dimer synthon formed by the carboxy group may be used to assemble a variety of supermolecules. For example, terephthalic acid forms a one-dimensional ribbon structure [61], trimesic acid with its threefold molecular symmetry forms a two-dimensional hydrogen-bonded sheet [62] and adamantane-1,3,5,7-tetracarboxylic acid with its tetrahedrally disposed carboxy functionality forms a diamondoid network [63].

Other common supramolecular synthons for solid-state supramolecular construction are formed via hydrogen bonding of amide groups [64]. Primary and secondary amides [65], pyridones [66] and 2-aminopyrimidines [67] form cyclic hydrogen-bonded dimers, as shown in Figure 5.

The complementary molecular functionality of substituted melamines and barbituric acids leads to efficient hydrogen bonding between such compounds. This pattern is shown in Figure 6 and has been used by Whitesides, Lehn and others to design various supramolecular structures such as a linear tape, a crinkled tape and a cyclic rosette [68, 69]. These structures are also illustrated in

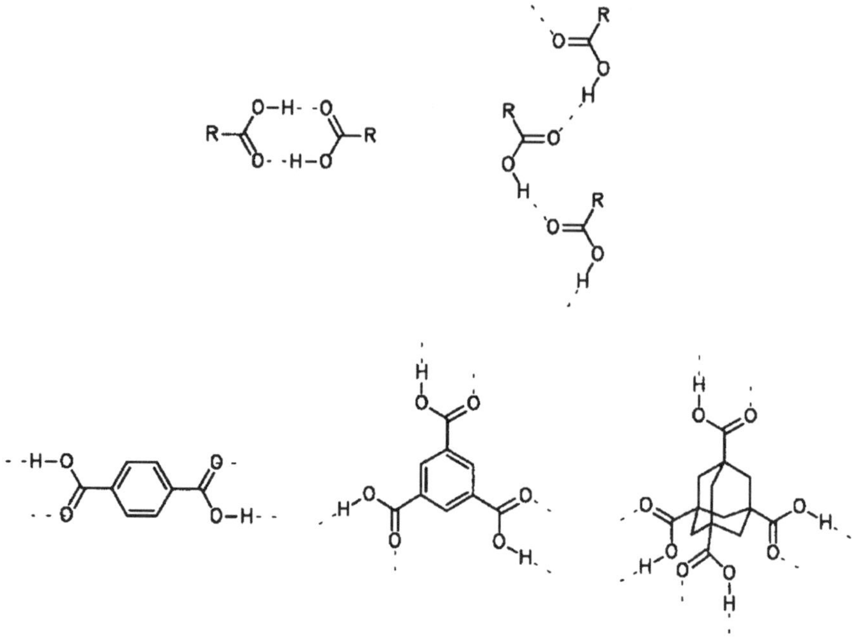

Figure 4 Some hydrogen-bonding patterns in crystalline carboxylic acids. The dimer motif is used in terephthalic acid, trimesic acid and adamantane-1,3,5,7-tetracarboxylic acid to obtain networks of increasing dimensionality

Figure 6. Whether a particular molecular complex adopts the tape or rosette structure depends on (what appear to be) relatively minor modifications in the molecular structures of the starting materials.

2-Aminopyridines and pyrimidines preferentially hydrogen bond to carboxylic acids. Tellado *et al.* showed that a unit containing two aminopyridine groups separated by a rigid spacer is a receptor for dicarboxylic acids with a correspondence between the length of the spacer and that of the carboxylic acid [70]. This is illustrated in Figure 7.

Most of the recognition strategies reported in the literature employ strong hydrogen bonds. The directional patterns of weak intermolecular interactions have not been extensively studied. These interactions are significant in themselves and also in their influence upon the stronger interactions in that they determine the secondary and tertiary features of crystal structures. The objective of crystal structure prediction often becomes elusive unless these interactions are also taken into account. With the aid of the CSD, it is possible to identify supramolecular synthons based on weak intermolecular interactions. Some of these synthons are shown in Figure 8 and their utility

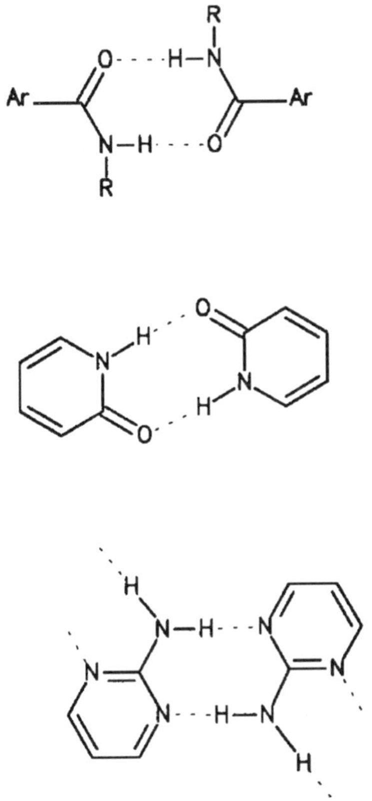

Figure 5 Some amide and amine recognition patterns

is reflected in Figure 9, which shows the C–H···O hydrogen-bonded ribbon structure formed in the crystal structure of *N,N*-dimethylnitramine. The same synthon is used to construct a ribbon in the 1:1 complex of 4-nitrobenzoic acid and 4-(*N,N*-dimethylamino)benzoic acid [71]. It may be mentioned in this connection that just as traditional organic synthesis progressed by the study of lesser-known and more subtle reactions, the use of some of these weaker intermolecular interactions in crystal engineering is expected to yield rich dividends in the future.

4. MOLECULAR RECOGNITION

The study of molecular recognition has evolved around the principles of biological functions, such as substrate binding to a receptor protein, enzyme reactions, assembly of protein–protein complexes, immunological antigen–

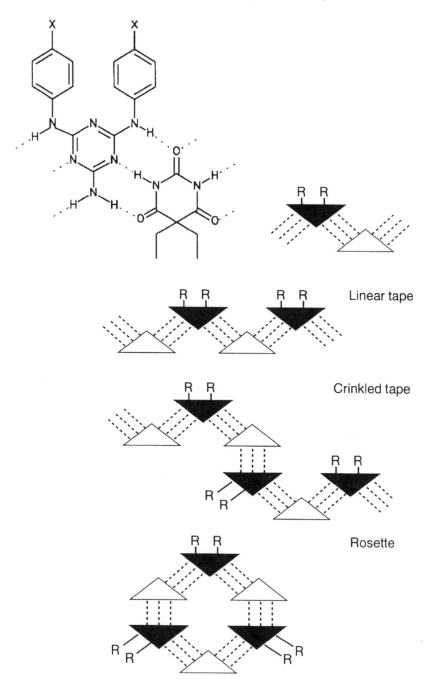

Linear tape

Crinkled tape

Rosette

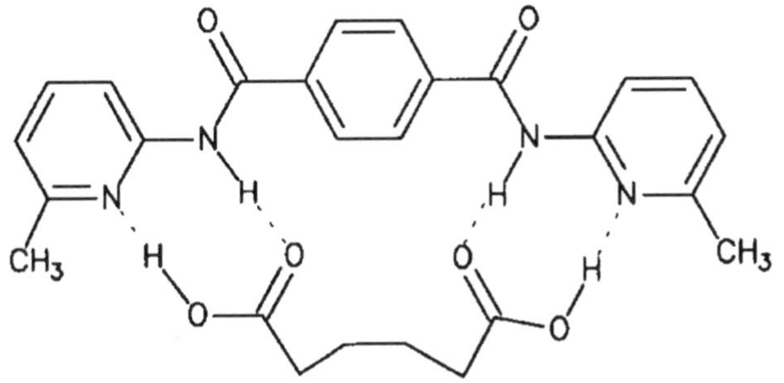

Figure 7 A receptor for dicarboxylic acids (see Tellado *et al.* [70].)

antibody association, intermolecular reading of the genetic code, signal induction by neurotransmitters and cellular recognition. Many of these functions can be manipulated by designing artificial (abiotic) receptors that can efficiently and selectively function like their biological counterparts. The design of artificial receptors requires a near-perfect match of the steric and electronic features of the noncovalent intermolecular forces between substrate and receptor. Thus molecular recognition has been defined as 'a process involving both binding and selection of a substrate by a given receptor molecule as well as possibly a specific function' [3, 72].

Receptors which can effect molecular recognition may be classified into two types based on their directional properties. In an exo-receptor (or divergent receptor), the intermolecular interactions which are responsible for the recognition event are directed outwards. In an endo-receptor (or convergent receptor), the interactions are directed inwards [73]. These are illustrated in Figure 10. Molecular size and shape are not so critical in exo-binding when compared to endo-binding because it is difficult to utilize fully the functional group capacity of substrate and receptor in this case. Recognition through exo-binding leads to the self-assembly of a molecular species. A molecular crystal is an example of an infinity of molecules, divergently bound. In contrast, the binding sites in endo-receptors are oriented into a molecular concavity, and most molecular recognition studies have been focused on such receptors. Most host–guest complexes are examples of such convergently/divergently bound species, and recognition occurs such that most of the molecular functionality is fully utilized.

Early work on molecular recognition was concerned mainly with a single-site strategy. In other words, either geometrical or chemical factors were

---

Figure 6 (*opposite*) Melamine–barbituric acid recognition pattern and a schematic depiction of the linear tape, crinkled tape and rosette structures derived from this pattern (see Zerkowski *et al.* [68].)

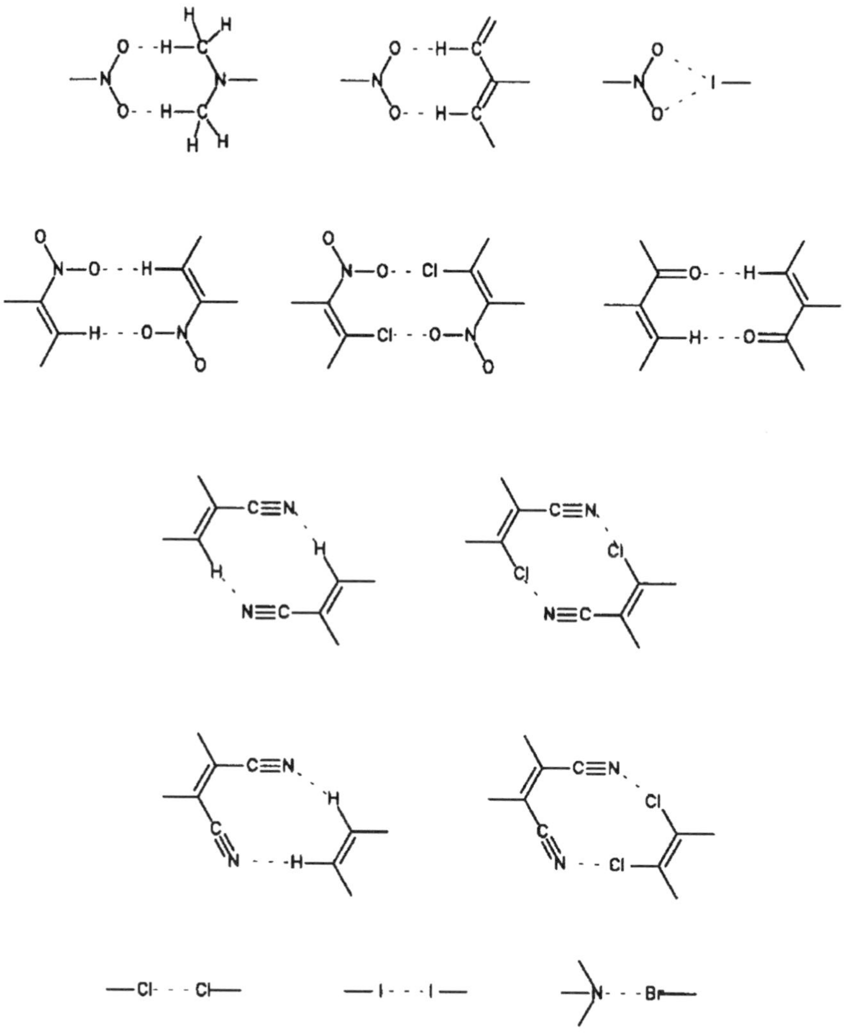

Figure 8   Representative supramolecular synthons

considered. Cyclodextrins, zeolites and crown ethers are examples of the first strategy wherein molecular shape and surface characteristics are paramount. Alternatively, the binding of hydrogen-bonded functional groups (carboxylic acids, amides) is considered. However, for most efficient manipulation, and especially if biological mimics are sought, a perfect match of size, shape and functionality between substrate and receptor must be obtained and the multiple matching of interactions is essential. A particularly appealing scheme for the

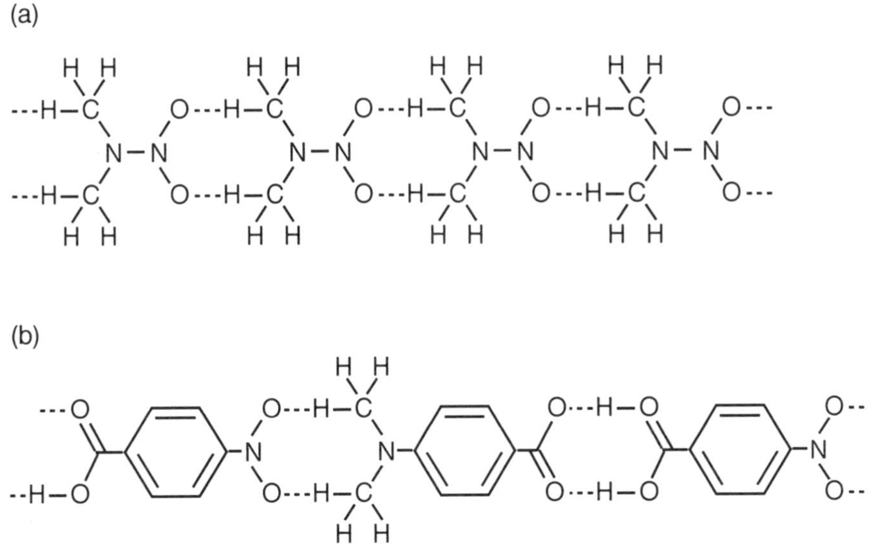

(a)

(b)

Figure 9 Use of the dimethylamino···nitro synthon in crystal engineering. Notice how the benzoic acid residues are introduced as spacers in the crystal structure of *N,N*-dimethylnitramine (a) to obtain the crystal structure of the 1:1 acid complex (b) (see Sharma *et al.* [71])

recognition of 1,4-benzoquinone has been recently developed by Hunter and is shown in Figure 11. The tetraamide host is able to capture the benzoquinone guest with a combination of four hydrogen bonds and four herringbone interactions. The methyl groups hold the isophthaloyl moieties perpendicular to the diarylmethane wall of the cavity, while the cyclohexyl groups provide solubility. The aesthetic appeal of these supermolecules belies their structural intricacy [74].

4.1 Self-Assembly

The spontaneous assembly of several molecules into a single, highly structured supramolecular aggregate is termed self-assembly [75]. The nature of the supramolecular species obtained will depend on the information stored in the components. A novel, irreversible self-assembling process has been developed by Philp and Stoddart for interlocking molecular components using $\pi\cdots\pi$ interactions and is shown in Figure 12 [76]. Such strategies may be used to make novel catenanes and rotaxanes. Recently, it has been shown by Wyler *et al.* that a tennis-ball-shaped molecular aggregate can be constructed by the self-assembly of a curved molecule with complementary hydrogen-bonding sites [77]. This is illustrated in Figure 13.

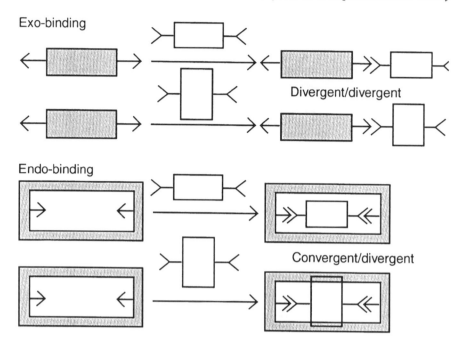

Figure 10   Binding in polycomponent crystals illustrated schematically

4.2   Molecular and Supramolecular Devices

In a general context, recognition implies the existence of information which is capable of being recognized. The availability or non-availability of information is an all-pervasive phenomenon in chemistry. For instance, a C–C single bond is information poor, whereas a C–X heteroatom bond is information rich because it can be recognized and attacked by a larger number of reagents. In the supramolecular context, an important application of molecular recognition is the design of molecular devices capable of processing information and signals at molecular and supramolecular levels. These devices have been defined by Lehn as 'structurally organized and functionally integrated chemical systems built into supramolecular architectures' [1]. A variety of complex functions can be performed by supramolecular devices with the integration of components which are specific to elementary functions (photoactive, electroactive, ionoactive). For example, the wavelength of incident light can be changed by a light conversion molecular device consisting of two components: a light collector and an emitter. A europium(III) cryptate of a macrocyclic ligand functions as a light converter [78].

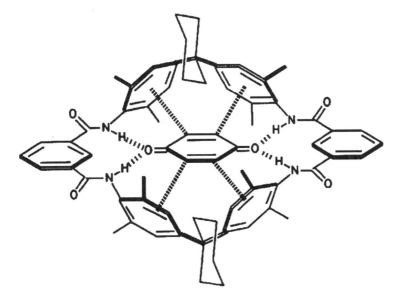

Figure 11 A scheme for recognition of 1,4-benzoquinone (see Hunter [74])

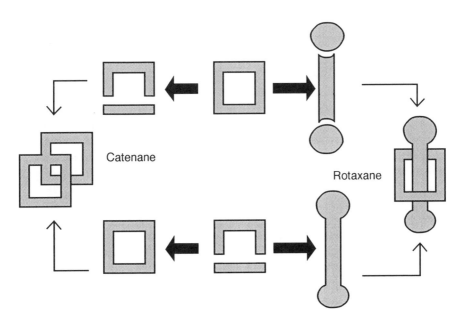

Figure 12 Supramolecularly mediated synthesis of catenanes and rotaxanes (after Philp and Stoddart [76])

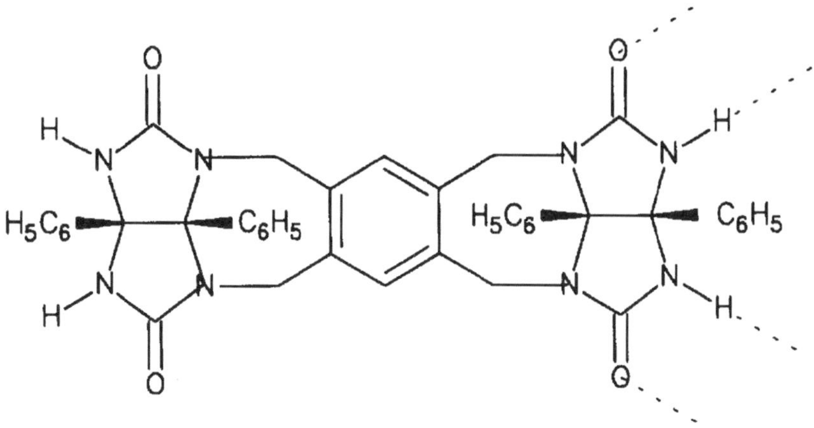

Figure 13 Assembly of a molecular 'tennis ball' via complementary hydrogen bonding. Notice that the molecule has a concave shape (see Wyler *et al.* [77])

5. HOST–GUEST COMPLEXES

Host–guest complexes result when the sizes, shapes and functional sites of the guest molecules are complementary to the cavities of the hosts. The host may be of two types: (1) a molecular host or cavitate containing a macrocyclic cavity; or (2) a supramolecular host or clathrate containing a multimolecular cavity. These are shown schematically in Figure 14. The host and guest can be defined as a chemical combination in which one component (the guest) fits into a cavity of the other (the host) or where the convergent binding partner of the molecular aggregate is specified as the host while the divergent species is called the guest. In practical terms, there are advantages and disadvantages associated with each of these two types of hosts. While it is tedious to synthesize chemical substances which act as macrocyclic hosts, the cavities of such hosts are well defined and suitable guests can be identified unambiguously. In effect, the guest plays a subsidiary role in these complexes. In contrast, supramolecular hosts can be formed by simpler host molecules, since the cavity is generated by aggregation of several small host molecules in the crystal lattice. But the cavity of these hosts is subject to variation and guest properties are critical in determining the stability of the complexes.

The formation of monomolecular host–guest complexes depends on intermolecular interactions between two molecular species only: the guest and the host which surrounds it completely. A vast amount of literature is

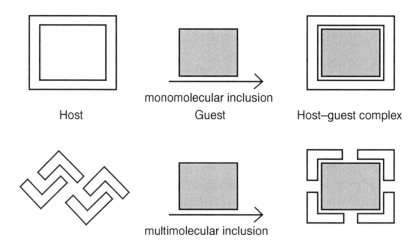

Figure 14   Monomolecular and multimolecular inclusions in host–guest complexes

available on these hosts and the initial ideas of molecular recognition, and indeed supramolecular chemistry itself, originated from this class of compounds [79, 80]. Variation in cavity sizes, shapes and functional sites causes the binding of a large number of different guest molecules. Some of the more popular host compounds are calixarenes, carcerands, cavitands, crowns, cryptands, cyclodextrins, podands, spherands and so on. Even molecules as large as C_{60} are capable of being encapsulated by the bowl-shaped calixarene shown in Figure 15, and such complexation constitutes a simple and economical way of purifying C_{60} from fullerene soot [81].

Very diverse classes of organic compounds can be used as supramolecular hosts. As these hosts are viable only in the solid state, the principles of crystal engineering have a significant bearing in the preparation of these host–guest complexes. In these cases, there are two levels of hierarchy in the supramolecular construction. Several host molecules are assembled to form a host framework which constitutes the first level of construction. The guest molecule or molecules then occupy the host voids to complete the second level of construction. A number of strategies are employed in the design of a molecule which assembles to give a host framework: (1) molecules having a bulky constitution and a rigid molecular conformation cannot close pack in the crystal, and consequently a cavity may be formed; (2) the requirement for strongly directional intermolecular interactions may be incompatible with close packing, resulting in cavity formation [82]; and (3) careful selection of the guest may induce host cavity formation. The last two options are illustrated in Figure 16.

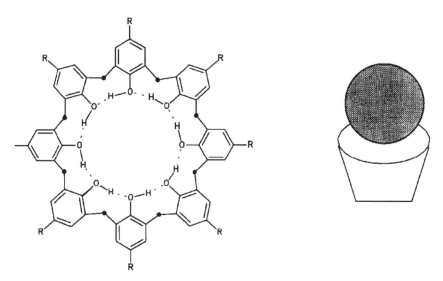

Figure 15 Encapsulation of a C_{60} molecule in a bowl-shaped calixarene host

The inclusion of toluene guest molecules in the supramolecular host cavity formed by 3-(3',5'-dinitrophenyl)-4-(2',5'-dimethoxyphenyl)cyclobutane-1,2-dicarboxylic acid highlights the earlier discussion [58]. Figure 17 shows a pair of host molecules which form a cavity by the centrosymmetric approach of two 1,2-diphenylethane moieties. The toluene molecule is enclosed by the cavity and is disordered around the same inversion centre, effectively appearing like a molecule of *p*-xylene. What is more significant is the orientation of the toluene guest in the cavity. Figure 17 shows that the toluene molecule is sandwiched between the two dinitrophenyl rings of the host while simultaneously making an angle of 74° with the dimethoxyphenyl rings. The electron-rich toluene stacks in a $\pi \cdots \pi$ fashion with the electron-deficient dinitrophenyl rings and makes herringbone or C–H$\cdots\pi$ interactions with the electron-rich dimethoxyphenyl rings. In effect, the orientation of the guest in the cavity is perfectly tuned electronically with the host cavity and is therefore predictable.

6. CRYSTAL ENGINEERING AS MOLECULAR RECOGNITION IN THE SOLID STATE

The concepts of crystal engineering and molecular recognition are exceedingly similar and both fields are concerned with the manipulation of intermolecular interactions in the architecture of supramolecular assemblies. Crystal engineering and molecular recognition are the supramolecular equivalents of

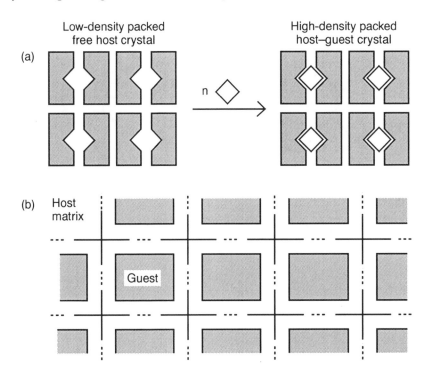

Figure 16 Schematic representation of host–guest interactions: (a) the host forms a well-defined cavity which can accommodate certain guests only; and (b) the guest induces the formation of the host cavity

organic synthesis, and, as in traditional synthesis, targets are defined in terms of aesthetics or utility. Generally, molecular recognition deals with convergent binding and crystal engineering mostly with the divergent binding of molecules. It is somewhat artificial to view these fields as distinct, and if there is some tendency for this rather superficial separation today, it might have arisen because crystal engineering and molecular recognition have been developed largely independently by researchers with quite distinct backgrounds and research interests. Crystal engineering has been developed by structural and physical chemists with a knowledge of crystallography and a view to designing novel materials and solid-state reactions [13, 55], while molecular recognition has been developed by physical organic chemists with a biological leaning and with an interest in mimicking biosynthetic processes [83]. The methodologies and goals of these two fields are summarized in Table 1.

Strategies for crystal engineering identify the characteristic binding features of various intermolecular interactions, and using this information they further

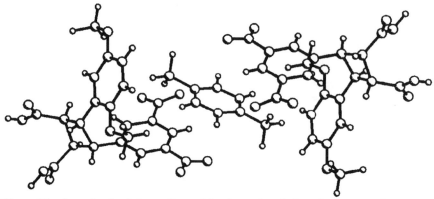

Figure 17 A cavity for toluene formed by four phenyl rings from two adjacent 1,2-diphenylcyclobutane molecules (see Sharma *et al.* [58])

Table 1 A comparison of crystal engineering and molecular recognition

Crystal engineering	Molecular recognition
(1) Concerned with the solid state	(1) Concerned chiefly with solution
(2) Convergent and divergent bindings of molecules are studied	(2) Convergent binding of molecules is mostly studied
(3) Intermolecular interactions are examined directly in terms of their geometrical features obtained from X-ray crystallography	(3) Intermolecular interactions are studied indirectly in terms of association constants obtained from spectroscopy (NMR, UV)
(4) Design strategies involve the control of the three-dimensional arrangement of molecules in the crystal. Such an arrangement ideally results in desired chemical and physical properties	(4) Design strategies are confined to the recognition of generally two species: the substrate and the receptor. Such recognition is expected to mimic some biological functionality
(5) Both strong and weak interactions are considered independently or jointly in the design or engineering strategy	(5) Only strong interactions such as hydrogen bonding are generally used for the recognition event
(6) The design may involve either single-component or multi-component species. A single-component molecular crystal is the best example of self-recognition	(6) The design usually involves two distinct species: the substrate and the receptor. Ideas concerning self-recognition are poorly developed
(7) In host–guest complexes, the host cavity is composed of several molecules, the synthesis of which may be fairly simple. The geometry and functionality of the guest molecules are often of significance in the complexation	(7) In host–guest complexes, the host cavity is formed by a single macrocyclic molecule, the synthesis of which is generally tedious. The host framework rather than the guest molecule plays a critical role in the complexation
(8) Systematic retrosynthetic pathways may be deduced with the CSD to design new recognition patterns using both strong and weak interactions	(8) There is no systematic set of protocols for the identification of new recognition patterns. Much depends on individual style and preferences

(a)   **(b)**

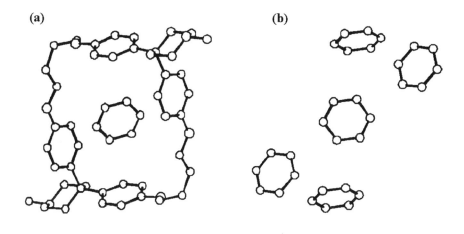

Cyclophane–Benzene   **Benzene**

Figure 18   Near identity of phenyl···phenyl interactions in a cyclophane–benzene host–guest complex (a) and in crystalline benzene (b) (see Klebe and Diederich [85])

consider if the interplay of all possible interactions for a given molecular skeleton will yield that predesired three-dimensional structure with particular physical and chemical properties. These efforts are not always fruitful because of our inability to control the weaker secondary and tertiary structural features. However, one is often successful in constructing well-defined primary structural patterns, i.e. in obtaining solid-state supermolecules consisting of a finite number of molecules (dimers, rosettes) or infinite one- and two-dimensional supermolecules (ribbons, sheets). Such efforts define an extremely attractive and competitive area of research combining molecular synthetic methodology with supramolecular ideas of molecular association [84]. In effect, crystal engineering is equivalent to solid-state molecular recognition, and the patterns of molecules found in crystals may well be extended to solution-state molecular recognition studies because the types of intermolecular interactions in the solid-state and solution are identical.

An important example which reveals the near identity of the fields of crystal engineering and molecular recognition is provided by the recent report by Klebe and Diederich of benzene inclusion in the substituted cyclophane shown in Figure 18 [85]. A molecule of benzene is enclosed by the molecular framework of the host but the orientation of the benzene guest relative to the phenyl rings of the host is very similar to the near-neighbour interactions in crystalline benzene itself. There is a close correspondence between the centre-to-centre distances and interplanar angles in the complex and the equivalent ones in benzene, showing that the recognition of the guest by the host in the complex is virtually identical to the self-recognition phenomenon in crystalline benzene. In reality there is hardly any difference between these two situations, and in the supramolecular context the aliphatic spacers between the phenyl rings in the host molecule are superfluous. This study is important because it asserts that a crystal is one of the best examples of a supramolecular entity and because it removes many unnecessary boundaries between solid-state and solution chemistry. Indeed, the distinction between single-component and polycomponent crystals is itself shown to be artificial for in the light of these results, it is not hard to perceive crystalline benzene as a host–guest complex of benzene in benzene!

In conclusion, crystal engineering helps in drawing into the realm of solution-state molecular recognition studies a keener perception of the properties of intermolecular interactions. The need for such a perception is greater for the weaker interactions which have still not been exploited fully in solution studies. A distinctive feature of supramolecular chemistry is its ability to cut across older barriers like the 'organic', 'inorganic' and 'physical' subdivisions of chemistry [86]. The present discussion shows that seemingly different research fields, i.e. 'solid-state' crystal engineering and 'solution-state' molecular recognition, are indeed one and the same in concept. A closer exchange of ideas between these fields is expected to enhance dramatically and synergistically the scope of supramolecular chemistry.

7. REFERENCES

1. J.-M. Lehn, *Angew. Chem., Int. Ed. Engl.*, **29**, 1304 (1990).
2. J.-M. Lehn, *Supramolecular Chemistry: Concepts and Perspectives*, VCH, Weinheim, 1995.
3. Y. Aoyama, in *Supramolecular Chemistry* (eds V. Balzani and L. De Cola), Kluwer, Dordrecht, 1992, p. 27.
4. This is because the geometrical preferences of intermolecular associations follow from chemical factors; see also A. Gavezzotti, *J. Am. Chem. Soc.*, **113**, 4622 (1991).
5. J. D. Dunitz, *Pure Appl. Chem*, **63**, 177 (1991).
6. A. I. Kitaigorodskii, *Molecular Crystals and Molecules*, Academic Press, New York, 1973.
7. F. H. Allen, J. E. Davies, J. J. Galloy, O. Johnson, O. Kennard, C. F. Macrae and D. G. Watson, *J. Chem. Inf. Comput. Sci.*, **31**, 187 (1991).
8. R. Taylor and F. H. Allen, in *Structure Correlation*, Vol. 1 (eds H.-B. Bürgi and J. D. Dunitz), VCH, Weinheim, 1994, p. 128.
9. G. R. Desiraju and R. Parthasarathy, *J. Am. Chem. Soc.*, **111**, 8725 (1989).
10. S. L. Price, A. J. Stone, J. Lucas, R. S. Rowland and A. Thornley, *J. Am. Chem. Soc.*, **116**, 4910 (1994).
11. V. R. Pedireddi, D. S. Reddy, B. S. Goud, D. C. Craig, A. D. Rae and G. R. Desiraju, *J. Chem. Soc., Perkin Trans.* **2**, 2353 (1994).

12. G. M.J. Schmidt, *Pure Appl. Chem.*, **27**, 647 (1971).
13. G. R. Desiraju, *Crystal Engineering: The Design of Organic Solids*, Elsevier, Amsterdam, 1989.
14. G. R. Desiraju and A. Gavezzotti, *J. Chem. Soc., Chem. Commun.*, 621 (1989).
15. G. R. Desiraju and A. Gavezzotti, *Acta Crystallogr., Part B*, **45**, 473 (1989).
16. C. A. Hunter and J. K. M. Sanders, *J. Am. Chem. Soc.*, **112**, 5525 (1990).
17. S. K. Burley and G. A Petsko, *J. Am. Chem. Soc.*, **108**, 7995 (1986).
18. T. Steiner, *J. Chem. Soc., Chem. Commun.*, 95 (1995).
19. F. H. Quina and D. G. Whitten, *J. Am. Chem. Soc.*, **97**, 1602 (1975).
20. C. Tanford, *The Hydrophobic Effect. Formation of Micelles and Biological Membranes*, Wiley, New York, 1980.
21. G. R. Desiraju and V. Nalini, *J. Mater. Chem.*, **1**, 201 (1991).
22. C. B. Aakeröy and K. R. Seddon, *Chem. Soc. Rev.*, **22**, 397 (1993).
23. L. Pauling, *The Nature of the Chemical Bond and the Structure of Molecules and Crystals—An Introduction to Modern Structural Chemistry*, Cornell University Press, Ithaca, NY, 1939, p. 449.
24. P. W. Atkins, *Physical Chemistry*, Oxford University Press, Oxford, 1987, p. 566.
25. G. C. Pimentel and A. L. McClellan, *The Hydrogen Bond*, Freeman, San Francisco, CA, 1960.
26. T. Steiner and W. Saenger, *J. Am. Chem. Soc.*, **115**, 4540 (1993).
27. G. A. Jeffrey and W. Saenger, *Hydrogen Bonding in Biological Structures*, Springer, Berlin, 1991.
28. R. Taylor and O. Kennard, *Acc. Chem. Res.*, **117**, 320 (1984).
29. D. Braga, F. Grepioni, P. Sabatino and G. R. Desiraju, *Organometallics*, **13**, 3532 (1994).
30. D. Braga, F. Grepioni, K. Biradha, V. R. Pedireddi and G. R. Desiraju, *J. Am. Chem. Soc.*, **117**, 3156 (1995).
31. G. Gilli, F. Bellucci, V. Ferretti and V. Bertolasi, *J. Am. Chem. Soc.*, **111**, 1023 (1989).
32. G. R. Desiraju, *Acc. Chem. Res.*, **24**, 290 (1991).
33. V. R. Pedireddi and G. R. Desiraju, *J. Chem. Soc., Chem. Commun.*, 988 (1992).
34. T. Steiner, *J. Chem. Soc., Chem. Commun.*, 2342 (1994).
35. K. Biradha, C. V. K. Sharma, K. Panneerselvam, L. Shimoni, H. L. Carrell, D. E. Zacharias and G. R. Desiraju, *J. Chem. Soc., Chem. Commun.*, 1473 (1993).
36. D. S. Reddy, B. S. Goud, K. Panneerselvam and G. R. Desiraju, *J. Chem. Soc., Chem. Commun.*, 663 (1993).
37. V. R. Thalladi, K. Panneerselvam, C. J. Carrell, H. L. Carrell and G. R. Desiraju, *J. Chem. Soc., Chem. Commun.*, 340 (1995).
38. J. L. Atwood, F. Hamada, K. D. Robinson, G. W. Orr and L. R. Vincent, *Nature (London)*, **349**, 683 (1991).
39. M. A. Viswamitra, R. Radhakrishnan, J. Bandekar and G. R. Desiraju, *J. Am. Chem. Soc.*, **115**, 4868 (1993).
40. R. Hunter, R. H. Haueisen and A. Irving, *Angew. Chem., Int. Ed. Engl.*, **33**, 566 (1994).
41. S. C. Nyburg and W. Wong-Ng, *Proc. R. Soc. London, Ser. A*, **367**, 29 (1979).
42. D. E. Williams and L.-Y. Hsu, *Acta Crystallogr., Part A*, **41**, 296 (1985).
43. F. H. Allen, B. S. Goud, V. J. Hoy, J. A. K. Howard and G. R. Desiraju *J. Chem. Soc., Chem. Commun.*, 2729 (1994).
44. D. S. Reddy, D. C. Craig, A. D. Rae and G. R. Desiraju, *J. Chem. Soc., Chem. Commun.*, 1737 (1993).
45. K. Xu, M. Ho and R. A. Pascal, *J. Am. Chem. Soc.*, **116**, 105 (1994).

46. M. R. Bryce and L. C. Murphy, *Nature (London)*, **309**, 119 (1984). For an update on the literature concerning crystal engineering of TTF–TCNQ-type materials, see M. D. Ward and M. D. Hollingsworth (eds), *Chem. Mater.*, **6**, 1087 (1994).
47. D. Braga and F. Grepioni, *Acc. Chem. Res.*, **27**, 51 (1994).
48. L. Brammer and D. Zhao, *Organometallics*, **13**, 1545 (1994).
49. S. S. Pathaneni and G. R. Desiraju, *J. Chem. Soc., Dalton Trans.*, 319 (1993).
50. J. M. Thomas, S. E. Morsi and J. P. Desvergne, *Adv. Phys. Org. Chem.*, **15**, 63 (1977).
51. D. E. Williams, *Acta Crystallogr., Part A*, **27**, 452 (1971).
52. G. Filippini and A. Gavezzotti, *Acta Crystallogr, Part B*, **49**, 868 (1993).
53. H. R. Karfunkel and R. Gdanitz, *J. Comput. Chem.*, **13**, 1171 (1992).
54. Cerius^2, BIOSYM/Molecular Simulations, San Diego, CA and Cambridge, U.K., 1995.
55. G. R. Desiraju. *Angew. Chem., Int. Ed. Engl.*, **34**, 2311 (1995).
56. G. R. Desiraju, in *Comprehensive Supramolecular Chemistry*, Vol. 6 (eds D. D. MacNicol, F. Toda and R. Bishop), Elsevier, Oxford, 1996, in press.
57. M. C. Etter, *Acc. Chem. Res.*, **23**, 120 (1990).
58. C. V. K. Sharma, K. Panneerselvam, L. Shimoni, H. Katz, H. L. Carrell and G. R. Desiraju, *Chem. Mater.*, **6**, 1282 (1994).
59. A. I. Kitaigorodskii, *Mixed Crystals*, Springer, Berlin, 1984, p. 214.
60. L. Leiserowitz, *Acta Crystallogr., Part B*, **32**, 775 (1976).
61. M. Bailey and C. J. Brown, *Acta Crystallogr.*, **22**, 387 (1967).
62. F. H. Herbstein, in *Topics in Current Chemistry* (ed. E. Weber), Springer, Berlin, 1987, p. 107.
63. O. Ermer, *J. Am. Chem. Soc.*, **110**, 3747 (1988).
64. J. C. MacDonald and G. M. Whitesides, *Chem. Rev.*, **94**, 2383 (1994).
65. L. Leiserowitz and G. M. J. Schmidt, *J. Chem. Soc.*, 2372 (1969).
66. Y. Ducharme and J. D. Wuest, *J. Org. Chem.*, **53**, 5787 (1988).
67. M. C. Etter and D. A. Adsmond, *J. Chem. Soc., Chem. Commun.*, 589 (1990).
68. J. A. Zerkowski, C. T. Seto and G. M. Whitesides. *J. Am. Chem. Soc.*, **114**, 5473 (1992).
69. J.-M. Lehn, M. Mascal, A. DeCian and J. Fischer, *J. Chem. Soc., Perkin Trans. 2*, 461 (1992).
70. F. G. Tellado, S. J. Geib, S. Goswami and A. D. Hamilton, *J. Am. Chem. Soc.*, **113**, 9265 (1991).
71. C. V. K. Sharma, K. Panneerselvam, T. Pilati and G. R. Desiraju, *J. Chem. Soc., Chem. Commun.*, 832 (1992).
72. J.-M. Lehn, *Angew. Chem., Int. Ed. Engl.*, **27**, 89 (1988).
73. E. Weber, *J. Mol. Graph.*, **7**, 12 (1989).
74. C. A. Hunter, *J. Chem. Soc., Chem. Commun.*, 749 (1991).
75. G. M. Whitesides, J. P. Mathias and C. T. Seto, *Science*, **254**, 1312 (1991).
76. D. Philp and J. F. Stoddart, *Synlett*, 445 (1991).
77. R. Wyler, J. de Mendoza and J. Rebek, *Angew. Chem., Int. Ed. Engl.*, **32**, 1699 (1993).
78. B. Alpha, J.-M. Lehn and G. Mathis, *Angew. Chem., Int. Ed. Engl.*, **26**, 266 (1987).
79. F. Vögtle and E. Weber (eds), *Host–Guest Complex Chemistry. Macrocycles*, Springer, Berlin, 1985.
80. F. Vögtle, *Supramolecular Chemistry*, Wiley, Chichester, 1991.
81. J. L. Atwood, G. A. Koutsantonis and C. L. Ratson, *Nature (London)*, **368**, 229 (1994).
82. I. Csöregh, M. Czugler, E. Weber and J. Ahrendt, *J. Incl. Phenom.*, **8**, 309 (1990).

83. A. D. Hamilton (ed.), *Tetrahedron Symp*, **51**, 343 (1995).
84. E. C. Constable, A. J. Edwards, P. R. Raithby and J. V. Walker, *Angew. Chem., Int. Ed. Engl.*, **32**, 1465 (1993); C. Dietrich-Buchecker, B. Frommberger, I. Lüer, J.-P. Sauvage and F. Vögtle, *Angew. Chem., Int. Ed. Engl.*, **32**, 1434 (1993); F. Vögtle, W. M. Müller, U. Müller, M. Bauer and K. Rissanen, *Angew. Chem., Int. Ed. Engl.*, **32**, 1295 (1993); E. A. Wintner, M. M. Conn and J. Rebek, *Acc. Chem. Res.*, **27**, 198 (1994); P. R. Ashton, D. Philp, N. Spencer, J. F. Stoddart and D. J. Williams, *J. Chem. Soc., Chem. Commun.*, 181 (1994); M. J. Gunter, D. C. R. Hockless, M. R. Johnston, B. W. Skelton and A. H. White, *J. Am. Chem. Soc.*, **116**, 4810 (1994); H. L. Anderson, A. Bashall, K. Henrick, M. McPartlin and J. K. M. Sanders, *Angew. Chem., Int. Ed. Engl.*, **33**, 429 (1994).
85. G. Klebe and F. Diederich, *Philos. Trans. R. Soc. London, Ser. A*, **345**, 37 (1993).
86. G. R. Desiraju, *J. Mol. Struct.*, **374**, 191 (1996).

Chapter 3

Molecular Shape as a Design Criterion in Crystal Engineering

RAYMOND E. DAVIS, JAMES K. WHITESELL, MAN-SHING WONG AND NING-LEH CHANG

The University of Texas at Austin, USA

1. INTRODUCTION

Central to all interpretations of the term 'supramolecular chemistry' is the concept of molecular recognition, which necessarily involves intermolecular interactions. The most orderly result of molecular recognition among organic molecules is the molecular crystal — a supramolecular array *par excellence*; such an array provides the opportunity for detailed study of the geometrical aspects of these intermolecular interactions, principally through the technique of X-ray crystallography. The well-known progress since the 1960s in experimental and computational equipment and methodology has resulted in the knowledge of the structures of tens of thousands of molecular crystals in varying levels of detail. However, rather few crystal structure reports have contained any significant analysis of (indeed, often no reference at all to) the molecular arrangement in the crystal [1], presumably because the principal motivation for many individual crystal structure studies was the elucidation of some molecular feature.

Fortunately, a knowledge of the (usually) available atomic coordinates together with the crystal unit cell and space group symmetry allows one to reconstruct fully the packing arrangement. The ready availability in the Cambridge Structural Database [2] of such data for tens of thousands of

The Crystal as a Supramolecular Entity. Edited by G. R. Desiraju
©1996 John Wiley & Sons Ltd

organic and organometallic crystal structures, together with continually evolving retrieval and analysis software, provides a largely untapped resource that is one focus of many active programs aimed at the elucidation and understanding of the organic solid state. Among the many important reviews in this area, of particular note is the superb recent article with the tantalizing title 'Towards a Grammar of Crystal Packing' by Brock and Dunitz [3]. The interplay between crystal structure studies and synthetic chemistry in the design of self-assembling molecular arrays has been discussed and nicely exemplified by Amabilino and Stoddart [4].

Despite recently invigorated interest in the study of crystal packing, there is yet no general approach to predicting, let alone controlling, molecular orientations in crystals. The rational design and preparation of crystalline and other supramolecular materials for a wide range of applications is a growing field known as crystal engineering (see the excellent book by Desiraju [5]). Development of these design methods has been hampered by insufficient knowledge of those factors that control packing. Whereas considerable attention has been paid to the development of substances with desired molecular properties, the details of assembly of single-crystal materials have until now been left mainly to chance. In the absence of general methods for controlling multimolecular arrays, the best we can do is to try to alter the odds significantly in favor of an array that includes a particular packing feature. Any approach that increases the chance of obtaining a usefully oriented assembly of molecules represents a tool for crystal engineering. Two recent papers that report the use of hydrogen bonding as the key design element in the design of periodic arrays of molecules, one by Whitesides *et al.* and another by Lauher *et al.*, appear, appropriately, in the special issue of *Chemistry of Materials* dedicated to the late Margaret C. Etter in recognition of her pioneering work in this area of crystal engineering [6]. With their extensive references, these articles provide a valuable entry into the literature of this area of crystal engineering.

Optimization of bulk forces requires tailoring of intermolecular orientations in addition to molecular properties [7]. For example, the optimal relative orientation of the polarization of the collinearly propagating waves and of molecular charge transfer axes for second-order nonlinear optics has been described [8], and these properties disappear for rigorously centrosymmetric structures. Methods for crystal engineering, especially of polar crystals [9] because of their unique physical properties (e.g. nonlinear optical effects, piezoelectricity), have received considerable attention from solid-state chemists. With this as partial motivation, we undertook the development of approaches to provide at least a bias away from centrosymmetry, and perhaps eventually more definite control of intermolecular relationships in the assembly of molecular crystals. Several research programs are actively pursuing the use of hydrogen bonding as the key design element for tailoring such molecular arrays [6]. We have chosen to base our approach on less specifically directional forces than hydrogen bonding, focusing instead on the role of molecular shape and approximate molecular symmetry as a possible crystal design tool. This chapter reports our early progress in this effort.

The bias for centrosymmetry is estimated to be greater than 10:1 for crystals of achiral, nonpolar organic molecules [10], and is about 4:1 for all structures (see Table 8 in Brock and Dunitz [3]). We have communicated our statistical study of the influence of molecular dipoles on crystal packing [11]; this study indicated that for molecules that lack strong hydrogen bonding, the magnitudes of the molecular dipoles have no significant correlation with centrosymmetry. We believe that molecular shape represents the dominant factor in determining the high preference for centrosymmetric alignments in most molecular crystals [12]. We suggest that the high propensity for centrosymmetry, whatever its physical origin, can be turned to advantage to provide a desirable alignment of molecules that, by virtue of the presence of groups of nearly identical shapes, possesses quasisymmetry either between pairs of molecules or within a single molecule. This quasisymmetry within a pair of molecules can then allow prediction, and hence tailoring, of desirable orientational aspects within that pair of molecules.

We have developed two approaches, which we discuss below in turn. In the first, two molecules that are otherwise enantiomers differ in some detail of element or group identity but retain their steric similarities; this can lead to crystallization as quasiracemic crystals, in which two separate molecular assemblies are intertwined in such a way that individual molecular features no longer rigorously cancel. In the other approach, the near-symmetry is built into a single molecule in such a way that the shape 'drive' (whatever that means!) for centrosymmetry can be satisfied, while still leaving the crystal not rigorously centrosymmetric in full detail. The idea that groups and molecules of similar shape and size can be substituted intentionally one for another in molecular crystals is not new. The application of quasiracemates for the assignment of absolute configuration dates from the early part of the twentieth century [13]. In a more recent application, 'structural mimicry' was utilized to produce solid solutions of two compounds in which solid-state photoreactivities were altered [14]; in this more recent case, however, enantiomerism (or near-enantiomerism) was not involved, as the objective was not the production of a noncentrosymmetric arrangement.

2. QUASIRACEMATES

An equimolar solution (or molten or gas-phase mixture) of two enantiomers, say (R)-A and (S)-A, can crystallize in one of two main ways.

(1) They can crystallize with the two enantiomers in separate crystals, i.e. some crystals consist entirely of (R)-A and (a presumably equal mass of) other

crystals consist entirely of (*S*)-A; this phenomenon is termed conglomerate crystallization. Each enantiomer in such a mixture can be considered as an impurity in the other, so the melting point (m.p.) diagram of such a mixture is characterized by equal melting points for the two pure crystalline forms and a sharp minimum melting point (eutectic) for the 1:1 molar ratio mixture (Figure 1a). The two crystalline forms are enantiomers of one another, so either pure or mixed in any proportion they would display the same X-ray powder diffraction pattern.

(2) They can crystallize with the two enantiomers present in equal amounts in the same crystal, nearly always with two enantiomeric molecules related by inversion symmetry; such a crystalline phase is termed a true racemate or, more simply, a racemate. In this case, there are three pure crystalline phases possible, resulting in three maxima (one for each pure enantiomeric phase and one for the racemic crystalline phase at a 1:1 molar ratio of enantiomers) and two eutectic points. The melting point of the racemic phase can be either lower (Figure 1b) or higher (Figure 1c) than that of either of the two enantiomeric phases. Thus, a sample that was crystallized from an equimolar mixture of enantiomers and that displays a higher melting point than either of the two enantiomeric phases can be assumed to be a racemic phase; however, one that displays a lower melting point than either of the two enantiomeric phases could result from either conglomerate (Figure 1a) or racemate (Figure 1b) crystallization. Because the packing pattern is different in the racemic crystal compared to either of the two pure enantiomers, its X-ray powder diffraction pattern would be (markedly) different from that of either enantiomeric phase.

Alternatively, they can form a solid solution (a pseudoracemate) in which one enantiomer replaces the other at random in the packing arrangement. This situation, extremely rare in practice, would be characterized by either a maximum or a minimum in the melting point diagram and by the same X-ray powder diffraction pattern over the entire range of concentrations [15].

Thus, melting points (even in the absence of a full melting point phase diagram) and X-ray powder diffraction patterns are useful criteria for assessing the results of such crystallizations; if suitable single crystals are formed, the more elaborate (but far more informative in terms of structural detail) method of single-crystal structure determination provides more definitive confirmation and structural descriptions of phase identities.

In 1899, Centnerszwer [16] observed that (+)-chlorosuccinic acid and (−)-bromosuccinic acid form what was then termed a 'molecular compound'. In fact, the same behavior had been noted by Pasteur in one of his earliest papers on optical activity [17]. In general, one would not expect cocrystallization of two different compounds. A and A′, unless there is a strong, specific interaction (e.g. hydrogen bond formation) between them [18]. However, if A and A′ are very similar in shape (isosteric), then special situations can arise. The like-handed compounds, e.g. (*R*)-A and (*R*)-A′, could

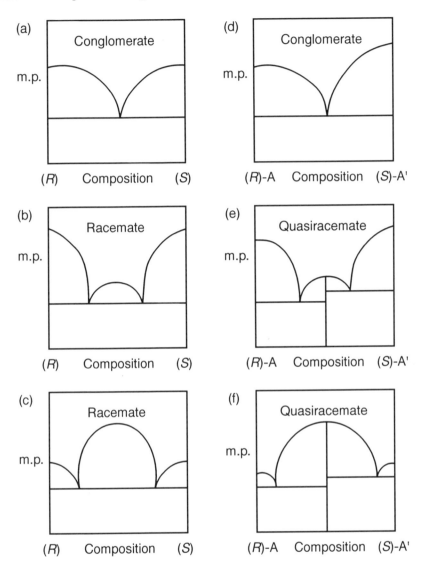

Figure 1 Melting point phase diagrams for (a–c) enantiomeric and (d–f) quasienantiomeric mixtures: (a) conglomerate crystallization by an enantiomeric pair; (b) true racemate formation (racemate melting point lower than that of enantiomeric crystal); (c) true racemate formation (racemate melting point higher than that of enantiomeric crystal); (d) conglomerate crystallization by a quasienantiomeric pair (i.e. separate crystals of (*R*)-A and (*S*)-A′); (e) quasiracemate formation (racemate melting point lower than that of either enantiomeric crystal); and (f) quasiracemate formation (racemate melting point higher than that of either enantiomeric crystal)

form solid solutions through the range of concentrations. Crystallization of an equimolar mixture of opposite near-enantiomers, say (*R*)-A and (*S*)-A', has possible outcomes analogous to those described above for true enantiomeric mixtures. They could form separate crystals (conglomerate crystallization), as in Figure 1(d), or they could form quasiracemic crystals with either lower (Figure 1e) or higher (Figure 1f) melting points than those of one or both of the pure substances. These melting point diagrams are similar to those for true enantiomers racemates (Figures 1a–1c), but they are no longer symmetrical. Conglomerate crystallization would lead to an X-ray powder pattern that is a composite of those for the two components, whereas quasiracemic crystals would give a pattern that does not match either pure enantiomer. A quasiracemic packing pattern that approximates either of the true racemates would lead to a corresponding similarity in their powder patterns. Once again, melting point and X-ray powder diffraction measurements can serve as criteria for assessing this behavior. One way of describing a quasiracemate is as a molecular crystal derived from true racemate by a not too extensive change in the structure of one of the two enantiomeric components, e.g. (*R*)-A and (*S*)-A', where A and A' are sterically similar molecules.

Quasiracemate formation is discussed extensively in the classic book by Jacques *et al.* [10, p. 100]. This cocrystallization phenomenon was utilized in the powerful quasiracemate method for establishing the absolute configurations of optically active substances. This method and its application have been described by Fredga [13], one of its leading practitioners. In the process of establishing their utility in the determination of absolute configuration, Fredga and his collaborators determined and published hundreds of melting point phase diagrams for such systems. Nevertheless, full crystal structural data on quasiracemates are sparse in the literature [19].

We should not leave this discussion, however, without issuing a caveat regarding the use of X-ray powder patterns as the sole criterion for assessing conglomerate/racemate/quasiracemate crystallization. Among the thousands of reported crystal structures, a number of enantiomer–racemate pairs have been reported [20]. Of these, a small but significant number have structures in which there is a great resemblance between the packing of the racemate and the corresponding enantiomers [21]. Interestingly, nearly all of these exhibit significant intermolecular hydrogen bonding in the molecular crystal. A pronounced similarity between enantiomeric and racemic crystals would lead to quite similar X-ray powder patterns and a possibly erroneous conclusion regarding conglomerate versus racemate crystallization. This ambiguity could be resolved in at least two ways. One is the determination of melting point data for mixtures other than for pure enantiomers and their 1:1 molar ratio, i.e. enough other mixtures to indicate whether the melting point of the 1:1 molar ratio is a maximum (racemate) or a minimum (conglomerate). This could involve synthesis and purification of both enantiomers of each compound under investigation, or it could be accomplished by mixing one available enantiomer in varying proportions with the racemic (or quasiracemic) equimolar mixture. Another, more definitive, experiment would be the complete single-crystal structure determination, provided, of course, that suitable single crystals were available. The lack of either of these definitive measurements leaves us with less than full satisfaction regarding certain of the results reported below. However, we believe that our success in the design and production of crystals with predictable relationships represents a step that tips the statistical scales of crystal engineering.

3. DESIGN OF QUASIRACEMIC CRYSTALS

As we have already indicated, our goal is to turn the strong drive for centrosymmetry to advantage in designing noncentrosymmetric crystals. Let us clarify this seemingly paradoxical goal. In a true racemate, molecules of the two enantiomers are related pairwise by an inversion center. Thus, any vectorial property in one molecule is exactly balanced by an equal and opposite vectorial property in the centrosymmetric mate, as illustrated in cartoon fashion in Figure 2(a). Our design approach is to introduce a subtle change in the structure of one type of molecule by substituting an isosteric group while altering some vectorial property of the molecule (say dipole moment or hyperpolarizability); if molecular shape is (and if it remains) a sufficiently dominant factor in determining crystal packing, the resulting arrangement would be quasicentrosymmetric by shape but perhaps not so in regard to other properties (Figure 2b).

In the course of developing quasiracemate formation as a tool for assessing chirality [13], Fredga used many isosteric pairs of chemical groups, some of which are noted in Table 1. Our initial work in designing pairs of molecules for quasiracemate formation used some of these isosteric sets (S, O, CH_2). However, our interest in designing types of materials that might exhibit nonlinear optical properties led us eventually to incorporate isosteric groups whose electronic properties such as their electron-donating/withdrawing nature differed more markedly.

Furthermore, much of the systematic study of quasiracemate formation by Fredga and by others involved species that were strongly linked by hydrogen bonds. Our desire to explore, and perhaps utilize, the role of molecular shape in crystal packing and crystal engineering has led us to avoid compounds with the potential for hydrogen bonding in favor of molecules that would form crystals that are (to borrow Desiraju's chapter title) 'based mostly on van der Waals forces' [5].

This work has focused on two classes of hyperpolarizable compounds, benzoates and sulfoxides, in which the carboxylate and sulfinyl functionalities

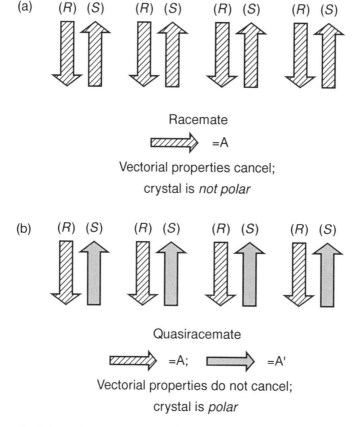

Figure 2   Schematic representation of racemic (a) and quasiracemic (b) arrangements

are somewhat electron withdrawing in nature (Figure 3). The skeletons of these molecules are composed of either an aromatic core or an extended conjugated stilbene framework, with an electron-donating group at the other end. The sulfoxide group is intrinsically chiral but the benzoate group is not, so for that series of compounds various groups were introduced into the molecule as a chiral handle; of those we attempted to use — 2-butyl, menthyl and *trans*-2-phenylcyclohexyl — only the benzoates that included the last of these showed a general propensity to crystallize.

3.1 Benzoates

It was suggested that suitable modification of an enantiomeric pair that form a stable true racemate could lead to a high tendency for the formation of a quasiracemate [14a]. In order to explore the possibility of using this concept in

Table 1 Some group replacements that have been used in quasiracemate formation

S	Se		
S	O	CH$_2$	NH
SH	Me		

crystal engineering, both isomers of each of the benzoates (**1a**), (**1b**) and (**1c**) were synthesized. Crystallization experiments were carried out with enantiomeric and with racemic solutions of each benzoate to examine whether or not these compounds have the tendency to form true racemate crystals. Melting points and X-ray powder patterns were used to characterize the resulting crystallization tendencies, with results as shown in Table 2(a). Powder patterns for the six crystalline samples appear in Figure 4. The virtual coincidence of the patterns in Figure 4 suggests that the crystallization from the racemic solution of (**1a**) did not form true racemate crystals, but rather a conglomerate mixture of (+)-crystals and (−)-crystals. The same result was found for (**1c**), from which we concluded that these compounds had not

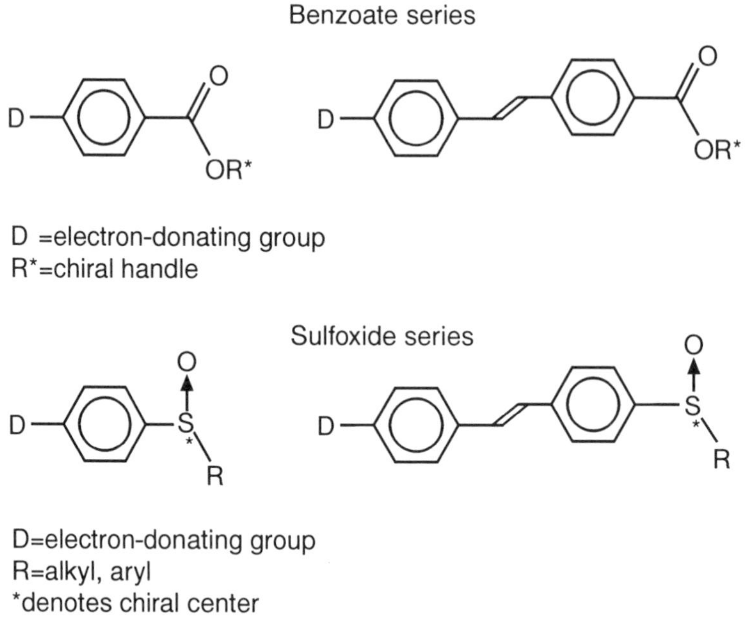

Benzoate series

D =electron-donating group
R*=chiral handle

Sulfoxide series

D=electron-donating group
R=alkyl, aryl
*denotes chiral center

Figure 3 General structures of the benzoate and sulfoxide series of compounds used in this study

formed racemic crystals in these crystallizations. Only for (**1b**) did comparisons of the powder patterns and the crystal morphologies indicate the formation of a true racemate. This evidence is not fully conclusive in that the melting point of the putative racemate is still lower than that of the enantiomer (see Figure 1b); it is still possible, though unlikely, that the presence of each enantiomer could have induced the formation by the other of a new polymorphic enantiomeric crystal form with a different melting point. The sharpness of the melting point of crystals from the mixture (+)-(**1b**)–(−)-(**1b**) mitigates against solid solution formation, but the melting point versus concentration data that would settle this question were not available.

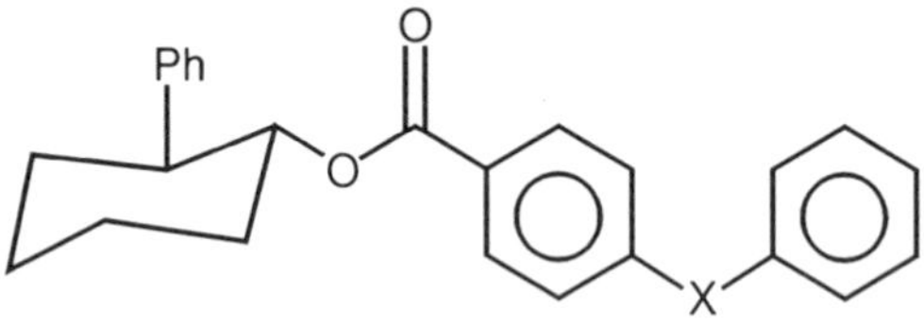

(**1a**) X=S; (**1b**) X=O; (**1c**) X=CH$_2$

Despite the evidence that two of these three benzoates failed to form true racemates, we carried out crystallization experiments to investigate the possible

Table 2(a) Crystal morphologies, melting points and comparisons of powder patterns for benzoates (**1a**), (**1b**) and (**1c**)

Crystal from	Morphology	Melting point (°C)	Powder patterns
(+)-(**1a**)	Rods	142	Same
(+)- and (−)-(**1a**)	Rods	121	(Figures 4a and 4b)
(+)-(**1b**)	Large plates	123	Different
(+)- and (−)-(**1b**)	Needles	116	(Figures 4c and 4d)
(+)-(**1c**)	Small rods	119	Same
(+)- and (−)-(**1c**)	Small rods	91	(Figures 4e and 4f)

formation of quasiracemates in this series. In each of these experiments, equimolar solutions of opposite enantiomers of pairs of these benzoates, e.g. (−)-(**1a**) and (+)-(**1b**), were prepared and crystallized. Again, crystal morphologies and X-ray powder patterns (Figure 5) were used to assess the outcome of the crystallization experiments. The results, summarized in Table 2(b), suggest only conglomerate crystallization, with no evidence of quasiracemate formation. Numerous attempts to produce crystals of suitable quality of (**1a**), (**1b**) and (**1c**) either as enantiomers or racemates, or in combination as quasiracemates, for single-crystal structural studies were unsuccessful, yielding only very thin needles or plates or poorly ordered rods.

Results with other derivatives of *trans*-2-phenylcyclohexyl benzoate esters were also equivocal owing to the tendency of many of these samples to form oils or very poorly formed crystalline samples, either from solution or from the melt.

3.2 Sulfoxides

Progress on the benzoates having been less than encouraging, we turned to another series of compounds based on diaryl sulfoxides, a chiral group for which we already had developed a considerable synthetic methodology [22]. This grouping has the advantage that the sulfoxide, when properly substituted, is intrinsically chiral, so that no separate chiral handle needs to be attached, but also the disadvantage that enantiomeric conversion at high temperatures might preclude both sample preparation by melting–solidification and characterization of phase behavior from melting point diagrams. The first compounds synthesized were the enantiomers and racemates of (**2a**) and (**2b**). The close similarity in peak positions and intensities in the two X-ray powder patterns in Figures 6(a) and 6(b) suggests that (**2a**) undergoes conglomerate crystallization. The patterns in Figures 6(c) and 6(d) show strong similarities in

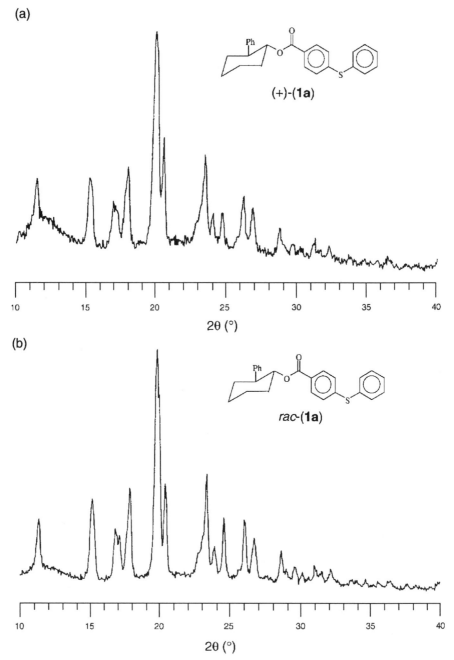

Figure 4   X-ray powder patterns of benzoates (**1**): (a) crystals from a solution of (+)-(**1a**); (b) crystals from a racemic solution of (**1a**)

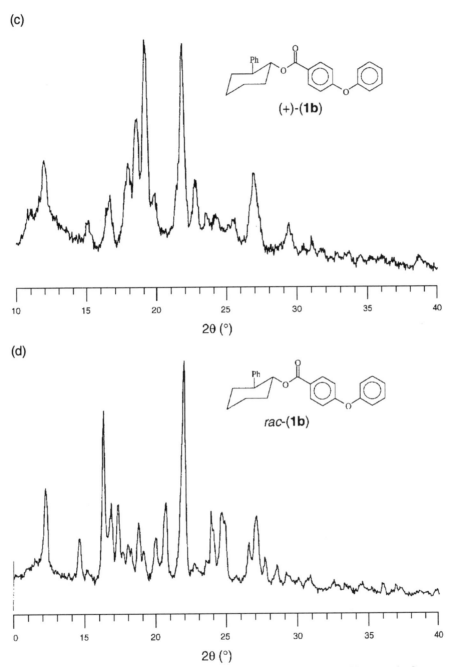

Figure 4 (*continued*) (c) crystals from a solution of (+)-(**1b**); (d) crystals from a racemic solution of (**1b**)

(e)

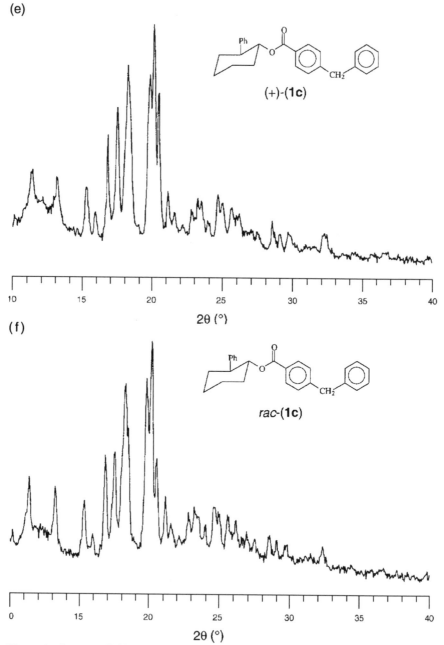

(+)-(**1c**)

2θ (°)

(f)

rac-(**1c**)

2θ (°)

Figure 4 (*continued*) (e) crystals from a solution of (+)-(**1c**); (f) crystals from a racemic solution of (**1c**). These and the following X-ray powder patterns were recorded with CuKα radiation

Table 2(b) Crystal morphologies and powder patterns of attempted quasiracemates involving (1a), (1b) and (1c)

Crystals from	Morphology	Powder pattern
(+)-(1b) and (−)-(1a)	Combination of A1 and B1	Combination of enantiomeric 1a and 1b (Figure 5a)
(+)-(1a) and (−)-(1c)	Combination of A1 and C1	Combination of enantiomeric 1a and 1c (Figure 5b)
(+)-(1b) and (−)-(1c)	Combination of B1 and C1	Combination of enantiomeric 1b and 1c (Figure 5c)

peak positions, but more variations in peak intensities; this could be attributed either to conglomerate crystallization with some unexplained experimental effects resulting from preferred orientation of sample grains, or to a strong resemblance between enantiomer and racemic packing arrangements [23b]. Attempted formation of quasiracemate crystals from an equimolar mixture of (+)-(2a) and (−)-(2b) gave a crystalline sample whose powder pattern (Figure 6e) is quite similar to those of the two enantiomers (Figures 6a and 6c), inconsistent with the formation of quasiracemates crystals.

(2a) X=CH$_2$; (2b) X=S

The next chemical modification was an attempt to 'bury' the isosteric modification more deeply within the molecule in order to lessen its effect on overall molecular shape, viz. (3a) and (3b) with *n*-heptyl groups in place of methyl. Comparison of the powder patterns in Figures 7(a)–7(d) clearly indicates four different crystal structures for enantiomers and racemates of (3a)

(a)

(b)

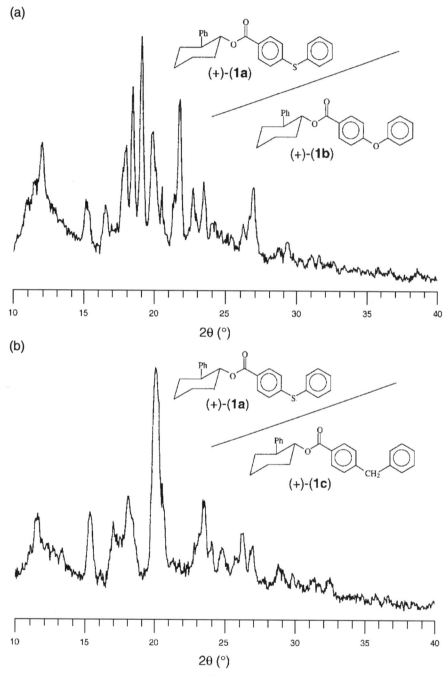

Figure 5

(c)

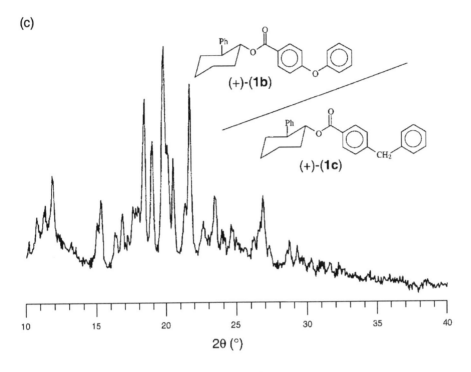

2θ (°)

Figure 5 (*continued*) X-ray powder patterns of crystals from equimolar solutions of benzoates (**1**): (a) (−)-(**1a**) and (+)-(**1b**); (b) (+)-(**1a**) and (−)-(**1c**) and (c) (+)-(**1b**) and (−)-(**1c**)

and (**3b**), leading to the strong likelihood that these two compounds form true racemates. Crystallization from an equimolar mixture of (*R*)-(**3a**) and (*S*)-(**3b**) in CH$_2$Cl$_2$ gave a crystalline sample (m.p. 94–95°C) that exhibited the X-ray powder pattern shown in Figure 7(e). The close similarity between this pattern and the one shown in Figure 7(d) indicates the formation of a quasiracemic phase with a structure very similar to that of *rac*-(**3b**), i.e. the quasiracemate can be described as being derived from the structure of *rac*-(**3b**) by replacing each molecule of (*R*)-(**3b**) with a molecule of (*R*)-(**3a**). The resulting structure cannot be strictly centrosymmetric as it is constituted entirely of chiral molecules that are not enantiomers, but we assume that it is quasicentro-symmetric in terms of shape relationships, resembling the presumably centrosymmetric structure of *rac*-(**3b**).

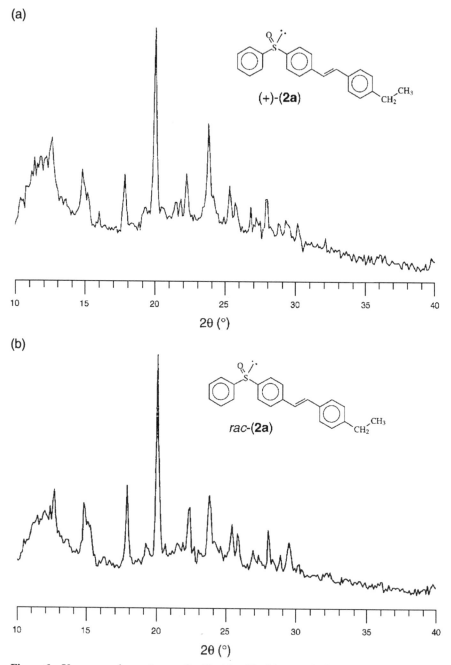

Figure 6 X-ray powder patterns of sulfoxides (**2**): (a) crystals from a solution of (+)-(**2a**); (b) crystals from a racemic solution of (**2a**)

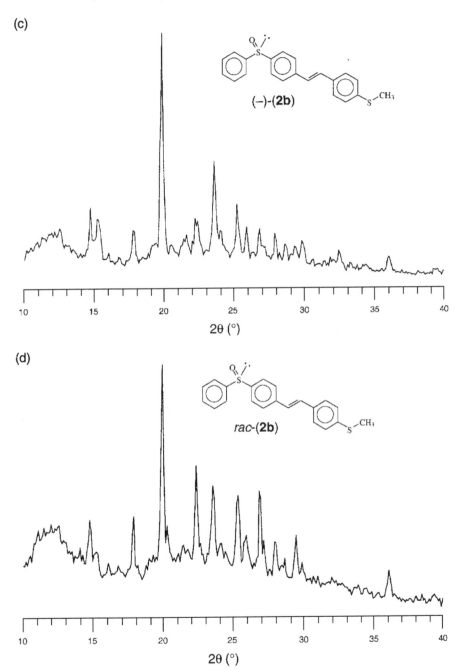

Figure 6 (*continued*) (c) crystals from a solution of (−)-(**2b**); (d) crystals from a racemic solution of (**2b**)

(e)

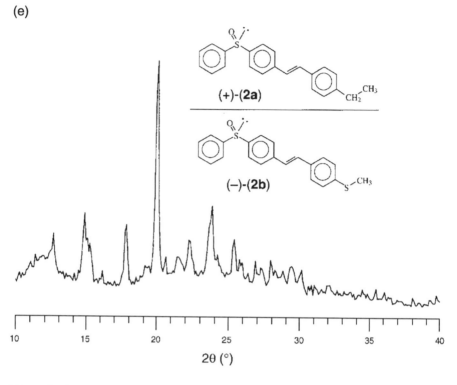

Figure 6 (*continued*) (e) crystals from an equimolar mixture of (+)-(**2a**) and (−)-(**2b**)

(**3a**) X=O; (**3b**) X=S

The same equimolar mixture of (*R*)-(**3a**) and (*S*)-(**3b**) when crystallized from ethyl acetate also gave a crystalline sample, but one whose X-ray pattern (Figure 7f) did not correspond to any of the four in Figures 7(a)–(d), or to a mixture of those patterns. This phase could correspond to a polymorphic

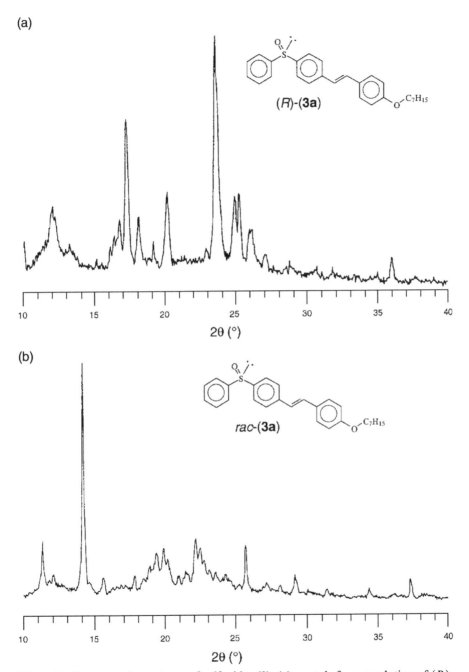

Figure 7 X-ray powder patterns of sulfoxides (3): (a) crystals from a solution of (*R*)-(**3a**); (b) crystals from a racemic solution of (**3a**)

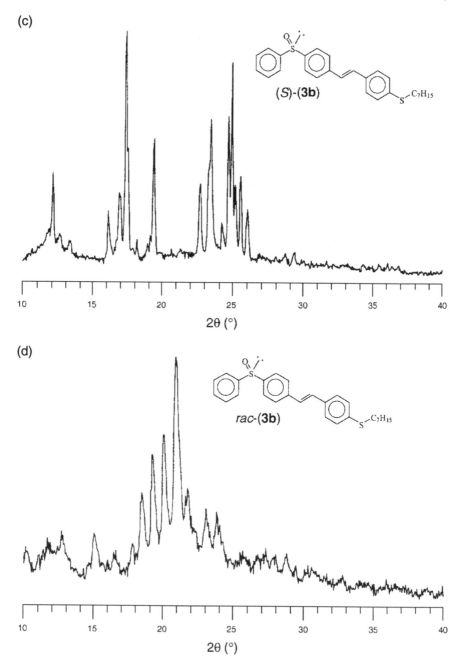

Figure 7 (*continued*) (c) crystals from a solution of (*S*)-(**3b**); (d) crystals from a racemic solution of (**3b**)

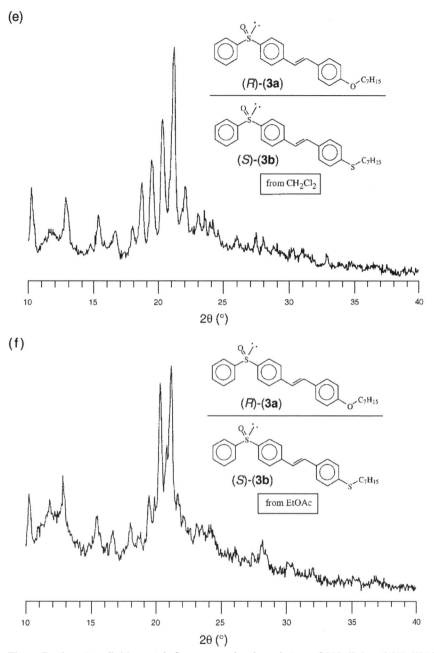

Figure 7 (*continued*) (e) crystals from an equimolar mixture of (*R*)-(**3a**) and (*S*)-(**3b**) in methylene chloride; (f) crystals from an equimolar mixture of (*R*)-(**3a**) and (*S*)-(**3b**) in ethyl acetate

quasiracemic crystalline phase different from that obtained from CH_2Cl_2, or it could represent conglomerate crystallization of two enantiomeric phases different from those represented in Figures 7(a) and 7(c). Regrettably, none of the enantiomeric, racemic or quasiracemic crystalline phases obtained from this system were of suitable quality for single-crystal analysis.

All of the isosteric substitutions described so far have involved substitutions of groups 'inside' the molecule, i.e. atoms or small groups of atoms that bridge two relatively larger portions of the molecule. We have been more successful with isosteric substitutions of pendant groups, provided those groups match well enough both in volume and in conformational relationship to the rest of the molecule. Table 3 gives van der Waals volumes and surface areas for substituents that might be considered as potential isosteric groups [23]. The pair isopropenyl and dimethylamino, (4a) and (4b), are especially well matched in several important respects: they are quite similar in volume and surface area; they would each be expected to adopt a conformation nearly coplanar with an aromatic ring; and they have quite different electron-donating/withdrawing properties, necessary for nonlinear optical target systems. The remaining systems discussed here utilize these two groups as isosteric substituents on aromatic rings.

(4a)   (4b)

Crystalline samples of enantiomeric and racemic sulfoxides (5a) and (5b) gave the melting points reported in Table 4; the observation for both sulfoxides that the racemate melted at a higher temperature than the corresponding enantiomer indicated the existence of true racemates. Crystallization of an equimolar mixture of (S)-(5a) and (R)-(5b) gave crystals with an even higher melting point, indicating that quasiracemate crystals had formed. Indeed, both racemic sulfoxides and the putative quasiracemate formed crystals of excellent quality for single-crystal X-ray structure determination [24]. Some pertinent details of the crystal structure studies appear in Table 5. The two racemate

Table 3 van der Waals' volumes and surface area calculations

Name	Formula	Volume ($Å^3$)	Area ($Å^2$)
Hydrogen	—H	2.3	6.7
Fluoro	—F	5.3	11.6
Methyl	—CH_3	22.4	33.3
Ethyl	—CH_2CH_3	39.4	56.3
Isopropyl	—CH(CH₃)CH₃	54.9	72.8
Isopropenyl	(CH₃)C=CH₂	50.4	64.1
Dimethylamino	—N(CH₃)CH₃	50.5	66.8
Nitro	—NO_2	23.2	36.1

crystals are each centrosymmetric, space group $P2_1/c$, with very similar metrical constants for a cell containing four molecules. The quasiracemate unit cell is essentially the same size and shape, with a volume intermediate between those of the two racemates; this unit cell contains two formula units, but now each formula unit consists of two molecules, (*S*)-(**5a**) and (*R*)-(**5b**).

(**5a**) R= ... ; (**5b**) R= ...

Table 4 Crystal morphologies and melting points of sulfoxides (**5a**) and (**5b**)

Crystals from	Morphology	Melting point (°C)
(*S*)-(**2a**)	Fine powdery residue	67
(*R*)- and (*S*)-(**2a**)	Rods	86
(*R*)-(**2b**)	Fine powdery residue	75
(*R*)- and (*S*)-(**2b**)	Rods	87
(*S*)-(**2a**) and (*R*)-(**2b**)	Rods	103–104

Table 5 Crystal data for *rac*-(**5a**), *rac*-(**5b**) and the quasiracemate (*S*)-(**5a**)–(*R*)-(**5b**)

	rac-(**5a**)	*rac*-(**5b**)	(*S*)-(**5a**)–(*R*)-(**5b**)
Formula	*rac*-$C_{15}H_{14}ONS$	*rac*-$C_{14}H_{15}ONS$	(*S*)-$C_{15}H_{14}ONS$ and (*R*)-$C_{14}H_{15}ONS$
Temperature (K)	208	208	193
Space group	$P2_1/c$	$P2_1/c$	$P2_1$
a (Å)	14.300 5 (12)	14.182 2 (27)	14.224 9 (44)
b (Å)	7.579 0 (7)	7.659 8 (17)	7.636 5 (26)
c (Å)	11.587 1	11.520 1 (24)	11.544 6 (36)
$\beta(°)$	101.063 (8)	101.114 (16)	101.159 (24)
V (Å³)	1232.51 (2 0)	1227.97 (4 3)	1230.36 (6 7)
Z	4	4	2 (i.e. four molecules per cell)

As shown in Figure 8, the packing arrangements in these three structures are virtually identical. Each racemate unit cell (Figures 8a and 8b) has two centrosymmetrically related pairs of molecules, designated (A,B) and (A′,B′), with these two pairs related by a twofold screw axis, e.g. A and A′. In the quasiracemate, two sulfoxide molecules of opposite chirality, one with the isopropenyl group and the other with the dimethylamino group, form a pair related by a pseudoinversion center. Two such pairs related by a screw axis make up the unit cell. The symmetry thus approximates $P2_1/c$, but is actually $P2_1$. Because the two molecules in the quasiracemic pair have different substituents on the phenyl ring, molecular properties such as the dipole moment or molecular polarizability need not cancel by symmetry, as they must in the true racemate. In fact, the charge transfer (polarizability) direction lies along the $N \rightarrow S$ direction of (**5b**). Fortuitously, this direction lies almost perpendicular to the 2_1 screw axis, which leads to an antiparallel arrangement of this charge transfer property between the two (*R*)-(**5b**) molecules in the unit cell. This unfortunate result points up important considerations associated with packing by the statistically likely pseudo-$P2_1/c$ symmetry (true $P2_1$

(a)

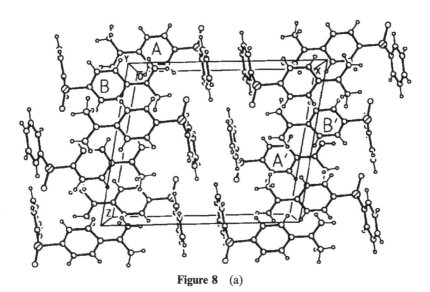

rac-(**5a**)

Figure 8 (a)

symmetry): any vectorial property parallel to the screw axis is reinforced by its screw axis relative (the most favorable case); and any vectorial property perpendicular to the screw axis is canceled by its screw axis relative (the least favorable case). The design features described here are based on quasicentrosymmetry; thus, they specifically address the relative relationships within a quasicentrosymmetric pair of molecules. They do not address the relationship between pairs of molecules where the effects of a second quasicentrosymmetric pair can cancel the effects of the first, as occurs in quasiracemic (**5a,5b**). (In practice, this may be an important consideration, as 38% of the 31 770 organic crystals studied by Brock and Dunitz [3] crystallized in space group $P2_1/c$, which has four generally equivalent positions.)

To test this strategy further, we prepared and cocrystallized another pair of sulfoxides, (*S*)-(**6a**) and (*R*)-(**6b**). These crystals were again suitable for single-crystal analysis. Crystal data appear in Table 6 and the packing diagram is

(b)

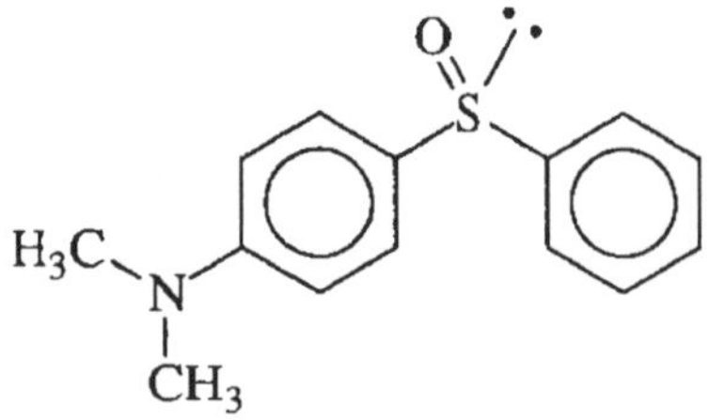

rac-(**5b**)

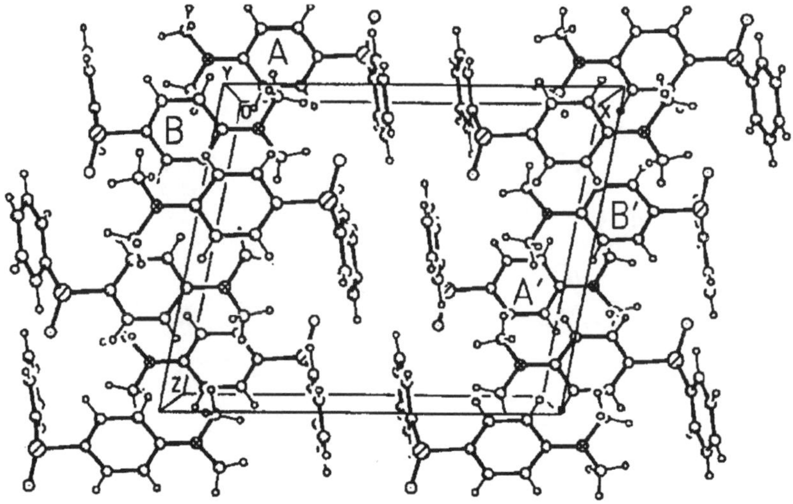

Figure 8 (*continued*)   (b)

shown in Figure 9. A strong similarity to the packing arrangements in Figure 8 is evident, with four molecules per unit cell in space group $P2_1$, approximately $P2_1/c$. This crystal structure establishes another successful production and analysis of a designed quasiracemic system.

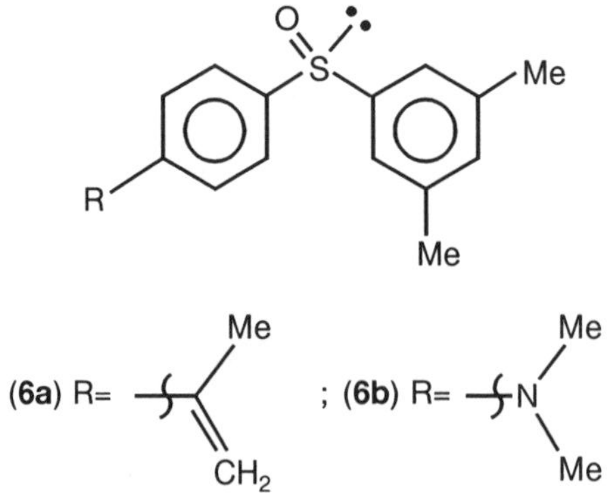

(**6a**) R=   ; (**6b**) R=

(c)

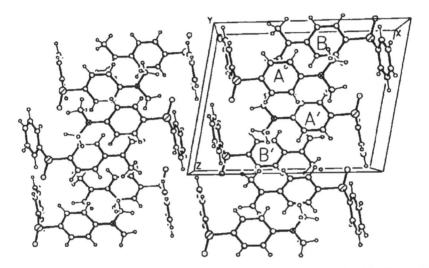

(S)-(5a) | (R)-(5b)

Figure 8 (*continued*)   Packing arrangements in crystals of sulfoxides (**5**): (a) *rac*-(**5a**); (b) *rac*-(**5b**); and (c) quasiracemic (*S*)-(**5a**)–(*R*)-(**5b**)

We have demonstrated that we can design pairs of chiral molecules with very similar size and shape but different donor or acceptor properties so that such pairs of molecules can cocrystallize to form a quasiracemic motif in which dipole moments or other physical properties need not cancel. This provides a rational strategy to design molecular crystals for practical applications. These successes provide further indication that molecular size and shape are controllable factors that direct how molecules arrange themselves in the

Table 6   Crystal data for the quasiracemate of (**6a**) and (**6b**)

	(*S*)-(**6a**)–(*R*)-(**6b**)
Formula	(*S*)-$C_{17}H_{18}ONS$ and (*R*)-$C_{17}H_{18}ONS$
Temperature (K)	193
Space group	$P2_1$
a (Å)	15.874 (2)
b (Å)	8.3514 (12)
c (Å)	11.2289 (13)
β (°)	97.293 (10)
V (Å³)	1476.6 (3)
Z	2
	(i.e. four molecules per cell)

crystal structure, without employing strongly directional intermolecular forces such as hydrogen bonding. The results discussed here indicate that chiral aryl sulfoxides constitute a viable series of compounds for testing our design ideas. Finally, they demonstrate the utility of the isopropenyl and dimethylamino groups as suitably interchangeable isosteric groups that possess substantially different electron-donating/withdrawing properties.

4.   SHAPE MIMICRY

The quasiracemate approach that we have described for producing polar, noncentrosymmetric crystals has some operational disadvantages: it requires the synthesis of two separate enantiomeric compounds; it requires cocrystallization of these two different compounds; and it provides little control over the relative orientations of effects due to the two isosteric groupings. To address these difficulties, we suggest that the high propensity for centrosymmetry can be turned to advantage to provide a desirable alignment of identical molecules that have internal quasisymmetry by virtue of the presence of isosteric groups within the same chiral molecule.

The rationale of this design method is shown in cartoon fashion in Figure 10. Each molecule depicted is chiral by virtue of the presence of four different groups (E, E', C and D) arranged about a tetrahedral center. In the centrosymmetric packing of such a unit with its enantiomer (a true racemate, Figure 10a), there is an antiparallel arrangement of all like vectors (e.g. X → D) between pairs of molecules. However, a single enantiomer cannot crystallize in such a centrosymmetric arrangement; yet to the extent that the groups E and E' are isosteric, a packing arrangement with the vector X → E of one molecule antiparallel to the X → E' vector of the other mimics centrosymmetry (Figure 10b). If the properties of E and E' differ in some way (e.g. polarizability), then

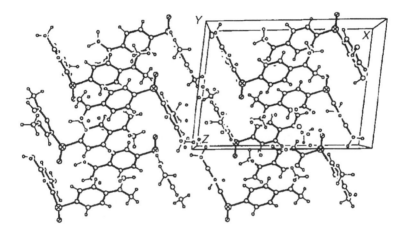

Figure 9 Packing arrangement in the quasiracemate crystal (*S*)-(**6a**)–(*R*)-(**6b**)

this quasicentrosymmetric packing can result in net additivity, instead of cancellation, of molecular properties. Indeed, in this quasisymmetric molecular packing, the spatial relationship of the vectorial properties of molecules so related is predictable. Within a quasicentrosymmetrically related pair, the vector X → E in one molecule is antiparallel to the X → E′ vector of the other (and vice versa); thus, the angle (β) subtended between the X → E vectors of the pair is equal to 180°-α, where α is the intramollecular E–X–E′ angle (Figure 11). For example, in the scenario shown in Figure 10 with ideal tetrahedral angles,

(a) 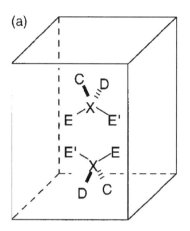   (b)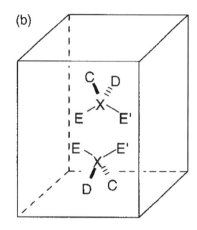

Figure 10 Schematic representation of the possible centrosymmetric packing for the racemate (a) and the quasicentrosymmetric packing for a single enantiomer (b) of a quasisymmetric chiral molecule

the angle β between the X–E bonds in the quasicentrosymmetric pair would be $180° - 109° = 71°$.

In the remainder of this chapter, we describe our attempts to implement this design strategy, which we have termed 'shape mimicry'. The sulfoxide (7) was designed to have internal quasisymmetry, with the isopropenyl and dimethylamino groups that we have already described serving as isosteric substituents. The hope was that (7) would mimic the shape of its own enantiomer by substituting one of these groups for the other, providing a packing arrangement with quasicentrosymmetry. In such a packing motif, pairs of molecules would be oriented about an approximate center of

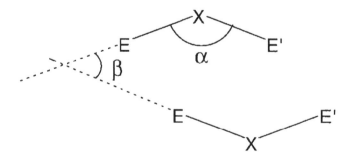

Figure 11 Vectorial relations in a quasicentrosymmetrically arranged pair of molecules

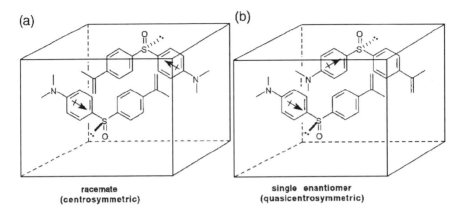

Figure 12 Schematic representation of the possible centrosymmetric packing for the racemate (a) and quasicentrosymmetric packing for a single enantiomer (b) of the nearly symmetric sulfoxide (**7**)

symmetry, as illustrated schematically in Figure 12. We note that this quasicentrosymmetric arrangement would not require any disorder in the structure. Indeed, the (*S*)-(**7**) enantiomer did form molecular crystals with approximate $P2_1/c$ symmetry, true space group $P2_1$, with quasicentrosymmetrically related pairs as shown in Figure 13 [25]. The packing arrangement (Figure 14) shows that the cell contains two such pairs, with the two molecules A and B related to each other by approximate centrosymmetry, as are A' and B'. Inverting the atomic positions of molecule A through the point (0.500,0.500,0.250) gives positions that range from 0.015 to 0.095 Å from the refined atomic positions of molecule B for atoms not included in the isopropenyl or dimethylamino groups and within 0.204 Å even for those atoms in the isopropenyl double bond; this quantifies the close approximation to centrosymmetry. The two molecular pairs (A,B) and (A',B') are related to one another by the screw axis symmetry operator that survives in this chiral crystal. The vectors N → S for B and B' are coincidentally nearly perpendicular (89.3°) to the screw axis, and are thus virtually antiparallel; however, those for molecules A and A' are inclined to the screw axis at 44.9°, so they combine to provide a net polar direction for the crystal.

We then obtained suitable crystal of racemic (**7**) and determined its single-crystal structure. Crystal data and a packing diagram for this structure appear in Table 7 and Figure 15(a), respectively. These crystals are also monoclinic, $P2_1/c$, but with eight molecules in a unit cell with twice the volume of that of (*S*)-(**7**). The general packing arrangements in these two structures are quite

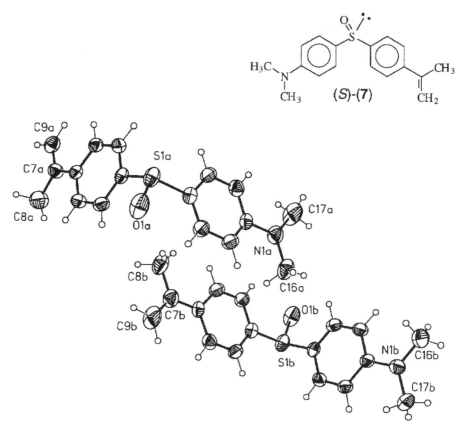

Figure 13 Two quasicentrosymmetrically related molecules in the single-crystal structure of (*S*)-(**7**). Thermal ellipsoids are shown at the 50% equiprobability level

Figure 14 (*opposite*) The four molecules per unit cell in the structure of (*S*)-(**7**). For clarity, N and S are shown as solid spheres. Molecules A and B are related by an approximate inversion center, as are A′ and B′; molecules A and A′ are related by the twofold screw axis of space group $P2^1$, as are B and B′; molecules A and B′ are approximately related by the pseudo-c-glide plane of space group $P2^1/c$, as are A′ and B. (a) Oblique view. (b) View along the x-axis. (c) View along the x-axis, with polarizability vectors N → S shown as arrows. Note that the N → S vectors of A and A′ have a net component parallel to the $+y$-axis, whereas the vectors of B and B′ are nearly antiparallel

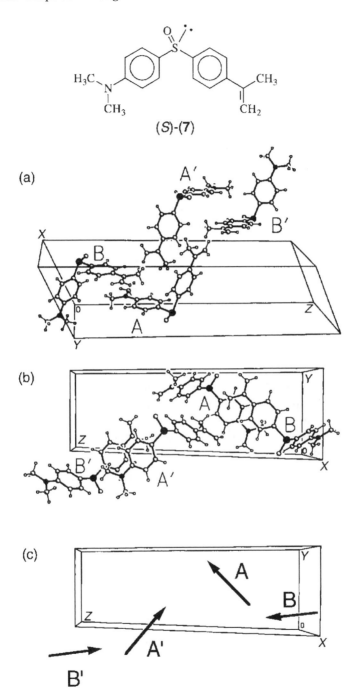

(*S*)-(**7**)

(a)

(b)

(c)

similar, except that the racemate crystal contains pairs of molecules approximately related by a noncrystallographic translation of approximately (0,0.05,0.25). Comparison of the powder patterns of (*S*)-(**7**) and *rac*-(**7**) shown in Figure 16 also emphasizes the general similarity of the two packing arrangements of this enantiomer–racemate pair. The (so far hypothetical) centrosymmetric racemate crystal structure that is mimicked by the quasicentrosymmetric structure of (*S*)-(**7**) is thus not the same structure we observe for the particular crystals obtained for *rac*-(**7**), but rather is a subset of that racemate; this does not detract from the prior success of the design of quasicentrosymmetric (*S*)-(**7**) crystals.

(**7**)

In the *rac*-(**7**) crystal structure, the sulfone oxygen and lone pair are disordered; interestingly, no other sulfoxide that we have studied [not even the shape mimicry enantiomer (*S*)-(**7**)] exhibits such disorder. The near-identity of this packing arrangement with that of the corresponding sulfone (**8**) (an achiral molecule) is obvious from a comparison of the crystal data in Table 7 and the packing diagrams in Figure 15.

Table 7 Crystal data for sulfoxides (*S*)-(**7**) and *rac*-(**7**) and achiral sulfone (**8**)

	(*S*)-(**7**)	*rac*-(**7**)	(**8**)
Formula	(*S*)-$C_{17}H_{19}ONS$	*rac*-$C_{17}H_{19}ONS$	$C_{17}H_{19}O_2NS$
Temperature (K)	193	293	193
Space group	$P2_1$	$P2_1/c$	$P2_1/c$
a (Å)	7.6716 (13)	16.517 (3)	16.755 (6)
b (Å)	8.250 (2)	8.3023 (13)	8.097 (6)
c (Å)	26.113 (5)	23.241 (5)	23.240 (5)
β (°)	114.716 (13)	103.49 (2)	104.74 (2)
V (Å^3)	1501.3 (5)	3099.2 (10)	3049.2 (10)
Z	4	8	8

(a)

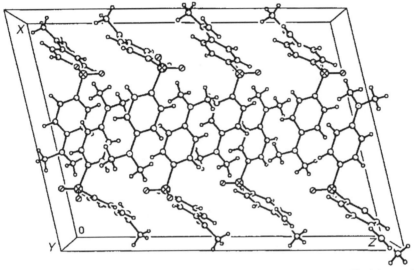

rac-(**7**)

Figure 15 (a) Packing arrangement in the crystal structure of the sulfoxide *rac*-(**7**)

To test our hypothesis further that the packing arrangement for (*S*)-(**7**) was directed by the molecular quasisymmetry, we prepared, crystallized and determined single-crystal structures (Table 8) of two additional enantiomeric sulfoxides, (**9a**) and (**9b**) [25]. In (**9a**) the isopropenyl group has been replaced by isopropyl, a somewhat more bulky group with larger volume and area, whereas in (**9b**) the isopropenyl is replaced by the much smaller methyl group.

(**8**)

(b)

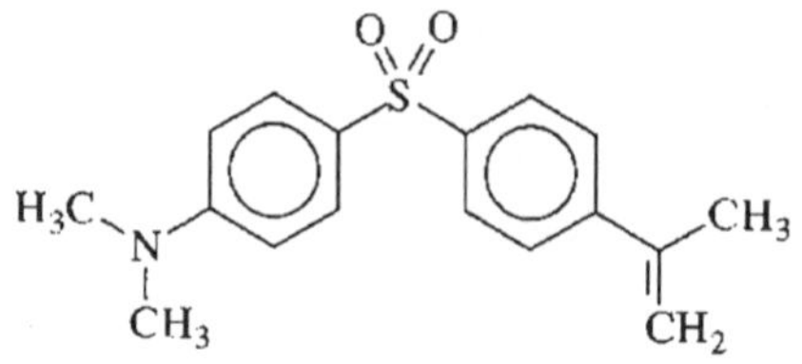

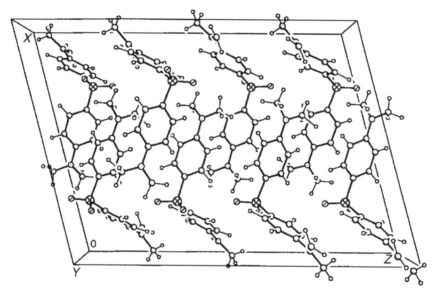

(8)

Figure 15 (*continued*)   (b) Packing arrangement in the crystal structure of the achiral sulfone (**8**)

Now we would expect the two ends of the molecule to be sufficiently different in their steric aspects that they would not pack in a quasicentrosymmetric fashion as did (*S*)-(**7**). The packing arrangements of these two sulfoxides are very similar to one another (Figure 17), and markedly different from that for (*S*)-(**7**). As for sulfoxide (*S*)-(**7**), the molecular crystals of enantiomeric (**9a**) and (**9b**) also have net polar directions, but now resulting from additivity of the sulfoxide moieties. Thus, arrangements based solely on electrostatic interactions cannot simultaneously rationalize all three structures.

Figure 16 (*opposite*) X-ray powder patterns of crystal forms of sulfoxide (**7**): (a) quasicentrosymmetric (*S*)-(**7**); and (b) *rac*-(**7**)

(a)

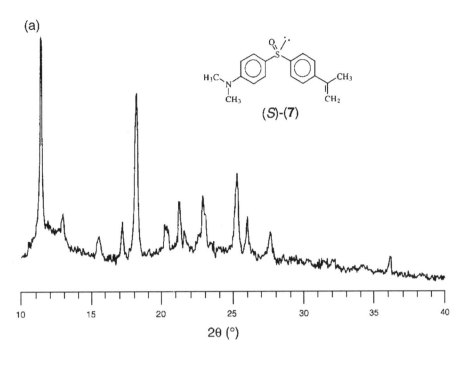

(S)-(7)

(b)

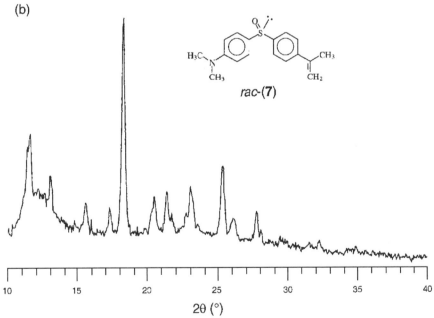

rac-(7)

Table 8 Crystal data for sulfoxides (*S*)-(**9a**) and (*S*)-(**9b**)

	(*S*)-(**9a**)	(*S*)-(**9b**)
Formula	(*S*)-$C_{17}H_{21}ONS$	(*S*)-$C_{15}H_{17}ONS$
Temperature (K)	293	293
Space group	$P2_1$	$P2_1$
a (Å)	7.803 0 (6)	7.752 3 (8)
b (Å)	6.035 5 (6)	5.986 9 (7)
c (Å)	17.037 (2)	14.813 (2)
β (°)	96.899 (7)	103.224 (8)
V (Å³)	796.56 (1 3)	669.29 (1 3)
Z	2	2

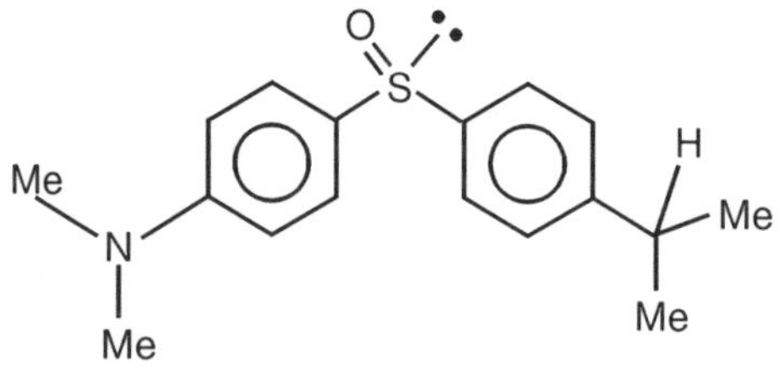

(9a)

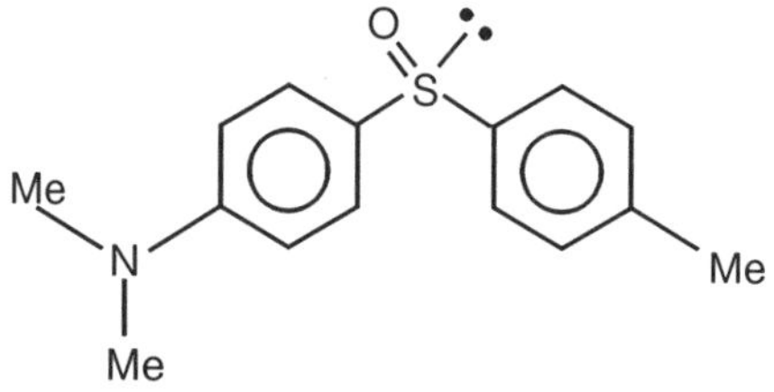

(9b)

5. SUMMARY

We have successfully designed molecular features that lead to predictable features in the packing motifs of molecular crystals, with net summation of a molecular property from the molecules that comprise a molecular crystal. The design features that we have utilized depend solely on molecular size and shape rather than on directional intermolecular interactions such as hydrogen bonding, emphasizing the important role of molecular shape as a determinative, tunable factor in molecular packing. This approach is based on the design of chiral molecules that crystallize to mimic centrosymmetric

(a)

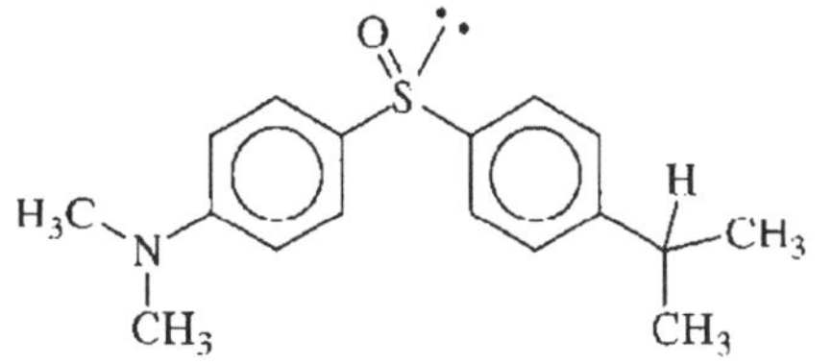

(*S*)-(**9a**)

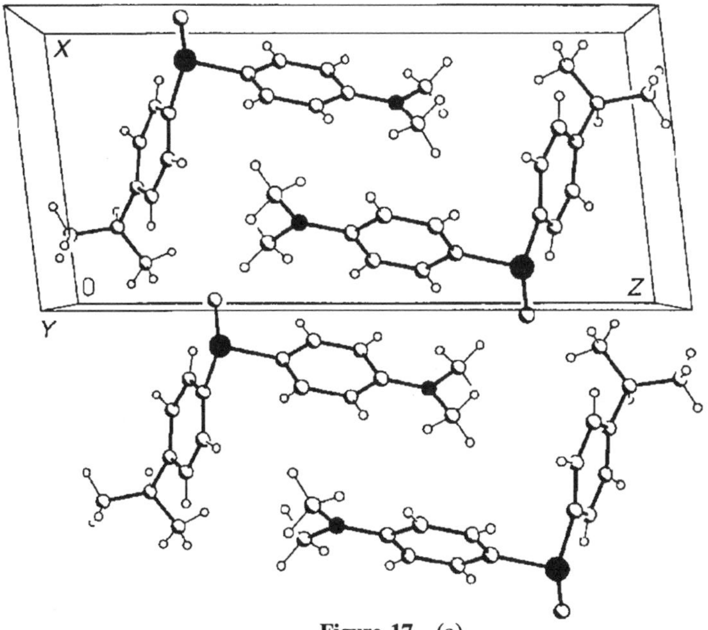

Figure 17 (a)

packing arrangements as a result of the high statistical preference for this symmetry. By proper design of molecular features, the resulting quasicentrosymmetric arrangement is in fact noncentrosymmetric and polar. Two approaches have been used: quasiracemate formation involving cocrystallization of two nearly enantiomeric molecules; and shape mimicry in which individual molecules are constructed so as to have internal quasisymmetry and thus can serve as their own quasienantiomers. The isopropenyl and dimethylamino substituents, with their very similar sizes, shapes and conformational preferences relative to an aromatic ring but with different electron-donating/withdrawing properties, have been found to be convenient, effective isosteric groups for this approach.

(b)

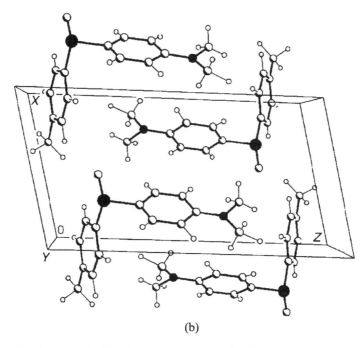

(S)-$(9b)$

(b)

Figure 17 (*continued*) Packing arrangements in the crystal structures of the enantiomeric sulfoxides (**9**): (a) (S)-(**9a**) and (b) (S)-(**9b**)

6. ACKNOWLEDGMENTS

Financial support from the donors of the Advanced Research Program of the Texas Higher Education Coordinating Board (grant 277 to J. K. W. and R. E. D.), the Robert A. Welch Foundation (grants F-233 to R. E. D. and F-626 to J. K. W.), the US National Science Foundation (grant DMR-9014026 to J. K. W.) and the Petroleum Research Fund, administered by the American Chemical Society (grant ACS-PRF AC-20714 to J. K. W.), is gratefully acknowledged.

7. NOTES AND REFERENCES

1. The reader is referred to the pioneering work in organic solid-state reactivity by Schmidt and his coworkers at the Weizmann Institute [e.g. G. M. J. Schmidt, *Pure Appl. Chem.*, **27**, 647 (1971)] and in crystal packing and crystal potential energy calculations by Kitaigorodskii (A. I. Kitaigorodskii, *Molecular Crystals and Molecules*, Academic Press, New York, 1973).

2. F. H. Allen, J. E. Davis, J. J. Galloy, O. Johnson, O. Kennard, C. F. Macrae, E. M. Mitchell, G. F. Mitchell, J. M. Smith and D. G. Watson, *J. Chem. Inf. Comput. Sci.*, **31**, 187 (1991).

3. C. P. Brock and J. D. Dunitz, *Chem. Mater.*, **6**, 1118 (1994).

4. D. B. Amabilino and J. F. Stoddart, *Chem. Mater.*, **6**, 1159 (1994).

5. G. R. Desiraju, *Crystal Engineering: The Design of Organic Solids*, Elsevier, Amsterdam, 1989.

6. (a) J. A. Zerkowski, J. C. MacDonald and G. M. Whitesides, *Chem. Mater.*, **6**, 1250 (1994); (b) L. M. Toledo, J. W. Lauher and F. W. Fowler, *Chem. Mater.*, **6**, 1222 (1994); (c) M. C. Etter, *J. Phys. Chem.*, **95**, 4601 (1991).

7. (a) T. Hahn (ed.), *International Tables for Crystallography*, Vol. A, Reidel, Dordrecht, 1983, Section 10-5; (b) J. F. Nye, *Physical Properties of Crystals— Their Representations by Tensors and Matrices*, Oxford Science Publications, Oxford, 1957 (reprinted by Dover, London, 1987); (c) T. S. Narasimhamurty, *Photoelastic and Electro-Optic Properties of Crystals*, Plenum Press, New York, 1981; (d) J. Zyss and G. Tsoucaris, in *Structure and Properties of Molecular Crystals* (ed. M. Pierrot), Elsevier, New York, 1990, p. 297.

8. (a) D. F. Eaton, *Science*, **253**, 281 (1991); (b) D. S. Chemla and J. Zyss (eds), *Non-Linear Optical Properties of Organic Molecules and Crystals*, Vols 1 and 2, Academic Press, New York, 1987; (c) D. J. Williams (ed.), *Non-Linear Optical Properties of Organic and Polymeric Materials*, American Chemical Society, Washington, DC, 1983; (d) D. J. Williams, *Angew. Chem., Int. Ed. Engl.*, **23**, 690 (1984).

9. D. Y. Curtin and I. C. Paul, *Chem. Rev.*, **81**, 525 (1981).

10. J. Jacques, A. Collet and S. H. Wilen, *Enantiomers, Racemates, and Resolutions*, Wiley-Interscience, New York, 1981.

11. J. K. Whitesell, R. E. Davis, L. L. Saunders, R. J. Wilson and J. P. Feagins, *J. Am. Chem. Soc.*, **113**, 3267 (1991).

12. The relationships of molecular shape and crystal packing are elegantly discussed in the classic text by Kitaigorodskii [1]. A recent study of packing energies has shown on a quantitative basis the relative importance of inversion among the symmetry operators in many organic molecular crystals: G. Filippini and A. Gavezzotti, *Acta Crystallogr., Part B*, **48**, 230 (1991).

13. For two extensive reviews of this application of quasiracemates, see (a) A. Fredga, *Tetrahedron*, **8**, 126 (1960) and (b) A. Fredga, *Bull. Soc. Chim. Fr.*, 174 (1973).

14. W. Jones, C. R. Theocharis, J. M. Thomas and G. R. Desiraju, *J. Chem. Soc., Chem. Commun.*, 1443 (1983).

15. Diffraction peaks that are most sensitive to the inevitable disorder in such structures would vary somewhat with concentration, but these are generally weak reflections with insignificant contributions to the powder pattern.

16. M. Centnerszwer, *Z. Physk. Chem.*, **29**, 715 (1899).

17. L. Pasteur, *Ann. Chim. Phys.*, *Ser. 3*, **38**, 437 (1853). The first examples described by Pasteur were the two pairs ammonium hydrogen $(+)$-tartrate–ammonium hydrogen $(-)$-malate and $(+)$-tartramide–$(-)$-malamide.

18. A cogent discussion of this point appears in Brock and Dunitz [3], p.1127. A common type of exception, not explicitly mentioned by Brock and Dunitz, would be the cocrystallization of solute and solvent to form solvated crystals, in which the solvent often plays only a space-filling role. See P. van der Sluis and J. Kroon, *J. Cryst. Growth*, **97**, 645 (1989).

19. (a) (+)-*m*-Methoxyphenoxypropionic acid–(−)-*m*-bromophenoxypropionic acid: I. Karle and J. Karle, *J. Am. Chem. Soc.*, **88**, 24 (1966); (b) (−)-Ormosamine–(−)-podopetaline: R. Misra, W. Wong-Ng, P.-T. Cheng, S. McLean and S. Nyburg, *J. Chem. Soc., < Chem. Commun.*, 659 (1980); (c) Octahedral valinatocobalt complexes: R. D. Gillard, N. C. Payne and D. C. Phillips, *J. Chem. Soc. A*, 973 (1968); (d) Octahedral salicylaldiminatochromium complexes: M. S. Bilton, *Cryst. Struct. Commun.*, **11**, 755 (1982).

20. C. P. Brock, W. B. Schweizer and J. D. Dunitz, *J. Am. Chem. Soc.*, **113**, 9811 (1991).

21. (a) A leading reference to some early reports of this phenomenon among amino acids and overcrowded halogenated compounds is B. DiBlasio, G. Napolitano and C. Pedone, *Acta Crystallogr., Part B*, **33**, 542 (1977); (b) A more recent discussion of the phenomenon appears in K. Marthi, S. Larsen, M. Ácks, J. Bálint and E. Fogassy, *Acta Crystallogr., Part B*, **50**, 762 (1994).

22. J. K. Whitesell and M.-S. Wong, *J. Org. Chem.*, **59**, 597 (1994).

23. These were calculated with the computer program OPEC: A. Gavezzotti, *J. Am. Chem. Soc.*, **107**, 962 (1985); A. Gavezzotti, *J. Am. Chem. Soc.*, **105**, 5220 (1983).

24. All structures described herein were solved and refined using Sheldrick's program SHELXS-93. Many of these structures are strongly pseudosymmetric, a situation that traditionally leads to troublesome refinement by conventional crystallographic least-squares methods versus $|F|$, owing to high parameter correlations. These difficulties were overcome with the use of a restrained refinement technique versus $|F|^2$ available in this program. Refinement by this approach was invariably well behaved, and the wild variations in bond lengths, bond angles and especially atomic displacement parameters that usually characterize the refinement of such highly correlated structures versus $|F|$ were entirely avoided.

25. J. K. Whitesell, R. E. Davis, M.-S. Wong and N.-L. Chang, *J. Am. Chem. Soc.*, **116**, 523 (1994).

Chapter 4

Molecular Engineering of Crystals by Electrostatic Templating

PAUL J. FAGAN

The Dupont Co., Wilmington, DE, USA

AND MICHAEL D. WARD

University of Minnesota, Minneapolis, MN, USA

1. INTRODUCTION

Crystalline molecular solids continue to be of fundamental and technological interest, primarily because of the potential for unprecedented control of solid-state properties through judicious choice of the molecular constituents comprising the crystal. However, it has become abundantly clear since the 1960s that solid-state properties and reactivity are not only dependent upon molecular structure, but also upon the arrangement of the molecules in the crystal, commonly referred to as the crystal packing, or within the theme of this volume as the supramolecular structure [1]. This was first demonstrated clearly by the pioneering work of Schmidt in the early 1960s. The solid-state photodimerization of *trans*-cinnamic acids was demonstrated to be dependent upon the proximity and orientation of molecules in the solid state [2]. This work can be considered as the origin of 'crystal engineering', an emerging discipline of organic solid-state science. The aim of this discipline is to elucidate the principles responsible for packing motifs in crystals, and to use these principles to develop guidelines for the design and synthesis of new materials

The Crystal as a Supramolecular Entity. Edited by G. R. Desiraju
©1996 John Wiley & Sons Ltd

with desired properties. Since the 1960s, numerous reports and reviews have been devoted to this issue, attempting to unravel the intermolecular interactions responsible for supramolecular motifs in crystals. This pursuit has been further motivated by the discoveries of organic crystals exhibiting electronic properties typically found in the domain of inorganic and elemental solids. These properties include metallic conductivity, semiconductivity, superconductivity, ferromagnetism and second-harmonic generation capability [3–7]. Crystal-engineering strategies can also be useful in the design of crystalline pharmaceutical compounds, for which morphology and polymorphism issues are crucial [8]. The rational design of stable, porous crystalline networks capable of entrapping guest molecules, while promising, remains a scientific and technological challenge.

While considerable advances in crystal engineering have been made since the 1960s, a unifying set of principles describing the intermolecular interactions that govern crystal packing has remained elusive. Rather, understanding of the critical factors responsible for crystal packing generally has been advanced by making small systematic changes within a limited set of structurally similar molecules [9]. The primary limitation in crystal engineering is the inability to predict crystal packing for compounds which differ significantly in shape, functionality and electronic structure. This limitation can be explained by the presence of numerous, but relatively weak, nonbonding interactions between molecules in the solid state. Common to all molecular crystals are van der Waals interactions, which also tend to be locally nondirectional, making the precise orientation of molecules in the solid state unpredictable. Computational studies illustrate the difficulties encountered in predicting crystal structures when only weak nonbonding forces are operative and the difference in the energies of different packing motifs is small [10–13]. However, recent computations of the structure of rigid molecules assembled into one-dimensional aggregates in which molecules were related by glide, screw and inversion operations gave insight into the relative importance of van der Waals and coulombic terms [14,15]. Many of these aggregates gave local energy minima which were substantially lower than the energy minima calculated for the aggregates in the actual crystal structures. This indicates that crystal-packing forces in the other dimensions are too important to ignore. This is not terribly surprising as organic crystals have dimensionalities, with respect to intermolecular bonding, which exceed one.

The lack of predictive power has been countered somewhat by building in interactions that provide directionality such as in halogenated aromatics, where halogen 'steering effects' direct molecular orientations[16]. Crystal-engineering strategies based on hydrogen bonding [17,18] which tends to be comparable in strength to but more directional than van der Waals bonding, show considerable promise. However, molecular crystals have a common characteristic: in spite of directing forces, molecules will pack in a manner that minimizes void space as described in Kitaigorodskii's 'close-packing principle' [19]. This principle is evidence of the importance of van der Waals interactions, illustrating that molecules will generally tend to find a way to maximize these. However, because the van der Waals interactions are locally nondirectional and therefore the energy differences between alternative molecular arrangements are small, the presence of functional groups with directing abilities can dramatically alter the structure. The 'shape' of a molecule and its attendant functional groups defines the topology of that molecule. In the absence of functional groups, van der Waals interactions play an important role, with packing motifs reflecting the tendency to maximize these interactions. For example, hexane and other aliphatics crystallize so that individual rod-shaped molecules pack parallel to each other [20], maximizing the van der Waals attractions between the hydrocarbon chains [Figure 1]. Adamantane crystallizes in either a face-centered ($Fm\bar{3}m$ space group) [21] or body-centered ($P\bar{4}2_1c$) [22] arrangement, as expected for a molecule with spherical topology. However, the effect of directing groups is nicely illustrated by the crystal structure of 1,3,5,7-tetracarboxyadamantane, in which the carboxylic acid groups define a tetrahedral topology [23]. This topology is evident in the solid-state structure, in which these molecules assemble into a diamond-like network as a result of intermolecular carboxylic acid hydrogen-bonded dimers. However, because of the large size of the molecules, considerable open space exists in this diamond-like network. This results in a solid-state structure that actually contains *five* interpenetrating networks, providing an elegant example of a directing influence such as hydrogen bonding and the tendency to maximize van der Waals interactions.

Our goal was to develop a new set of ionic molecular building blocks with well-defined topology which would serve as 'electrostatic templates' for the self-assembly of molecules into crystalline molecular networks. Polycations and polyanions would be ideally suited for this purpose, as the directing topology of the template would be defined by the spatial arrangement of the ionic centers within a given molecule (Figure 2). If the positive and negative charges are locked rigidly in the polycations and polyanions, the number of possible geometric arrangements within a crystal will be limited. For example, one could fabricate a diamond-like structure by cocrystallizing a tetrahedral tetracation and a rod-like dianion (Figure 2). This network would be topologically similar to that of 1,3,5,7-tetracarboxyadamantane. Solids possessing these packing characteristics may have unique properties resulting from nanopores, whose size could be controlled by adjusting the sizes of the cations and anions. This could lead to new materials which would be organic analogs of inorganic zeolites. One may therefore expect a new class of materials capable of small-molecule separations, encapsulation of molecular species with interesting optical properties and catalysis.

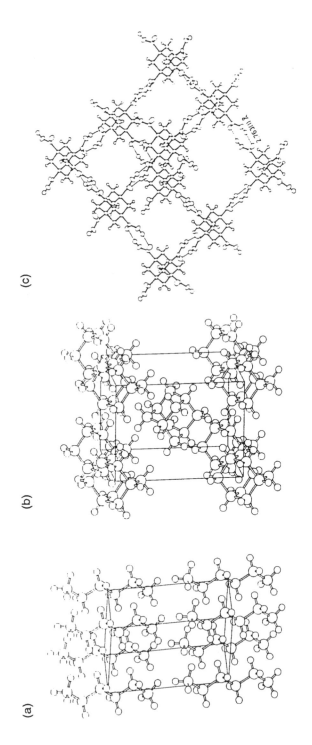

Figure 1 Crystal structures of (a) hexane, (b) body-centered cubic adamantane and (c) 1,3,5,7-tetracarboxyadamantane. Only one of the diamond networks is shown for (c)

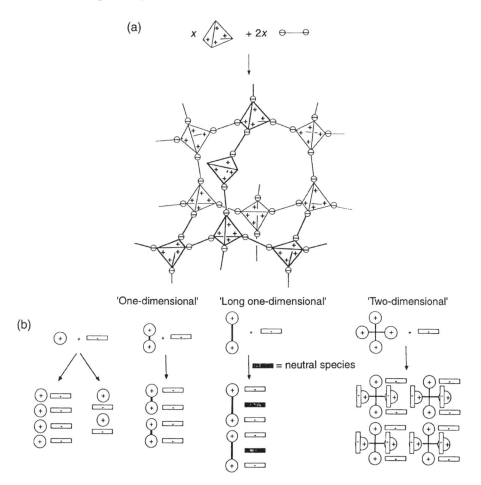

Figure 2 (a) A schematic representation of a diamond-like network formed from tetrahedral tetracations and linear dianions. The volume of the polycation and the length of the dianion will play a crucial role in determining the size of the nanometer-sized voids present in the network. (b) A schematic representation of electrostatic enforcement of anion networks by polycations. For convenience, the concept is portrayed in two-dimensional space. The topology of each cation is defined by the spatial arrangement of the individual cationic centers

The electrostatic templating principle would not be limited to anions capable of only electrostatic interactions. Intermolecular interactions between anions may predispose the anions to form extended networks. For example, anion networks may form as a consequence of charge transfer or hydrogen bonding, which both tend to be anisotropic. In these circumstances, the electrostatic template would direct the spatial orientation of the anion networks, instead of simply the anions. The presence of the secondary interactions responsible for the formation of the anion networks can serve to increase the ability to predict the crystal packing of rather complicated molecules. In particular, we were interested in the synthesis of crystals with controllable electronic properties such as conductivity and magnetism. Such materials generally contain planar, open-shell organic cations or anions that form ansiotropic one-dimensional stacks in which the ions are associated by π–π interactions. The overlap of the π-wavefunctions along the stack, which are partially filled owing to the open-shell nature of the ions, results in a partially filled quasi-one-dimensional electronic band capable of supporting electron transport.

This topological enforcement of anionic networks can be visualized in the following way. In the case of polycyanoanions, these species are predisposed to form extended π–π structures. Nucleation and growth of crystals therefore would involve the self-organization of linear aggregates of anions around the polycations in a manner that maximized the van der Waals attractive forces (i.e. shape) while also maximizing coulombic attraction between the ionic centers (Figures 2 and 3). Furthermore, the distance traversed by the cationic centers within a given template molecule would dictate the number of negative charges required for electroneutrality, established by the number of electrons per molecule in the anionic aggregate. Since it has been well established that polycyanoanion stacks can contain a fractional number of charges per molecule (this is equivalent to a stack containing neutral molecules as well as anions), the distance between cationic centers in the template could, in principle, govern the number of electrons per molecule, and hence the band filling in the solid state. This strategy therefore promises a high degree of control over the electronic properties of the crystalline materials realized with these templates.

The formation of nanoporous molecular networks can be illustrated by an example in which tetrahedral tetracations are linked by coulombic forces to linear dianions. This may lead to a diamond-like network not unlike the aforementioned 1,3,5,7-tetracarboxyadamantane structure. It is our goal here to describe the directing influence of electrostatic templates on anionic networks in crystalline materials, with a focus primarily on the structural aspects of these materials. Complete detailed descriptions of their synthesis, structure and properties are available elsewhere [24–28].

2. CONSTRUCTION OF ORGANOMETALLIC BUILDING BLOCKS

Several considerations went into designing electrostatic building blocks with which to test the ideas described above. We wanted the charges within the molecules held rigidly in place (as much as possible) and localized in a small

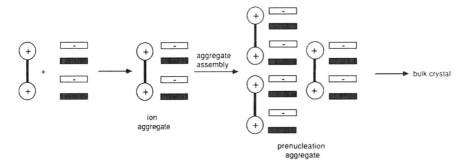

Figure 3 Mechanism for self-assembly of segregated anionic stacks directed by a generic dication, portrayed in two-dimensional space. The initial step involves formation of an ion aggregate, which is followed by further assembly into a prenucleation aggregate that structurally resembles the macroscopic crystal structure. The coulombic interaction between the cation centers and the anions, coupled with π-interactions that favor the formation of anion stacks, results in anion networks whose topologies mimic those of the polycations. The distance between the cations in the molecules will dictate the number of charges within the anion stack. Here the anions are depicted as white rectangles, while the shaded rectangles represent their neutral forms

region of space (Figure 4). A flexible molecule with charge delocalized over its framework would give rise to less predictable behavior upon crystallization. We thought the charges should be near the outer dimensions of the molecule to maximize the interactions with the oppositely charged counterions, although this may not be necessary in all cases. We also wanted to be able to change the spacing between charges within a particular framework, thus allowing us to expand or contract a particular lattice and incrementally change the physical properties of the crystal. The ability to tune the distance between charges in a molecule would also allow us to optimize spatial matches in fitting the anions and cations into a predesigned crystal architecture. Finally, we thought it would be advantageous to fill in the space within the molecules as much as possible to avoid penetration of one ion into another.

With these principles in mind, a number of polycationic templates were constructed using synthetic procedures established in our laboratories. These templates consisted of $[Cp^*Ru]^+$ fragments $(Cp^* = \eta^5\text{-}C_5Me_5)$ which were attached, in an η^6-mode, to benzene rings of polyaromatic molecules or molecules containing single or multiple arene rings. Previous work in our laboratory [29–33] as well as others, had already shown that organometallic complexes, particularly 'sandwich complexes' such as $[(\eta^6\text{-}C_6Me_6)_2Ru]^{2+}$ and $[Cp^*_2Fe]^{2+}$, were capable of forming charge transfer salts with a variety of polycyanoanions with a variety of interesting electronic properties. The

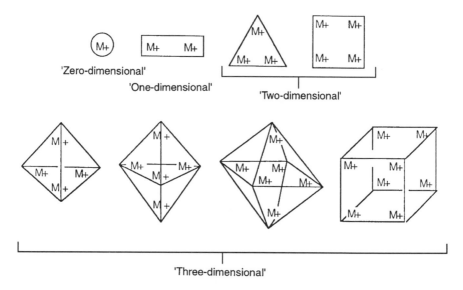

Figure 4 Schematic representation of different topologies for which polycations have been synthesized

crystal-packing motifs of these salts were fairly predictable owing to the simple topology of the organometallic cations in which the metal atom was effectively buried within the organic ligand matrix (Figure 5). Thus the shape of the molecules, defined by the ligand shells, was amenable to close packing via van der Waals interactions, while the ionic charge directed the packing in which the role of coulombic interactions was evident. Consequently, the choice to design organometallic reagents as templates seemed a reasonable strategy.

The ability of the $[CpM]^+$ ($Cp = \eta^5\text{-}C_5H_5$; $M = Fe$, Ru) fragment to bind arenes is well known [34,35]. In particular, we had observed a very strong interaction of the $[Cp*Ru]^+$ fragment with aromatic hydrocarbons. We

$$\tag{1}$$

Arene = Benzene, Toluene, Mesitylene, Hexamethylbenzene, etc.

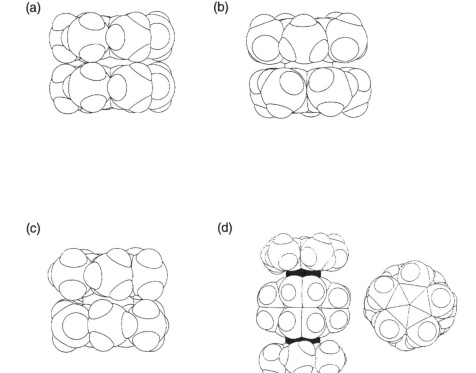

Figure 5 (a) A space-filling model of $[(\eta^6\text{-}C_6Me_6)_2Ru]^{2+}$. (b) A space-filling model of $[Cp_2^* Fe]^+$. (c) A space-filling model of $[Cp^*Ru(C_6Me_6)]^+$. (d) A space-filling model of $[(Cp^*Ru)_2(\text{paracyclophane})]^{2+}$. Two views are depicted in (d) to illustrate the one-dimensional topology of the dication. It can be seen readily that the metal ions are effectively buried within the hydrocarbon ligand matrix

developed a convenient synthesis of the reagent $[Cp^*Ru(MeCN)_3]OTf$ $(OTf = O_3SCF_3)$ which reacts with aromatic hydrocarbons to form arene complexes in high yield under relatively mild conditions, and which is tolerant of a variety of functional groups (equation 1). Numerous polycations can be prepared by the synthetic methodology of equation (1), with the topology of each defined by the arrangement of the cations, which in turn was dictated by the topology of the arene rings.

The zero-dimensional topology (based on the spatial arrangements of point charges) is exemplified by the simple adducts of $[Cp^*Ru]^+$ with benzene or other functionalized benzene molecules. Higher-dimensional topologies were realized by the synthesis of polycations. For example, binding of multiple [Cp*Ru]$^+$ fragments to arene ligands constrained to a linear array gave a one-dimensional topology. This was achieved by using paracyclophane to link two [Cp*Ru]$^+$ fragments to form [(Cp*Ru)$_2$(paracyclophane)]$^{2+}$ (equation 2), which has a one-dimensional array of two positive charges. This could be extended to polycations with different magnitudes of charge at each point along the array (by changing the metal valency), as in [(Cp*Ru)(paracyclophane)(CoCp*)]$^{3+}$, or to polycations with different lengths and charge, as in [{Cp*Ru(paracyclophane)}$_2$Ru]$^{4+}$.

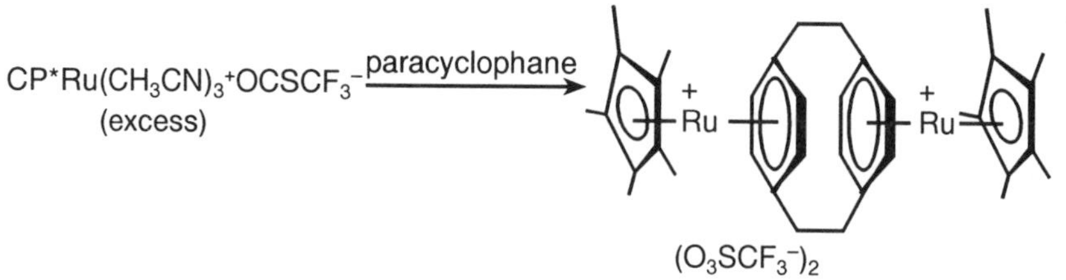

$$CP*Ru(CH_3CN)_3{}^+OCSCF_3{}^- \xrightarrow{\text{paracyclophane}} \qquad (2)$$

(excess)

$(O_3SCF_3{}^-)_2$

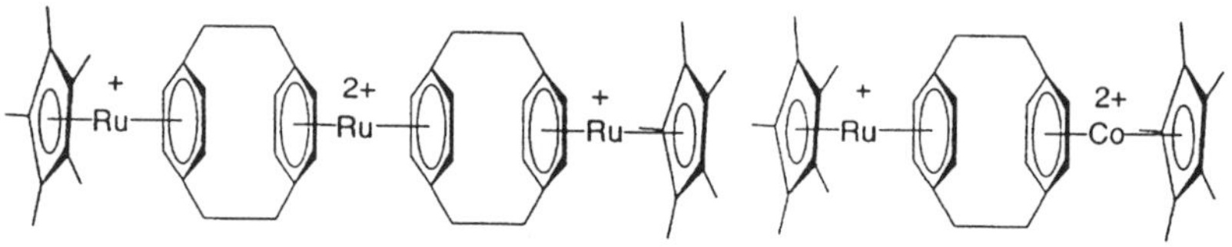

[(Cp*Ru-paracyclophane)$_2$Ru]$^{4+}$ (Cp*Ru)(paracyclophane)(CoCp*)$^{3+}$

A two-dimensional topology was realized with *p*-quaterphenyl and *p*-sexiphenyl as arene ligands, which provided for zigzag planar arrays of four and six positive charges, respectively. Triptycene provided a building block with a trigonal array of three positive charges, namely [(Cp*Ru)$_3$(triptycene)]$^{3+}$. This chemistry would be suitable for the preparation of square arrays of four, pentagonal arrays of five and hexagonal arrays of six positive charges.

Extending this synthetic strategy to three-dimensional arrays of charges presented a larger challenge. Tetrahedral tetracations were realized with tetraphenyl group 14 derivatives as ligands, resulting in a series of cations [{Cp*Ru(η^6-C$_6$H$_5$)}$_4$E]$^{4+}$ (E = C, Si, Ge, Sn, Pb). These provide a nearly tetrahedral array of four positive charges. For E = C and Si, the geometry is distorted to nearly perfect D_{2d} point group symmetry with respect to the positions of the four ruthenium atoms in the molecule. Overall, the molecules have almost perfect S_4 point group symmetry. However, for E = Ge, the X-ray structure shows deviation from S_4 symmetry due to the relaxation of steric constraints within the molecule. The increased germanium–phenyl bond length

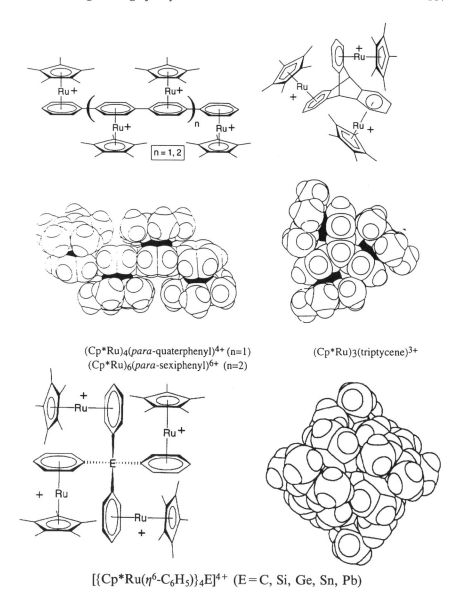

(Cp*Ru)₄(*para*-quaterphenyl)⁴⁺ (n=1)
(Cp*Ru)₆(*para*-sexiphenyl)⁶⁺ (n=2)

(Cp*Ru)₃(triptycene)³⁺

$[\{Cp^*Ru(\eta^6\text{-}C_6H_5)\}_4E]^{4+}$ (E = C, Si, Ge, Sn, Pb)

allows freer rotation about these bonds. This series is particularly interesting as the distance between the charges within these molecules can be systematically varied by simply changing the central element E. It was also possible to prepare the molecule $[\{Cp^*Ru(\eta^6\text{-}C_6H_5)\}_4B]^{3+}$. This molecule has the same shape as the group 14 cases, but has an overall charge of only 3^+. Therefore, one can

(a)   (b)

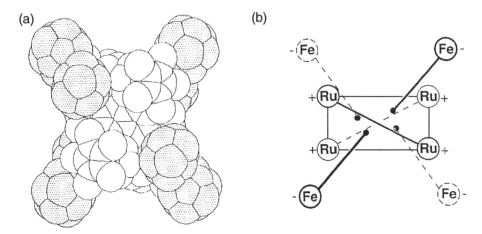

Figure 6 (a) The structure of a supramolecular aggregate contained in crystalline [(Cp*Ru)$_4$(rubrene)][(B$_9$C$_2$H$_{11}$)$_2$Fe]$_4$. It is readily apparent that the anions in this structure are in proximity to the cationic centers of the template cation. The hydrogen atoms have been removed for clarity. (b) Schematic representation of the packing of the [(B$_9$C$_2$H$_{11}$)$_2$Fe]$^-$ anions on the trigonal faces of the pseudotetrahedral array of charges defined by the ruthenium atoms of [Cp*Ru]$_4$(rubrene)]$^{4+}$

think of using it as a p-type dopant in a lattice composed primarily of the aforementioned tetracations.

An even more distorted tetrahedral array of charges was encountered in the novel molecule [(Cp*Ru)]$_4$(rubrene)]$^{4+}$. The geometry in this case arises from the severe distortions forced on the rubrene molecule by coordination of the bulky [Cp*Ru]$^+$ groups onto the four phenyl groups. A novel helical twist is induced on the central tetracene ring system. The single-crystal X-ray structure of [(Cp*Ru)$_4$(rubrene)][B$_9$C$_2$H$_{11)2}$Fe]$_4$ is actually quite revealing in the context

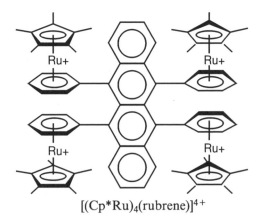

[(Cp*Ru)$_4$(rubrene)]$^{4+}$

of electrostatic enforcement (Figure 6). It is readily apparent that the anions in this structure are in proximity to the cationic centers of the template cation, and tend to pack on the trigonal faces of the pseudotetrahedral array of charges defined by the ruthenium atoms.

An octahedral array of charges was realized by reaction of the template hexa(*p*-methoxyphenoxy)benzene. Previous X-ray analyses of this class of aromatic compounds demonstrated that the outer six phenyl groups are arranged roughly in a trigonal antiprism, i.e. a pseudooctahedral array [36]. Reaction of slightly greater than six equivalents of the triflate salt of $[Cp^*Ru(MeCN)_3]^+$ with hexa(*p*-methoxyphenoxy)benzene in CH_2Cl_2 results in complete addition of six $[Cp^*Ru]^+$ groups onto the six *p*-methoxyphenoxy moieties to yield the complex $[\{Cp^*Ru(p\text{-}MeO\text{-}\eta^6\text{-}C_6H_4O)\}_6C_6](OTf)_6$. This hexasubstituted ruthenium complex has a distorted octahedron of six positive charges. In this case, the molecule is not as rigid as in some of the previously discussed cations.

$$[Cp^*Ru(p\text{-}CH_3O\text{-}\eta\text{-}C_6H_4\text{-}O)]_6C_6\}^{6+}$$

The final building blocks we will describe are two molecules that contain eight positive charges arranged in an almost cubic geometry. Reaction of $[Cp^*Ru(MeCN)_3]OTf$ with $(p\text{-}MeC_6H_4)_8Si_8O_{12}$ gave the triflate salt of the template polycation $[\{Cp^*Ru(\eta^6\text{-}p\text{-}MeC_6H_4)\}_8Si_8O_{12}]^{8+}$ whose single-crystal X-ray structure revealed a cubic siloxane cage with tolyl groups protruding from the corners [37]. The complete addition of all eight ruthenium atoms on this template required forcing conditions (heating in the presence of an excess of the ruthenium reagent, and repeated removal of acetonitrile). Spectroscopic results were in accord with the addition of ruthenium onto all eight of the tolyl groups. From a preliminary X-ray structural analysis, the ruthenium octacation produced in this manner has an arrangement of the ruthenium atoms that is halfway between a cubic and a square antiprismatic geometry. Many of the anions and included solvents in this structure are severely disordered, preventing a fully satisfactory modeling of the entire crystal structure. In our experience, this is a common problem with extremely large molecules that tend to include both solvent and small anions that are prone to disorder such as OTf^-. Consequently, the reader should view this X-ray structure as only a preliminary result.

$$[\{Cp^*Ru(\eta^6\text{-}p\text{-}MeC_6H_4)\}_8Si_8O_{12}]^{8+}$$

A second example of a distorted cubic arrangement of eight positive charges is found in the structure of the very unusual octacation $[\{Cp^*Ru(\eta^6\text{-}C_6H_5)\}_4\{(porphyrin)FeOFe(porphyrin)\}\{Cp^*Ru(\eta^6\text{-}C_6H_5)\}_4]^{8+}$, prepared by reaction of $[Cp^*Ru]^+$ with *meso*-tetraphenylporphyriniron(III)μ-oxo dimer. One might expect there to be free rotation around the central FeOFe unit in this molecule, and that the two porphyrin rings would adopt a staggered conformation (2) rather than the eclipsed conformation (1). However, it was found from an X-ray structure that this octacation adopts the eclipsed conformation (1), with the four $[Cp^*Ru(\eta^6\text{-}C_6H_5)]^+$ groups attached to one porphyrin ring directly overlying the four $[Cp^*Ru(\eta^6\text{-}C_6H_5)]^+$ groups of the

8 Cp*Ru(CH₃CN)₃⁺ + meso-tetraphenylporphyriniron μ-oxo dimer

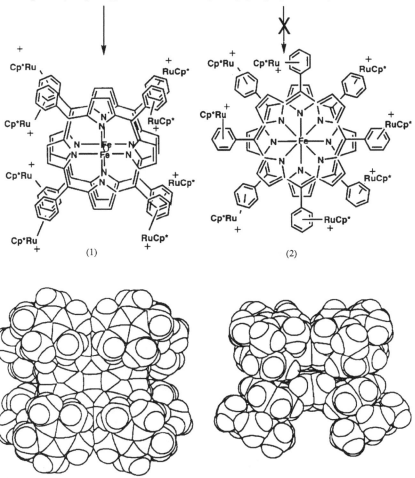

$[\{Cp^*Ru(\eta^6\text{-}C_6H_5)\}_4\{(porphyrin)FeOFe(porphyrin)\}\{Cp^*Ru(\eta^6\text{-}C_6H_5)\}_4]^{8+}$
(two views)

other porphyrin ring. This presumably arises from steric distortions induced by placing the four [Cp*Ru]⁺ groups on the porphyrin phenyl rings. The resulting distortion of each porphyrin ring into a saddle shape is commonly observed for porphyrin rings. For steric reasons, the two porphyrin saddles stack upon one another in an eclipsed geometry, forcing the [Cp*Ru(η⁶-C₆H₅)]⁺ groups also to be eclipsed. The overall result is a nearly square prismatic arrangement of positively charged ruthenium centers.

3. CONSTRUCTION OF ANIONIC NETWORKS WITH ORGANOMETALLIC ELECTROSTATIC TEMPLATES

In order to demonstrate the effectiveness of the organometallic templates in crystal engineering, we examined the solid-state packing of crystals formed from solutions containing the cations and different polycyanoanions. As stated earlier, these open-shell radical anions commonly form aggregates in the solid state by charge transfer interactions that result from π–π overlap. As a result, these anions form either dimers in which the anions each possess a single negative charge or infinite stacks in which the anions are arranged face to face. It is important to remember that in the case of the infinite stacks, each anion is capable of possessing either integral or nonintegral negative charge. We surmised, based on the electroneutrality principle, that the charge per anion site would be governed by the positive charge-to-distance values along a specific linear axis of a polycation template, as depicted in Figure 2. That is, electrostatic enforcement due to coulombic interactions would control the topology of the anion networks, although van der Waals interactions would still play an important role in dictating close-packing arrangements. The latter are short-range forces, suggesting that the molecules in the crystal will pack as closely as possible, but in a motif initially dictated by the electrostatic template.

Charge transfer complexes are commonly prepared *in situ* by electron transfer between donor and acceptor molecules. However, the organometallic templates used in our work can be described as 'preoxidized' in that they are already cationic and are not capable of donating electrons. The template complexes are very difficult to reduce, as evidenced by the highly negative reduction potentials ($E < -2.5$ V versus Ag/AgCl). This characteristic, which is partly a consequence of the 18-electron configurations of the Ru^+ centers, enables synthesis by either of two different methods. A very simple approach employs metathesis, in which the polycyanoanion is added slowly to a solution containing the cation (or polycation). However, we found that in some cases this route afforded crystals of poor quality. In these cases, we used electrochemical crystallization methods, which are convenient for the synthesis of high-quality crystals of poorly conducting as well as conducting materials. [38] The electrochemical approach involves the reduction of a redox-active acceptor at an electrode surface in the presence of the appropriate polycation. An example of such an acceptor is tetracyanoquinodimethane (TCNQ), which readily forms one-dimensional charge transfer stacks and is ubiquitous in low-dimensional conducting molecular crystals. Single crystals can be grown slowly at the electrode surface in this manner, with the rate of crystal growth governed by the concentrations of the acceptor and the cation and the current at the electrode surface.

As an illustration of this crystallization method, reduction of TCNQ ($E^0_{(TCNQ/TCNQ^-)} = +0.22$ V versus an Ag/AgCl reference electrode) at platinum

(a)

(b)

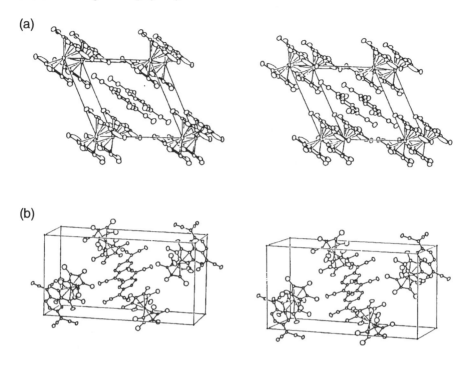

Figure 7 Stereoviews of the unit cells of the one-dimensional [Cp*Ru(C$_6$Me$_6$)]TCNQ green phase (a) and the herringbone dimer purple phase (b). Both are drawn with 20% ellipsoids

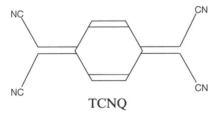

TCNQ

electrodes at an applied potential of +0.3 V in acetonitrile containing [Cp*Ru(C$_6$Me$_6$)]$^+$ afforded green crystals at the electrode surface. These crystals were of sufficient quality for X-ray analysis, whereas those grown by metathesis were fibrous and polycrystalline. However, metathesis methods in which TCNQ$^-$ was simply introduced into solutions of the cation did afford high-quality crystals of a purple phase. The green and purple phases both had the same stoichiometry, [Cp*Ru(C$_6$Me$_6$)]TCNQ, but single-crystal X-ray analysis revealed that their packing motifs were dramatically different.

The green phase crystallized in the $P\bar{1}$ space group and possessed one-dimensional mixed stacks of $[Cp*Ru(C_6Me_6)]^+$ and $TCNQ^-$ in a $\cdots D^+A^-D^+A^-D^+A^-\cdots$ motif (Figure 7). This structure is similar to other phases reported for organometallic TCNQ charge transfer complexes such as $[Cp*_2Fe]TCNQ$ [39]. In addition to the mixed stack motif, the structure revealed alternating 'ribbons' of anions and cations along the a-axis. However, the anions are flanked by the cations along the other two principal axes, suggesting favorable coulombic interactions. The purple phase, crystallizing in the $P2_1/c$ space group, possessed discrete D^+A^- A^-D^+ fragments arranged in a herringbone motif. The structure revealed that in each unit the two C_6Me_6 rings of the cations are parallel to and stacked face to face about the $(TCNQ)_2^{2-}$ dimer so that one of the methylidene carbon atoms (where most of the negative charge resides) on the $TCNQ^-$ anion lies nearly in the center of the C_6 ring. Furthermore, only the C_6Me_6 ring faces the $TCNQ^-$ anion. This may be because of lower repulsion compared to an arrangement in which the anion faces the formally negative Cp* ring. Additionally, it may be associated with the shorter distance between the Ru^+ center and the anion in the observed arrangement owing to the shorter ruthenium–arene bond length compared to the Ru–Cp* bond length. We concluded from these two examples, as well as our previous work with other $[M(arene)_2]^{2+}$ (polycyanoanion) salts, that crystal packing is a consequence of the 'zero-dimensionality' of the template cation. The structures, with the exception of some details, appear to maximize coulombic interactions by forming motifs not unlike those observed for simple inorganic salt structures.

Electrochemical growth methods were then used for crystallization of charge transfer complexes of $[(Cp*Ru)_2(paracyclophane)]^{2+}$ and $TCNQ^-$. Interestingly, two phases were also observed for crystals containing these components, but the selectivity toward a given phase could be controlled by adjusting the electrode potential during growth. At very negative potentials, large, high-quality, purple crystals with the composition $[(Cp*Ru)_2(paracyclophane)]$ $(TCNQ)_2$ formed at the electrode surface. However, at applied potentials positive of the E^0 value for $TCNQ/TCNQ^-$, large, black parallelepipeds with the composition $[(Cp*Ru)_2(paracyclophane)]$ $(TCNQ)_4$ grew exclusively. Since the charge on the organometallic template was constant $(2+)$, the composition of the black phase reflected an average nonintegral charge of 0.5^- on each TCNQ molecule. Infrared analysis, which can be used to determine the amount of charge on $TCNQ^-$, was consistent with this conclusion $(v(CN) = 2200\,cm^{-1})$ [25]. In addition, substantial infrared absorption was observed above $1800\,cm^{-1}$, indicative of free carrier absorption generally associated with metallic behavior.

Single-crystal X-ray analysis revealed that the purple phase $[(Cp*Ru)_2(paracyclophane)]$ $(TCNQ)_2$ crystallized in the $P\bar{1}$ space group and contained one-dimensional mixed stacks of $[(Cp*Ru)_2(paracyclophane)]^{2+}$ and

(a)

(b)

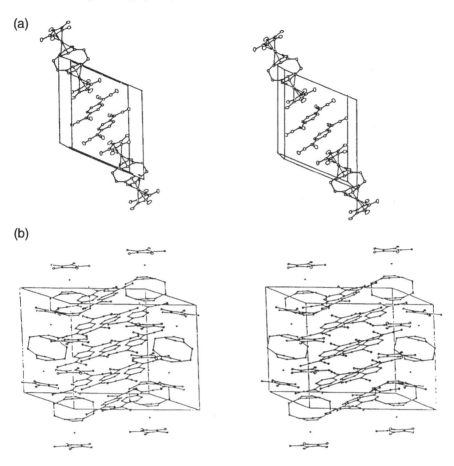

Figure 8 Stereoviews of the unit cells of the purple [(Cp*Ru)₂(para-cyclophane)](TCNQ)₂ phase (a) and the black [(Cp*Ru)₂(paracyclophane)](TCNQ)₄ phase (b). The one-dimensional stacks in each structure coincide with the long axis of the dication

$(TCNQ)_2^{2-}$ dimers in a $\cdots D^+ D^+ A^- A^- D^+ D^+ A^- A^- D^+ D^+ A^- A^- \cdots$ motif (Figure 8). Once again, ribbons of cations and anions are observed along the *a*-axis, but along the *b*-axis and *c*-axis the dications and dimeric dianions form an alternating network. This suggests favorable coulombic interactions in the bulk. The effect of the cojoined cations is evident when one compares this

structure to that of the [Cp*Ru(C_6Me_6)]TCNQ phases, in which a
$\cdots D^+ D^+ A^- A^- D^+ D^+ A^- A^- D^+ D^+ A^- A^- \cdots$ motif is not feasible owing to
the absence of the cyclophane linker which buffers the repulsive interactions
between cations in the chain.

In contrast, the single-crystal X-ray structure of the black
[(Cp*Ru)$_2$(paracyclophane)](TCNQ)$_4$ revealed a motif which was clearly a
consequence of the electrostatic template effect. This material also crystallizes
in the $P\bar{1}$ space group, but now the structure consists of segregated stacks of
cations and anions aligned approximately along the *a*-axis. The cations and
anions form segregated layers in the *bc* plane, whereas the *ac* plane contains
alternating stacks of cations and anions. The TCNQ$^-$ anions are arranged face
to face in the stacks, indicative of π–π interactions. However, the compound
actually contains two crystallographically independent TCNQ stacks. These
stacks differ with respect to the interplanar spacings between the anions and
their relative orientations with respect to each other in the stacks. More
importantly, the TCNQ stacks do not have uniform interplanar spacings,
which reflects the formation of a Mott–Hubbard semiconducting state. Indeed,
crystals of this material were found to be semiconducting, not metallic. This
property is typically associated with either dimerization or tetramerization of
anions within the stack. This is actually evident from the structure here, as each
stack is tetramerized in a $\cdots(AAAA)^{2-}(AAAA)^{2-}\cdots$ motif. These tetramers are
even evident in scanning tunneling microscopy (STM) images of the *ac* and *ab*
faces of the black phase (Figure 9). The activation energy for electron transport
that gives rise to semiconductivity can be conceptually understood as the
electron–electron repulsion associated with the formation of
$\cdots(AAAA)^-(AAAA)^{3-}\cdots$ sites during electron 'hopping'. Magnetic
susceptibility and electron paramagnetic resonance (EPR) measurements of
the black phase revealed interesting magnetic behavior associated with mobile
triplet-spin excitons, which single-crystal EPR analysis showed to be likely
associated with two spins residing in each tetramer.

The electrostatic enforcement provided by the one-dimensional
[Cp*Ru)$_2$(paracyclophane)]$^{2+}$ is clearly evident from the crystal structure of
the black phase. The concatenation of charge along the one-dimensional axis
of the dication induces the alignment of the anions, which have a proclivity to
form extended stacks parallel to the dication stacks. Van der Waals
interactions are also evident, as they are responsible for stacking of the
dications and the close packing between the cation and anion stacks. Most
importantly, the effect of template length and charge is demonstrated by the
nonintegral electron occupancy per TCNQ site observed for the black phase.
The average values of the interplanar separations in the two independent
TCNQ stacks are 3.29 Å and 3.27 Å. These afford average lengths of 9.87 Å
and 9.81 Å for the (TCNQ)$_4$$^{2-}$ units in each stack, respectively. These distances
are well matched to the distance between Cp* planes (9.96 Å) in the dication.

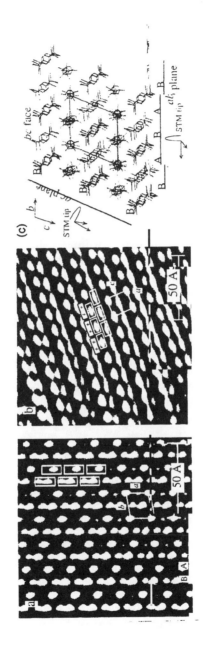

Figure 9 Scanning tunneling microscopy (STM) images of (a) the *ab* face and (b) the *ac* face of the black [(Cp*Ru)$_2$(paracyclophane)](TCNQ)$_4$ phase. A view down the stacking axis illustrating the TCNQ$^-$ anion layers on the *ab* face and the alternating cation and anion stacks on the *ac* face is depicted in (c). The *ab* and *ac* unit cells are drawn on the STM images. The rectangles on the STM images indicate the areas occupied by (TCNQ)$_4^{2-}$ tetramers. The cations are also displayed on the STM image of the *ac* face. These cations, which do not have an appreciable density of states near the Fermi level, are presumed to reside in the dark regions of the STM image where the tunneling current is negligible. The 'double rows' of tunneling current on the *ac* face may be due to tunneling to anions in the second layer below the surface

Therefore, the axial length of the dication accommodates four TCNQ acceptor molecules with the interplanar spacing required for intermolecular π–π overlap. Furthermore, the presence of two positive charges along the molecular axis of the dication dictates that only two negative charges reside over the length of the $(TCNQ)_4^{2-}$ unit. In summary, the charge-to-length ratio of the electrostatic polycation template governs the charge per anion site in the adjacent stack, thereby playing a crucial role in its electronic properties.

It was also discovered that electrostatic enforcement by $[(Cp^*Ru)_2(para-cyclophane)]^{2+}$ could be realized on other anions such as hexacyanotris(methylidene)propanide (HCTMP). Slow cooling of acetonitrile solutions of HCTMP$^-$ and $[(Cp^*Ru)_2(paracyclophane)]^{2+}$ resulted in the formation of long purple needles with the composition $[Cp^*Ru)_2(para-cyclophane)]$ (HCTMP)$_2$. Single-crystal X-ray analysis revealed that this compound crystallized in the $P2_1/a$ space group and possessed segregated one-dimensional stacks of $[(Cp^*Ru)_2(paracyclophane)]^{2+}$ cations and HCTMP$^-$ anions (Figure 10). These anion stacks form tetramers, with the anions strongly dimerized within the tetramers in an $\cdots A^- A^- \cdots A'^- A'^- \cdots A^- A^- \cdots A'^- A'^- \cdots$ motif. The distance between the Cp* rings was 3.76 Å, consistent with van der Waals packing along the stack. Views down the stack over several unit cells also reveal that the anion stacks are surrounded by cation stacks, signifying favorable coulombic interactions. It is important to note that Hückel calculations indicate that the negative charge on the HCTMP$^-$ anion resides mainly on the methylidene carbon atoms. Therefore, in the solid-state structure this negative charge is pointing toward the cation centers of the template. This example once again illustrates that the tethering of positive charge along a linear axis directs the formation of linear anion networks which are parallel to the axis.

HCTMP

Having established that linear networks of HCTMP$^-$ anions form as a result of templating by $[Cp^*Ru]_2(paracyclophane)]^{2+}$, we sought to examine the effect of the higher dimensionality present in $[\{Cp^*Ru(\eta^6 - C_6H_5)\}_4E]^{4+}$ (E = C, Si, Ge, Sn, Pb). These tetracations had several interesting structural features

129

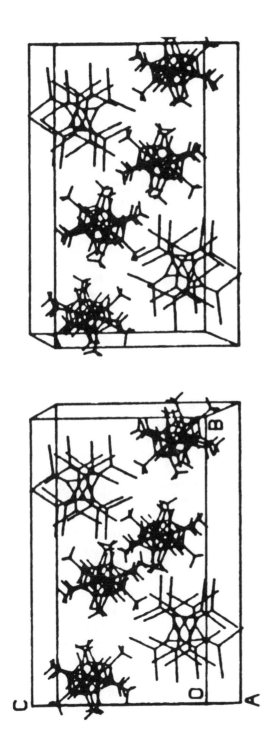

Figure 10 Stereoview of the unit cell of [Cp*Ru]₂(paracyclophane)](HCTMP)₂

which we envisaged would provide fine control over the assembly of anion networks. The S_4 molecular symmetry implied the presence of a C_2 rotation axis, which meant that the topology of positive charges was equivalent to two orthogonal axes of dipositive charge. We anticipated that this feature would induce the formation of orthogonal networks of polycyanoanions. Furthermore, the use of different central atoms E would provide subtle adjustments of the charge-to-distance ratio while maintaining the general symmetry.

The solid-state motifs observed in 1:4 complexes of the tetracations with HCTMP$^-$ anions, grown as large, block-shaped crystals from nitromethane, differed significantly from those observed for the one-dimensional $[(Cp^*Ru)_2(paracyclophane)]^{2+}$ phases. Single-crystal X-ray structural analysis revealed that $[\{Cp^*Ru(\eta^6\text{-}C_6H_5)\}_4C](HCTMP)_4 \cdot 6MeNO_2$ crystallized in the $C2/c$ space group, and possessed mutually orthogonal networks of HCTMP$^-$ anions. A similar motif was observed for $[\{Cp^*Ru(\eta^6\text{-}C_6H_5)\}_4Si]$ $(HCTMP)_4 \cdot 4MeNO_2$, which crystallizes in the lower-symmetry $P\bar{1}$ space group. The unit cells of these structures are too complicated to present here, so two alternative representations are depicted in Figure 11. In Figure 11(a), a section of the packing motif is shown which emphasizes the electrostatic enforcement axes of dipositive charge. This topology directs the formation of the anion network into orthogonal stacks such that the anion stacks are aligned along the mutually orthogonal axes on the tetracation.

The electrostatic template principle is not limited to the structural enforcement of π–π charge transfer networks. For example, we have determined that hydrogen-bonded chains of anions can also be directed to conform to the organometallic template topology in $[(Cp^*Ru)_2(paracyclophane)][HO_2C(C_6H_4)CO_2] \cdot 2MeOH$ (Figure 12). This salt crystallized as large, faint-yellow crystals from methanol in the $P2_1/n$ space group. The single-crystal X-ray structure revealed segregated linear chains of the anions and cations, with the anion chain direction coincident with that of the dication chains. This structure therefore bears a topological similarity to $[(Cp^*Ru)_2(paracyclophane)](TCNQ)_4$. While the direction of the chains is apparently enforced by the topology of cationic charges in the cation, the role of van der Waals interactions and the tendency for close packing are revealed by the corrugation of the anion chains, which buckle toward the cation chains in the cyclophane region. Conversely, the anion chains buckle away from the cation chains in the Cp* regions owing to steric interactions with the methyl groups. This structure illustrates that other intermolecular forces are compatible with the electrostatic templates. In this case it is intermolecular hydrogen bonding of the type $CO_2^- \cdots HO_2C$, in which the proton (donor) of a carboxylic acid group is hydrogen bonded to a carboxylate (acceptor) of the next molecule in the chain. These hydrogen bonds, commonly referred to as Speakman salts [40], are known to be exceptionally strong with respect to other types of hydrogen bonds. However, hydrogen bonds are also known to have shallow potential profiles with respect to

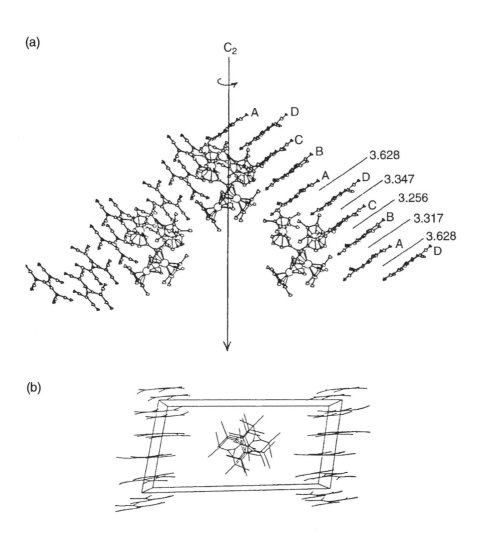

Figure 11 (a) Representation of two orthogonal HCTMP⁻ anion stacks with their nearest [{Cp*Ru(η^6-C$_6$H$_5$)}$_4$C]$^{4+}$ tetracations, viewed nearly normal to the *ab* plane. (b) Unit cell of [{Cp*Ru(η^6-C$_6$H$_5$)}$_4$Si](HCTMP)$_4$·MeNO$_2$ depicting the two unique HCTMP⁻ stacks. The tetracations have been removed for clarity so that the mutually orthogonal relationship of the HCTMP⁻ stacks can be discerned

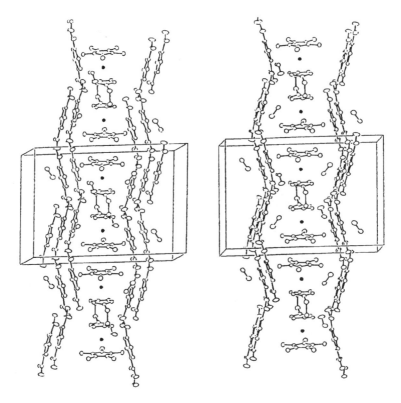

Figure 12 Stereoview of the unit cell of [(Cp*Ru)$_2$(paracyclophane)][HO$_2$C (C$_6$H$_4$)CO$_2$]·2MeOH. Ellipsoids are drawn at 20% probability

the O–H···O angles, which probably accounts for the ability of the chains to pucker about the dications.

4. NEW DIRECTIONS

We have described the use of organometallic templates which can enforce, by electrostatic interactions, the organization of anion networks in the crystalline solid state. The primary barrier to crystal engineering is the inability to predict confidently the crystal packing even when only small changes are made to the molecular components. The observation of similar electrostatic enforcement for both charge transfer and hydrogen-bonded anion networks indicates that this strategy can be applied to a diverse variety of molecular species, and that gross molecular structural changes can be tolerated. Extension of this work to systems which contain polycations *and* polyanions is particularly intriguing. For example, the condensation of a tetrahedral tetracation with a linear dianion described earlier represents a particularly attractive strategy for synthesizing crystalline networks having voids of controlled size. Indeed, the essential principles of topological control of self-assembly in crystallization are being practiced increasingly, as evident from recent reports of materials synthesized from organometallic building blocks. Three-dimensional polymeric networks have been reported in which Cu^+ ions link 4,4',4'',4'''-tetracyanotetraphenylmethane molecules via $CN\cdots Cu\cdots CN$ bonds [41,42]. The tetrahedral topology of this aromatic species directs the formation of an adamantane-like supercage. Similarly, three-dimensional polymeric networks based on 5,10,15,20-tetra-4-pyridyl-21H,23H-porphinatopalladium anions bridged by Cd^{2+} ions have been reported [43]. These examples further illustrate the design flexibility made possible by organometallic templates. Related covalent networks based on tetrahedral $Ge_4S_{10}^{2-}$ units linked by Mn^{2+} ions have been reported recently, demonstrating that design by topological principles is not limited to organometallic building blocks [44]. Indeed, significant contributions are also expected from design strategies based on rigid organic molecular building blocks [45]. We anticipate that the molecular approach to 'designer materials' will be a vital component of materials research for the foreseeable future.

5. ACKNOWLEDGMENTS

The authors gratefully acknowledge the support of Dupont (P.J.F.) and the Office of Naval Research (M.D.W.).

6. REFERENCES

1. J.-M. Lehn, *Angew. Chem., Int. Ed. Engl.*, **27**, 89 (1988).
2. G. M. Schmidt, *J. Pure Appl. Chem.*, **27**, 647 (1971).
3. J. S. Miller, J. C. Calabrese, H. Rommelmann, S. R. Chittpedi, J. H. Zhang, W. M. Reiff and A. J. Epstein, *J. Am. Chem. Soc.*, **109**, 769 (1987).
4. S. Chittipeddi, K. R. Cromack, J. S. Miller, A. J. Epstein, *Phys. Rev. Lett.*, **58**, 2695 (1987).
5. G. T. Yee, J. M. Manriquez, D. A. Dixon, R. S. McLean, D. M. Groski, R. B. Flippen, K. S. Narayan, A. J. Epstein and J. S. Miller, *Adv. Mater.*, **3**, 309 (1991).
6. W. E. Broderick, J. A. Thompson, E. P. Day and B. M. Hoffman, *Science*, **249**, 401 (1990).
7. D. S. Chemla and J. Zyss (eds), *Nonlinear Optical Properties of Organic Molecules and Crystals*, Vol. 1, Acadmic Press, Orlando, FL. 1987.
8. S. R. Byrn, *Solid State Chemistry of Drugs*, Academic Press, New York, 1982.

9. G. R. Desiraju, *Crystal Engineering: The Design of Organic Solids*, Elsevier, Amsterdam, 1989.
10. J. Maddox, *Nature* (London), **335**, 201 (1988).
11. H. R. Karfunkel and R. J. Gdanitz, *J. Comput. Chem.*, **13**, 1171 (1992).
12. J. R. Holden, Z. Du and H. L. Ammon, *J. Comput. Chem.*, **14**, 422 (1993).
13. A. Gavezzotti, *J. Am. Chem. Soc.*, **113**, 4622 (1991).
14. J. Perlstein, *J. Am. Chem. Soc.*, **114**, 1955 (1992).
15. J. Perlstein, *J. Am. Chem. Soc.*, **116**, 455 (1994).
16. J. A. R. P. Sarma and G. R. Desiraju, *Acc. Chem. Res.*, **19**, 222 (1986).
17. (a) M. C. Etter, *J. Phys. Chem.*, **95**, 4601 (1991); (b) M. C. Etter, *Acc. Chem. Res.*, **23**, 120 (1990).
18. (a) J.-M. Lehn, M. Mascal, A. DeCian and J. Fischer, *J. Chem. Soc., Perkin Trans. 2*, 461 (1992); (b) J.-M. Lehn, M. Mascal, A. DeCian and J. Fischer, *J. Chem. Soc., Chem. Commun.* 479 (1990); (c) C.Fouquey, J.-M. Lehn, and A.-M. Levelut, *Adv. Mater.*, **2**, 254 (1990); (d) J. A. Zerowski, C. T. Seto, G. M. Whitesides, *J. Am. Chem. Soc.*, **114**, 5473 (1992); (e) C. T. Seto and G. M. Whitesides, *J. Am. Chem. Soc.*, **113**, 712 (1991); (f) G. M. Whitesides, J. P. Mathias, and C. T. Seto, *Science*, **254**, 1312, (1991); (g) J. A. Zekowski, C. T. Seto, D. A. Wierda, and G. M. Whitesides *J. Am. Chem. Soc.*, **112**, 9025 (1990); (h) C. T. Seto and G. M. Whitesides, *J. Am. Chem. Soc.*, **112**, 6409 (1990); (i) F. Persico, and J. D. Wuest, *J. Org. Chem.*, **58**, 95 (1993); (j) M. Simard, D. Su and J. D. Wuest, *J. Am. Chem. Soc.*, **113**, 4696 (1991); (k) Y. Ducharme and J. D. Wuest, *J. Org. Chem.*, **53**, 5787 (1988); (l) M. Gallant, M. T. P. Biet, J. D. Wuest, *J. Org. Chem.*, **56**, 2284 (1991). (m) M. Scoponi, E. Polo, F. Pradella, V. Bertolasi, V. Carassiti and P. Goberti *J. Chem. Soc., Perkin Trans. 2*, 1127 (1992); (n) Y.-L. Chang, M.-A. West, F. W. Fowler, and J. W. Lauher *J. Am. Chem. Soc.*, **115**, 5991 (1993); (o) J. W. Lauher, Y.-L. Chang, F. W. Fowler and J. W. Lauher, *J. Am. Chem. Soc.*, **112**, 6627 (1990); (p) X. Zho, Y.-L. Chang, F. W. Fowler and J. W. Lauher, *J. Am. Chem. Soc.*, **112**, 6627 (1990); (q) M. D. Hollingsworth, B. D. Santansiero, H. Oumar-Mahamat and C. J. Nichols, *Chem. Mater.*, **3**, 23 (1991); (r) F. Garcia-Tellado, S. J. Geib, S. Goswami, and A. D. Hamilton, *J. Am. Chem. Soc.*, **113**, 9265 (1991).
19. A. I. Kitaigorodskii, *Molecular Crystals and Molecules*, Academic Press, New York, 1973.
20. N. Norman and H. Mathisen, *Acta Chem. Scand.* **15**, 1755 (1961).
21. (a) C. E. Nordman and D. J. Schmitkons, *Acta Crystallogr.*, **18**, 764 (1965); (b) J. P. Amoureux, M. Bee and J. C. Damien, *Acta Crystallogr., Part B*, **36**, 2633 (1980).
22. (a) J. Donohue and S. H. Goodman, *Acta Crystallogr.*, **22**, 352 (1967); (b) J. P. Amoureux and M. Foulon, *Acta Crystallogr., Part B*, **43**, 470 (1987).
23. O. Ermer, *J. Am. Chem. Soc.*, **110**, 3747 (1988).
24. P. J. Fagan, M. D. Ward, J. C. Calabrese, *J. Am. Chem. Soc.*, **111**, 1698 (1989).
25. M. D. Ward, P. J. Fagan, J. C. Calabrese, and D. C. Johnson, *J. Am. Chem. Soc.*, **111**, 1719 (1989).
26. P. J. Fagan, M. D. Ward, J. V. Caspar, J. C. Calabrese and P. J. Krusic, *J. Am. Chem. Soc.*, **110**, 2981 (1988).
27. P. J. Fagan, and M. D. Ward, *Sci. Am.*, **267**, 48 (1992).
28. J. R. Morton, K. F. Preston, M. D. Ward and P. J. Fagan, *J. Chem. Phys.*, **90**, 2148 (1989).
29. M. D. Ward, *Organometallics*, **6**, 754 (1987).
30. M. D. Ward and D. C. Johnson, *Inorg. Chem.*, **26**, 4213 (1987).
31. M. D. Ward and J. C. Calabrese, *Organometallics*, **8**, 593 (1989).

Molecular Engineering of Crystals   135

32. J. R. Morton, K. F. Preston, Y. Le Page, A. J. Williams and M. D. Ward, *J. Chem. Phys.*, **93**, 2222 (1990).
33. J. S. Miller, M. D. Ward, J. H. Zhang and W. M. Reiff, *Inorg. Chem.*, **29**, 4063 (1990).
34. (a) M. Lacoste, F. Varret, L. Toupet and D. Astruc, *J. Am. Chem. Soc.*, **109**, 650 (1987); (b) V. Guerchais and D. Astruc, *J. Organomet. Chem.*, **312**, 97 (1986); (c) R. G. Sutherland, M. Iqbal and A. Piorko, *J. Organomet. Chem.*, **302**, 307 (1986); (d) I. W. Robertson, T. A. Stephenson and D. A. Tocher, *J. Organomet. Chem.*, **228**, 171 (1982) and references therein; (e) J.-R. Hamon, D. Astruc, *Organometallics*, **7**, 1036 (1988).
35. (a) J. L. Schrenk, A. M. McNair, F. B McCormick and K. R. Mann, *Inorg. Chem.*, **25**, 3501 (1986); (b) A. M. McNair and K. R. Mann, *Inorg. Chem.*, **25**, 2519 (1986); (c) A. M. McNair, J. L. Schrenck, and K. R. Mann, *Inorg. Chem.*, **23**, 2633 (1984); (d) T. P. Gill and K. R. Mann, *Organometallics*, **1**, 485 (1982); (e) V. S. Kaganovich, A. R. Kudinov and M. I. Rybinskaya *Izv. Akad. Nauk SSSR, Ser. Khim.*, 492 (1986); (f) J. A. Segal, *J. Chem. Soc., Chem. Commun.*, 1338 (1985); (g) N. A. Vol'kenau, I. N. Bolesova, L. S. Shul'pina, A. N. Kitaigorodskii and D. N. Kravtsov, *J. Organomet. Chem.*, **288**, 341 (1985); (h) N. A. Vol'kenau, I. N. Bolesova, L. S. Shul'pina and A. N. Kitaigorodskii, *J. Organomet. Chem.*, **267**, (1984) 313; (i) R. M. Moriarty, Y.-Y. Ku and U. S. Gill, *Organometallics*, **7**, 660 (1988).
36. C. J. Gilmore, D. D. MacNicol, A. Murphy and M. A. Russell, *Tetrahedron Lett.*, **24**, 3269 (1983).
37. Sample obtained from Prof. F. Feher and Dr. T. Budczhuski, University of California at Irvine.
38. M. D. Ward, in *Electroanalytical Chemistry*, Vol. 16 (ed. A. J. Bard), Dekker, New York, 1989, p. 181.
39. J. S. Miller, J. A. H. Reis Jr, E. Gebert, J. J. Ritsko, W. R. Salaneck, L. Kovnat, T. Cape, T. W. and R. P. van Duyne, *J. Am. Chem. Soc.*, **101**, 7111 (1979).
40. (a) J. C. Speakman, *Struct. Bond.* **12**, 141 (1972); (b) D. R. McGregor and J. C. Speakman, *J. Chem. Soc., Perkin Trans. 2*, 1740 (1977).
41. B. F. Hoskins and R. Robson, *J. Am. Chem. Soc.*, **112**, 1546 (1990).
42. B. F. Hoskins and R. Robson, *J. Am. Chem. Soc.* **111**, 5962 (1989).
43. B. F. Abrahams, B. F. Hoskins and R. Robson. *J. Am. Chem. Soc.*, **113**, 3606 (1991).
44. O. M. Yaghi, Z. Sun, D. A. Richardson and T. L. Groy, *J. Am. Chem. Soc.*, **116**, 807 (1994).
45. (a) P. Kaszynski, A. C. Friedli and J. Michl, *J. Am. Chem. Soc.*, **114**, 601 (1992); (b) K. Hassenrück, G. S. Murphy, V. M. Lynch, J. Michl, *J. Org. Chem.*, **55**, 1013 (1990).

Chapter 5

Supramolecular Inorganic Chemistry

IAN DANCE

University of New South Wales, Sydney, Australia

1. INTRODUCTION

Supramolecular chemistry encompasses all of the extramolecular attractions and associations which influence the condensed phases of molecular compounds. Supramolecularity is also described by Lehn as 'the designed chemistry of the intermolecular bond' [1]. While supramolecular chemistry originated with biomolecules (under the name 'molecular recognition') and has been developed further for organic molecules, its purview is not restricted to light-atom systems. All elements form molecules, and therefore possess a supramolecular dimension to their chemistry. The only requirement for supramolecularity is structural molecularity — the existence of the molecule. Indeed, inorganic [2] compounds demonstrate a diversity of molecules far beyond that of organic or biomolecules, and a corresponding diversity of supramolecular chemistry should be expected.

The conventional classes of molecular compounds containing metal atoms, i.e. coordination complexes and organometallic compounds, are included. As well as extending over all the metallic elements throughout the periodic table, these compounds frequently have associated with them elements from groups 14, 15, 16 and 17. Classification of compounds as coordination complex, organometallic or main group is hardly necessary: what is more useful is the occurrence of various elements in different domains of the compound, particularly at the periphery of the molecule, as is described below. The scope of this article is intentionally general.

The Crystal as a Supramolecular Entity. Edited by G. R. Desiraju
©1996 John Wiley & Sons Ltd

Some clarifications of terminology are needed. In supramolecular chemistry there is frequent reference to the non-bonded interactions between molecules, but the adjective 'non bonded' is unfortunate, and it really means weakly attractive or weakly repulsive, or not strongly bonding. Also, some explanation and definition of the terms 'molecular', 'intermolecular' and 'non-molecular' is needed in the context of the more complicated structures of inorganic compounds. A molecular structure possesses a group of atoms strongly bonded to each other, and interacting only very weakly with neighbouring atoms. Structural molecularity, a geometric property, is defined in terms of the placements of non-bonding boundaries, with four broad categories. Compounds are described as molecular in three dimensions, two dimensions, one dimension or zero dimensions according to the placement of non-bonding boundaries in three, two, one or zero dimensions, respectively. A conventional molecule is surrounded by non-bonding boundaries in three dimensions, while an infinite chain is surrounded by non-bonding boundaries in the two dimensions perpendicular to the non-molecular extension, and an infinite layer (or surface) is molecular in the one dimension perpendicular to the layer. The intermolecular interactions at the focus of supramolecular chemistry traverse the non-bonding boundaries which define the structural molecularity.

In this context three-dimensionally non-molecular structures would normally not have a supramolecular chemistry. However, there is a significant number of crystalline inorganic compounds which are technically non-molecular (in the sense that finite groups of atoms defining the molecule cannot be identified), but which paradoxically possess intermolecular interactions, and thus supramolecularity. This occurs because the number density of bonds establishing the non-molecularity is low, and the connections are usually quite elongated. Therefore in these crystals there are substantial regions of contiguous non-bonded groups which are supramolecular. In effect, these are like molecules turned inside out, with the periphery on the inside, and this will be referred to as internal supramolecularity in the following. The concept and importance of invaginated molecules are well known in molecular biology. Instances of internal supramolecularity occur in structures with at least two three-dimensionally non-molecular lattices which mutually interpenetrate without bonding (see Section 17), in compounds $Cd(SAr)_2$ (Section 12) and in nanoporous oxide lattices (zeolites and related) containing small molecules.

In this article the term supermolecule will be restricted to large molecules, often oligomers, totally connected by conventional bonding: it will not mean supramolecular assemblies or non-molecular structures.

Lehn [1] has also envisaged a wonderful array (literally) of supramolecular systems containing photoactive, electroactive and ionactive components. These systems can be designed and created to provide functions and properties such as light conversion by energy transfer, non-linear optics, molecular wires, ion-responsive monolayers, molecular channels through membranes, self-organizing helicates and liquid crystal phases. Lehn looks forward to chemionics, 'the chemistry of recognition-directed, self-organized, and functional entities of a supramolecular nature which may eventually be capable of replication'.

Metal and polymetal complexes will be valuable components of supramolecular devices because they are photoactive [3], electroactive [4], magnetoactive [5], bistable [6], and tunable. In addition, metal atoms usually engage a considerable number of bonds, and in this way they provide foci for structural organization.

In order to incorporate these attributes in the design and fabrication of supramolecular entities, it is necessary to develop a sound understanding of supramolecular forces and energies for inorganic compounds. Much of this understanding comes from observation of structure and patterns of structure, coupled with the theory and the computation of intermolecular interaction energies. The data on molecular and intermolecular structure come predominantly from crystal structures; the data are voluminous and valuable, and can be analysed specifically and statistically. Thus the objective of this article is to advance the understanding of supramolecular interactions in inorganic compounds by providing valuable viewpoints and perspectives on the theme of this volume: the crystal as a supramolecular entity.

This article is not a comprehensive review of previous research. As an expression of new perspectives, it is intended to be more a prescription for continuing research.

2. SCOPE

The scope and the themes of inorganic supramolecularity can be illustrated by the following features of inorganic chemistry.

2.1 Solubility, Crystal Growth and the Phenyl Factor

It is well known to synthetic chemists that compounds with exposed aromatic and heterocyclic sections, or covered with aryl groups, have relatively low solubility. Thus tetraphenylporphyrin and its complexes are substantially less soluble then etioporphyrin and its complexes: the alkyl substituents (Me_4Et_4) of etioporphyrin disrupt the directed supramolecular interactions which increase the lattice energy and decrease the solubility of tetraphenylporphyrin. Amongst the innumerable metal complexes with phosphines as ligands, the phenylphosphines, particularly Ph_3P, $Ph_2PCH_2CH_2PPh_2$ and many relatives [7], occur frequently. This is not only for reasons of cost and air stability, but the simple fact that compounds with these ligands crystallize first, and often well. Formation of adducts of metal complexes with fullerenes has been aided by the attractive interactions of phosphine phenyl substituents with the fullerene surface. Again, while ferrocene is soluble in many solvents, decaphenylferrocene is insoluble in all solvents [8]. Metal complexes with benzenethiolate are easier to crystallize than homologous complexes with alkanethiolates. The isolation and crystallization of anionic metal complexes from synthetic soups are frequently aided by use of phenylated cations such as Ph_4P^+. All of these phenomena are supramolecular chemistry in action, and all of them are due to the special supramolecular influence of the phenyl group: this phenomenon, the phenyl factor, will be developed later in this article.

2.2   The Crystal Field and Molecular Structure

As already mentioned, a very large amount of information about molecular structure now comes from the crystal structure, which is routinely determined for most new compounds that can be satisfactorily crystallized. The literature is replete with diagrams of molecular structure extracted from the crystal structure. But these descriptions of molecular structure come with the significant but tacit assumption that the crystal environment has not distorted the molecular structure. The assumption that molecular geometry in crystals is equivalent to molecular structure is now implicit in the pages of most current chemical journals: some authors genuflect to crystal packing, but very few analyse it. Many years ago Kitaigorodskii considered this, and concluded that the crystal field would have a negligible effect on relatively rigid hydrocarbons [9].

But coordination complexes and organometallic compounds are not always rigid. The number of ligand atoms bonded to a metal can vary in related compounds, and the concept of a larger number of weaker bonds or a smaller number of stronger bonds is fundamental for metallamolecules. Variable stereochemistry at metal atoms can yield structural isomers, and allow variable configurations and conformations. One of the distinctive characteristics of inorganic compounds is their geometric and electronic plasticity, which is important for their catalytic and biological activities.

Therefore in inorganic supramolecular chemistry there is a new meaning for the term crystal field (originally developed for metal ions doped in crystal lattices). The supramolecular crystal field is the total influence of the crystal surroundings on a target molecule: the magnitude, symmetry and anisotropy of the crystal field are important. We must continually ask to what extent the structures of molecular inorganic compounds are different in solution phases and crystal lattices, and seek quantification of this crystal field and its influences.

2.3 Nitrogenase, Metalloenzymes, Inorganic Molecular Biology and Biomineralization

The crystal structure of the Fe–Mo protein of the enzyme nitrogenase [10] has recently been determined [11], revealing the unexpected (cysteine-S)Fe(μ_3-S)$_3$Fe$_3$(μ-S)$_3$Fe$_3$(μ_3-S)$_3$Mo(homocitrate)(histidine) cluster, which is the site of the binding and complete reduction of one of the most recalcitrant molecules of chemistry, N$_2$. The cluster and its tantalizing catalytic ability depend on the protein surrounds, again in an unusual fashion because the cluster is bound to the protein through only two amino acid residues at opposite poles of the cluster. Some of the iron atoms in the cluster appear to be undercoordinated, and yet the relationship between these atoms and the protein surrounds is non-bonding, i.e. supramolecular. A proposed model for the binding and reduction of N$_2$ depends critically on the protein side chains surrounding the equator of the cluster, on hydrogen bonding to the sulfur atoms of the cluster, and on torsional motion of the cluster relative to the protein [12]. It is clear that the supramolecular effects important in all metalloenzymes are accentuated and special for this iron–molybdenum sulfide cluster, and need to be understood in detail.

This introduces an important theme of supramolecular chemistry, namely that nature — evolved molecular biology — is the supreme supramolecular chemist. There is much supramolecular inorganic chemistry to be learned from molecular biology. This dictum is illustrated by a variety of metalloenzymes such as nitrogenase, and also in the topical field of biomineralization. Current research is uncovering the variety of sophisticated inorganic materials formed under biological control, with special properties relating to mechanical strength, chemical resistance and morphology [13]. Biominerals are also involved in the processes of metal storage and metal detoxification [14].

2.4 Metal Complexes and DNA

Non-bonded associations of metal complexes and the helices of DNA are vital [15]. These phenomena include the 'zinc finger' proteins which bind specifically to DNA [16], the metalloregulatory proteins involving Hg, Fe and Cu which trigger gene function [17], the planar metal chelate systems which intercalate the stacked base pairs at the major groove, and the association with chiral chelate complexes such as [Ru(phen)$_3$]$^{+2}$ (phen = 1,10-phenanthroline) in the minor groove with enantioselectivity involving only hydrophobic effects [15]. Applications of these supramolecular assemblies include metallopharmaceuticals [18] and metallocomplexes as probes of DNA protein assemblies through photo-footprinting scission analyses of DNA and RNA [15, 19].

2.5   Metallomesogens, Ordered Arrays and Cooperative Physical Properties

Ordered (and partially ordered) arrays of metal sites and complexes enable the cooperation of their special electronic, magnetic and optical properties. Such materials have long been sought for their expected physical properties and applications in optics, electrooptics, superconductivity and sensors. The ordering can be by various mechanisms, such as adsorption on surfaces, intercalation into layered structures, formation of mesomorphic structures and liquid crystals, and adoption of specific crystal-packing motifs, all of which are supramolecular phenomena. Organic liquid crystals and their applications are now commonplace, and in recent years the incorporation of metal atoms into mesogenic molecules has demonstrated the occurrence of similar metallomesophases [20].

2.6   Condensation of Gas-Phase Species

Gas-phase inorganic chemistry is undergoing a renaissance, mostly through the use of laser ablation and plasma techniques for synthesis, coupled with ion-trap mass spectrometry for separation, monitoring and reactivity studies of the products [21]. Laser ablation of metal oxides, carbides, nitrides, phosphides, sulfides, selenides and tellurides has yielded many hundreds of new ions, which are molecular structures for compounds which until recently were known only as solids with non-molecular structure. The significant challenge for this research is condensation of the newly discovered gas-phase species, a process which imposes supramolecular influences on isolated molecules. For example, a number of cyanocomplexes of Zn and Cd, with the general formula $[M_n(CN)_{2n+1}]^-$ ($n \leqslant 27$) have been generated, and are proposed to contain unique helical structures (such as that shown in Figure 1) dependent on non-bonding interactions between contiguous strands [22]. The challenge is how to crystallize these structures without reverting to the different inclusion lattices already known in the crystalline derivatives of $Zn(CN)_2$ and $Cd(CN)_2$ [23]. The issues raised in the condensation of all new pristine binary molecules are the essence of supramolecular inorganic chemistry.

2.7   Electronic and Magnetic Properties of Metallamolecules

The original use of the description 'crystal field' was more specialized, referring to the influence of a crystal on the electronic and magnetic states of a transition metal ion doped into the lattice. This meaning and the more general supramolecular crystal field defined above come together in the phenomenon of spin-state variation in metal complexes [3b, 5a, 5b, 6b, 24]. In these compounds where the strength of the ligand field (replacing the original crystal field) places the molecule on the borderline between high-spin and low-spin

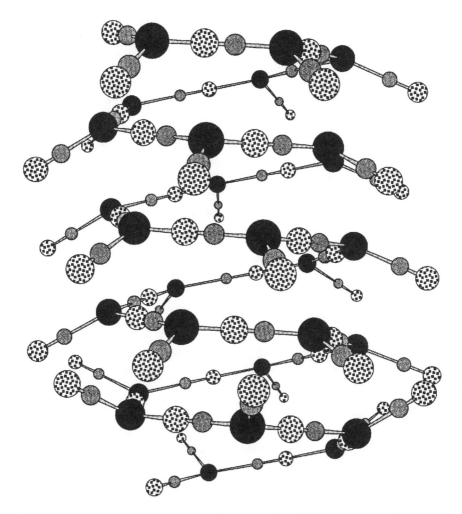

Figure 1 The proposed helical structure for the ion $[Zn_{27}(CN)_{55}]^-$ synthesized in the gas phase. Dark circles are trigonal planar Zn atoms, speckled circles N, grey circles C. The helix is maintained by coulombic attractions, parallel to the axis of the helix, between non bonded $Zn^{\delta+}$ and $CN^{\delta-}$

states, it is often found that the magnetic properties of solids are lattice dependent and affected by variations in counterion or solvent in the lattice. These small variations in the lattice crystal field are sufficient to change the electronic crystal field at the metal centre. In addition, the crystal surroundings of the magnetically active molecule affect the manner in which the spin state changes when induced thermally or by hydrostatic pressure: the spin transition can be (1) gradual and continuous over a wide temperature range, (2) discontinuous over a narrow temperature range, or (3) discontinuous with hysteresis. The change in spin state involves substantial changes in metal–ligand bond lengths (ca.0.2 Å for $Fe^{II} - N$) and in molecular volume (ca. 5%).

A demonstration of the subtleties of the interaction of a metal complex and its environment is found in a detailed examination of the spin-state crossover in two related crystalline compounds $Fe(L-L)_2(NCS)_2$, where $L-L = 1,10$-phenanthroline (phen) and 2,2'-bis(4,5-dihydrothiazine) (bdhtz) [25]. Both crystalline compounds undergo a thermal spin-state change which is abrupt for the phen compound but gradual for the bdhtz compound. Both compounds have the same crystal packing, and both undergo the same intramolecular changes in geometry associated with the spin transition. Nevertheless, the subtle coupling and cooperation between the intramolecular and inter-molecular electronic and geometric structures are different in the two compounds.

An understanding of the mutual supramolecular interactions between the magnetically active molecule and its crystal environment is crucial for progress with these intriguing and potentially valuable phenomena.

2.8 Supramolecular Supermolecules

One active area of research in coordination chemistry is the synthesis of supermolecules, which provide multiple coordination sites linked through bridging ligands. The class of metallomacrocycles has been reviewed recently [26]: these are metal complexes elaborated with additional ligands and able to coordinate further metal atoms into oligometal complexes. The construction precepts of 'complexes as metals' and 'complexes as ligands' have been developed to generate polymetallic molecules designed specifically for amplification of photophysical phenomena [27]. These oligomeric systems depend on bridging ligands or bridging metal atoms, not only on non-bonded interactions, and thus are not strictly supramolecular. However, the number of bonding interations is small, and the supramolecular chemistry of these systems is important. Supermolecules of these types relate also to an objective in contemporary chemistry to replicate larger metal-containing structures of molecular biology, such as the photon absorption and electron transfer centres of photosynthetic organisms. There is vibrant research on supermolecules containing metal centres and undergoing photoinduced charge separation and long-range vectorial migration of energy or electrons [3a, 28].

Sauvage *et al.* have elaborated special molecules with the topologies of catenated rings and knots [29]. The syntheses of these depend on metal coordination sites which direct the assembly of the components, but in general these catenanes and knots have few bonded connections. They are special because non-bonded sections are topologically trapped to coexist, thereby creating new opportunities for investigation of supramolecular relationships. A catenated dimer of a cyclic trimer of a zinc porphyrin involves π–π supramolecular interactions [30].

2.9 Inorganic Host–Guest Inclusion, Guest–Host Templation and Nanostructures

The diversity of the inclusion of molecules of any type into host lattices which are distinctly inorganic is now becoming apparent [31]. Apart from the oxidic zeolites, aluminosilicates and derivatives with other tetrahedral sites, there are the layered silicates and minerals, pillared clays and numerous other intercalation and inclusion systems, including graphite, metal sulfides and metal cyanides [23, 32, 33].

A related rapidly developing field of research uses organic and biological mesophases as frameworks for the construction of inorganic compounds. The mesophases themselves are supramolecular assemblies, usually of amphiphiles, which adopt micellar, layered or tubular morphology. When inorganic mixtures undergo crystallization or precipitation in the presence of these mesophases, exciting new inorganic nanostructures can be formed [34]. This connects with biominerals, mentioned above. The ability of molecular biology to grow SiO_2 skeletons with amazing morphologies [35], to shape singly crystalline $CaCO_3$, and to form magnetosomes containing Fe_3O_4, inspires biomimetic strategies for the development of inorganic materials [34c].

This is supramolecular chemistry because it appears that in general there are relatively weak interactions between the mesophase, or adsorbed solute molecules, and the growing inorganic solid. Further, the species that control the formation of the inorganic material are contained in the product only at very low concentrations, if at all. The supramolecular and biomimetic approaches to the preparation of inorganic materials and nanostructures are mild, low-energy processes, and contrast strongly with the high-energy, high-vacuum techniques of sputtering, vapour deposition and implantation. The more gentle, 'natural' processes have a philosophical attraction, and possibly a reliability, not possessed by the higher-energy ('military') techniques.

3. CONCEPTUAL AND THEORETICAL FRAMEWORK

3.1 Definition of Domains

Discussion of the supramolecularity of metal-containing molecules is conveniently organized through the definition of the distinguishable domains of molecules and their surroundings, illustrated in Figure 2. The metal domain $\mathbb{M}$, often at or close to the core of the molecule, contains the metal atom(s). Surrounding this are the ligand donor atoms and the framework which connects them, collectively labelled domain $\mathbb{L}$. The domain $\mathbb{P}$ contains the

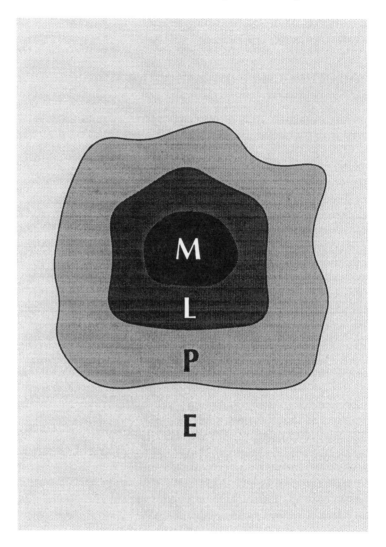

Figure 2   The domains of inorganic supramolecular chemistry. The molecule comprises domains M, L and P within its environment E

atoms on the periphery of the molecule, able to contact the molecular surrounds, domain E. All atoms surrounding the molecule are in domain E: E can be the solvation sheath, the crystal lattice, an amorphous solid array, or E can be absent as in gas-phase chemistry. An alternative concept of the enclosure shell (ES) of surrounding molecules in contact with the target has been used [36], but is restrictive in dealing with the longer-range coulombic interactions.

Recognition of domain E is important because most practical chemistry is effected in the presence of E, the cumulative influences of which can be substantial. These influences of E are often neglected in molecular mechanics computations of molecular structure, and frequently ignored in interpretations of molecular structure in crystals. However, inorganic molecular biology demonstrates that E can make sophisticated contributions to chemistry.

These domains help to organize the plethora of inorganic compound types for which the supramolecularity is to be analysed. This is because supramolecular interactions are not restricted to contiguous atoms (as are chemical bonds), and can be transmitted through domains: for instance, there can be coulombic interactions between partially charged metal atoms in M and partially charged atoms in domain E, traversing domains L and P. Thus, while first views of supramolecularity tend to focus on the $P–E$ interface, and particularly on the shape of the molecule and this interface, the other domains of inorganic molecules can have strong direct or indirect influences. The whole molecule needs to be considered: modifications of bonding and structure in M can readily cause large modifications in P and interact with E. The variability of inorganic molecular stereochemistry is much larger than that of organic molecules, and the influences of E on the stereochemistry of M, L and P need to be understood because they are significant.

The domains of inorganic supramolecularity are valuable also as a classification to accommodate substantial chemical diversity. While it is often true in inorganic systems that the peripheral domain P is heavily populated with organic substituents (jocularly dubbed the 'spinach'), such that the molecules look organic from the outside, in fact the intrigue of inorganic chemistry is that most elements can occur in any of the domains, and the supramolecular interactions and considerations can be much more variable than for conventional organic molecules. The three molecular domains may constitute very different proportions of the molecular volume.

A sizeable number of inorganic molecules possess an open domain P which is penetrable by other molecules. An example is the complex $[Fe(SPh)_4]^{2-}$, shown in Figure 3(a) as it occurs when crystallized with Ph_4P^+ [37]. This complex and others like it have another property relevant to the supramolecularity, namely substantial conformational flexibility. All four ligands can rotate around Fe–S bonds and S–C bonds, while maintaining tetrahedral angles at Fe and S, allowing numerous molecular conformations. These conformations (one of which is illustrated in Figure 3b) have different intramolecular interactions between phenyl substituents. For compounds like this there is a substantial intramolecular aspect for the supramolecularity. A metalloporphyrin is anisotropic in shape, with a markedly penetrable domain P and domain M exposed, but in contrast it has little conformational flexibility.

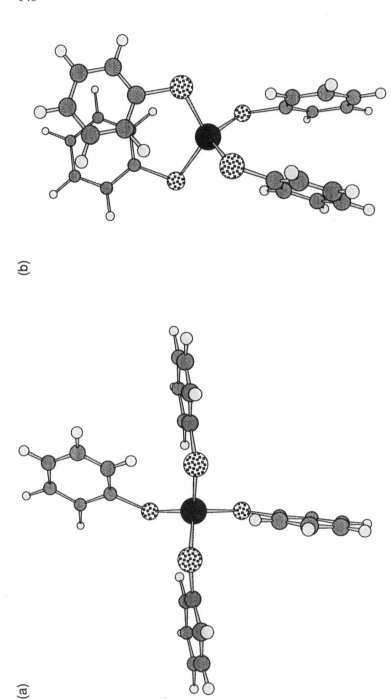

(b)

(a)

Figure 3 (a) $[Fe(SPh)_4]^{2-}$ as it occurs when crystallized with Ph_4P^+, showing the open periphery of the molecule. (b) An alternative conformation for $[Fe(SPh)_4]^{2-}$, by rotation about Fe–S and S–C, with favourable attractive interactions between Ph substituents

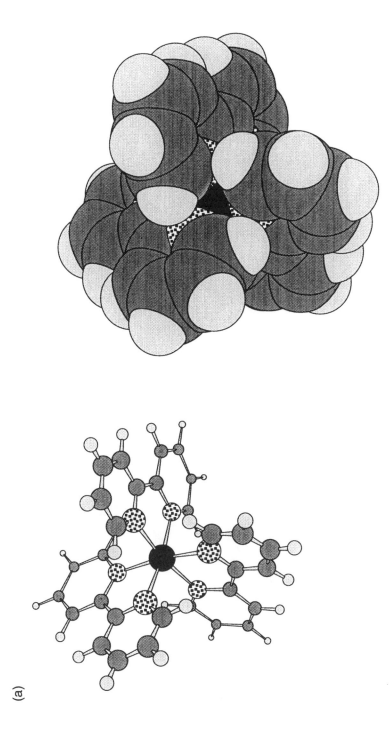

Figure 4 Representative inorganic molecules and their domains for supramolecular purposes: (a) $[M(bipy)_3]^{2+}$. In each case the metal domain $\mathbb{M}$ is shaded dark, the donor atoms (domain $\mathbb{L}$) are speckled, and C and H atoms (in domain $\mathbb{L}$ or $\mathbb{P}$) are grey

150

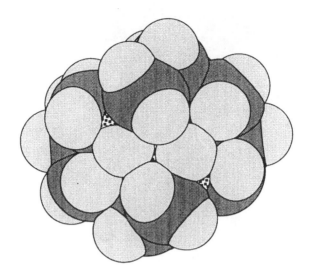

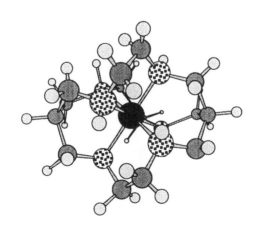

Figure 4 (*continued*)   (b) a metal sepulcrate or cryptate

(b)

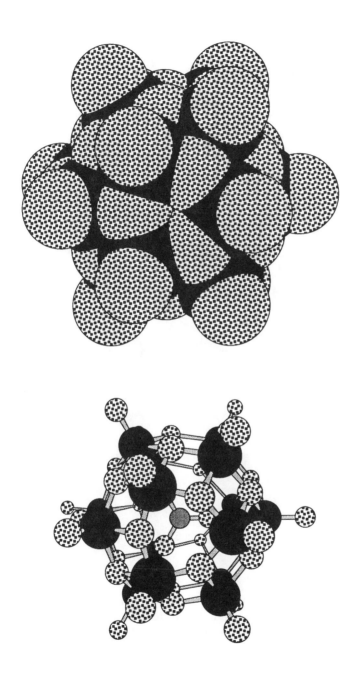

Figure 4 (*continued*)   (c) the centred oxovanadate $[(CO)_3V_{15}O_{36}]^-$

(c)

(d) i

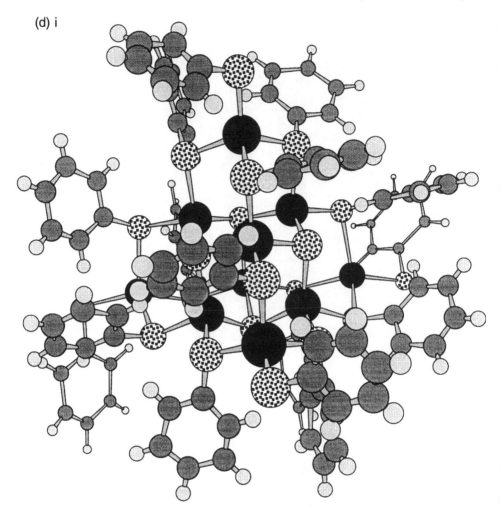

Figure 4 (*continued*)   (d) i: $[S_4Cd_{10}(SPh)_{16}]^{4-}$

3.2 Illustration of Domains

The compounds pictured in Figure 4 exemplify the scope of inorganic
molecules and their domains, and illustrate the ambit of inorganic
supramolecular interactions. The variable shape of the peripheral surface
and the exposure of the metal and donor atoms are evident in the space-filling
representations.

In this variety of compounds — a mosaic of the periodic table — the
distinction between bonding and non-bonding interactions is sometimes

(d) ii

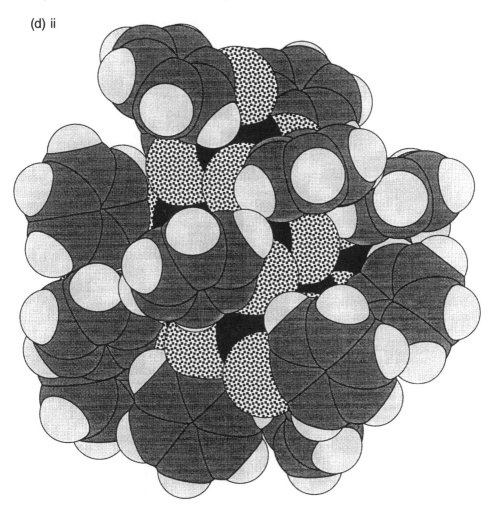

Figure 4 (*continued*)   (d) ii: $[S_4Cd_{10}(SPh)_{16}]^{4-}$

blurred. There is very much more variability of molecular chemistry, and of supramolecular chemistry, than occurs in the more familiar but restricted domains of organic compounds and most biological compounds. It is important to examine and define what are meant by bonding and non-bonding interactions, and to establish the applicable conceptual and theoretical framework. The objective of the following sections is to raise and develop the relevant questions and approaches.

3.3   Primary Bonding, Secondary Bonding, Metal–Metal Bonding, Non-Bonding

Primary bonding of a metal or other high-Z atom involves concepts of oxidation state, coordination number, coordination stereochemistry, bond length and strength, and bond polarity, all of which may vary quite widely. To exemplify the variability, how does Mo in $[Mo(CO)_6]$ compare with Mo in $[MoS_4]^{2-}$, or with Mo in $[Mo_7O_{24}]^{6-}$, or with the Mo atom in the active cluster of nitrogenase? These are very different Mo atoms, and it would be reasonable to expect that at least their partial charges (and thus their contributions to coulombic supramolecular energies) could be quite different. Similarly, what about Fe in $[Fe(\eta^5\text{-}C_5H_5)_2]$, in $[FeCl_4]^-$, and in haemoglobin? The phosphorus atom in PPh_3 is different from the phosphorus atom in the phosphotungstate anion $[PW_{12}O_{40}]^{3-}$. Such chemical diversity must induce diversity of supramolecularity. While concepts and theory for strong primary bonding are well developed for multifarious compounds of the same element, concepts of weakly bonding interactions are still nascent.

(e) i

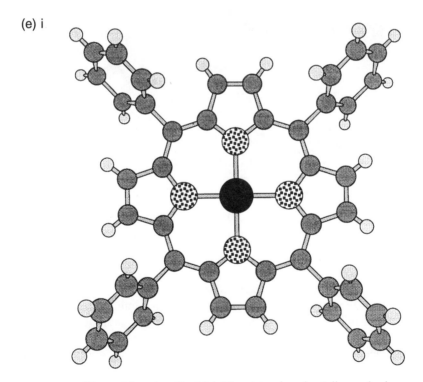

Figure 4 (*continued*)   (e) i: Mesotetraphenylmetalloporphyrin

One complication is the variability of coordination number and stereochemistry around a metal atom. It is defensible to hold a concept of metal coordination in which metal coordination requirements can be satisfied by a larger number of weaker (longer) bonds or a smaller number of stronger (shorter) bonds, or any intermediate collection. This notion follows from the well-established correlations of bond length and coordination number [38]. A metal such as zinc can be six-coordinate or five-, four- or three-coordinate, usually with softer anionic ligands or bulkier ligands supporting the lower coordination numbers. Again, this variability of primary coordination will introduce a variability of supramolecular effects, for instance through changes in the partial atomic charges on metal and ligand atoms.

There is a further complication. Some metals, especially the late transition and post-transition metals, engage long bonds — secondary coordination — in addition to the regular primary bonds. This is prevalent in mercury thiolate

(e) ii

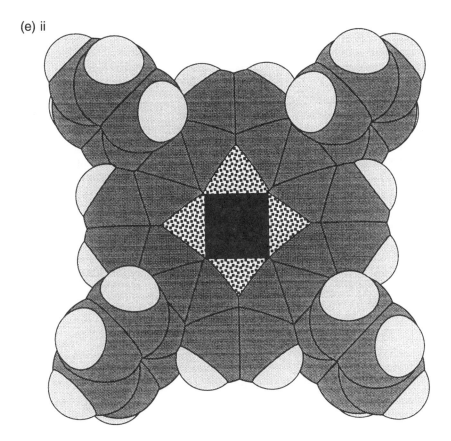

Figure 4 (*continued*)   (e) ii: Mesotetraphenylmetalloporphyrin

compounds; for example, crystalline $Hg(SPr^i)_2$ [39] has two Hg–S bonds of length 2.33 Å, one at 3.07 Å, one at 3.29 Å and one at 3.52 Å, all less than the sum of the van der Waals contact radii; in crystalline $Hg(SMe)_2$ the bonds around Hg are 2.36 Å ($\times 2$) and 3.25 Å ($\times 3$) [40]; and in $[Hg_2(SMe)_6]^{2-}$ the Hg–S bond lengths are 2.44 Å, 2.47 Å, 2.63 Å, and 2.71 Å [41]. Continuously variable elongation of axial bonds in Cu^{II} coordination is well known. While coordination number and stereochemistry at Pb^{II} are widely variable, a common motif is three primary bonds, approximately orthogonal, and two or three bonds ca.1 Å longer. This blurs the distinction between bonded and non-bonded interactions. Where does supramolecular chemistry begin in such compounds?

For conventional molecules the conceptual basis for intramolecular and intermolecular interactions involves two types of distance–energy curves, as illustrated in Figure 5. One curve (A, typically a Morse function) with a deep minimum describes the bonding interaction, while a second (B, typically a Lennard–Jones or equivalent function, see below) with a much shallower minimum at longer distances describes the non-bonded van der Waals attractions and repulsions. While curves A and B appear to be in conflict in the distance regime X, with one curve expressing attraction and the other repulsion, it must be remembered that curve B represents the energy for approach to a coordinatively saturated central atom. This is, when the central atom has a full complement of bonds at distance d_A, and all bonding (and probably non-bonding) molecular orbitals filled, approach by an additional atom will follow the energy of curve B.

Metallocompounds with variable coordination numbers and plastic coordination stereochemistries may not behave in this conventional way: the energy–distance curves for any one bond could vary with the number and lengths of the other bonds. With this variability of the primary bonding, the residual, more weakly bonded interactions could be expected to be even more variable. Thus an open mind is required in developing parameters for weakly bonded interactions for metallocompounds. Further development of fundamental theory for long, weak interactions to metal atoms is required.

Late transition metals and post-transition metals, especially Ag, Au and Hg, commonly have very low coordination numbers, exposing the metal atom. In such compounds domain $\mathbb{M}$ penetrates domains $\mathbb{L}$ and $\mathbb{P}$ to $\mathbb{E}$. Bonding and non bonding concepts are challenged further in silver and gold compounds, where there are questions about weak bonding between contiguous, exposed metal atoms. Consider the crystal structure of $AgSCMeEt_2$ [42], shown in Figure 6. There are two infinite strands, each $-(\mu\text{-}SR)\text{-}Ag\text{-}$, which are entwined but with no interstrand bonds. This double-helical structure is coated with the thiolate substituents, and the crystal structure is two-dimensionally molecular, with consequent supramolecularity. But what is the relationship between the strands of the chain? The S–Ag–S segments are approximately linear, but note

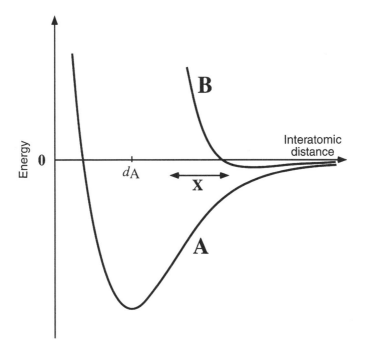

Figure 5 Comparison of the Morse curve (A) for chemical bonding and the van der Waals energy curve (B): see text for explanation

that the Ag atoms of one chain are not adjacent to S atoms of the other (as would occur for secondary coordination, or for coulombic interactions). Instead, Ag atoms are adjacent to Ag atoms. The question of the relationship between the stands translates to: is there Ag–Ag bonding?

The closest contact between the strands of the chain is Ag1···Ag2 at a distance of 2.89 Å, which is equal to the interatomic separation in silver metal. However, my interpretation of this contact is that it represents the onset of a repulsive interaction between the silver atoms. In this structure the linear geometry at silver indicates that its bonding is fulfilled by the two thiolates, and there is no requirement for secondary Ag···S bonding. The slight bending of Ag1 and Ag2 outside their S···S lines, away from each other, is indicative of weak repulsion, not attraction, between them. It is believed that the two zigzag strands of the chain approach each other until limited by the internal repulsion between opposite silver atoms. A corollary of this interpretation is a strong

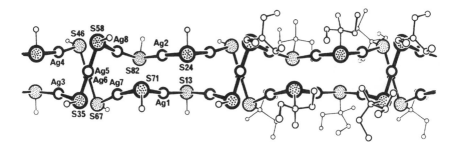

Figure 6 The double-stranded chain of AgSCMeEt$_2$, in which the zigzag strands of alternating Ag and S atoms are intertwined but not bonded: note the small outward displacements of opposing Ag atoms on different strands, consistent with a weak repulsive interaction. Reproduced with permission from *Inorg. Chem.*, **22**, 3785 (1983)

dependence of both the capacity and the distance for Ag$\cdots$Ag bonding on the coordination number and nature of the other ligands. Thus, an Ag–Ag separation of 2.88 Å is normal bonding when the Ag atoms are 12-coordinate (in the metal), but the same separation is nonbonding when the Ag atoms are coordinated by two anionic thiolate ligands.

There are now a large number of cystalline compounds which contain Au atoms separated by ca. 3.0 Å (±0.2 Å) in a weakly bonding interaction [43]. Many of these compounds have been designed, principally by Schmidbaur, following recognition of this aurophilicity of Au atoms; pronounced relativistic effects in gold underlie this aurophilicity. Some of the aurophilic compounds are described in Section 10 in another supramolecular context. As a reference point, strong Au–Au bonding occurs in the Au$_2$ molecule, with a bond length of 2.5 Å and a dissociation energy of ca. 230 kJ mol^{-1}. A correlation of bond distance with the Au–Au stretching force constant has been established [44]. An important datum is the energy of the general Au–Au attraction at 3.0 Å. Through ingenious experiments in which the Au–Au interaction competes with a known torsional barrier in the ligand, Schmidbaur estimates that Au–Au attraction at 3.0 Å is not greater than 24–32 kJ mol^{-1} [43a, 43b, 45]. *Ab initio* theory for this interaction calculates it to be about 15–25 kJ mol^{-1} [46], consistent with the experimental data. The occurrence of this aurophilic interaction in many crystalline compounds is important because it provides a comparison energy datum for other ligand-based supramolecular interactions. Further illustration of the variability of bonding versus nonbonding interactions at gold is found in the compounds of the type $Au_2(P–P)_2X_2$, where P–P is a chelating ligand and X is a halide or pseudohalide [47].

The aurophilic interaction appears to be an important paradigm in inorganic supramolecular chemistry because it calibrates intermolecular interactions between domains M. However, because 'gold is different' (for relativistic reasons) [43a, 43b], extension to other metals is ambiguous.

It is clear that supramolecular inorganic chemistry requires expansion of the conceptual base conventionally used for organic and biochemical molecules. Molecules with elements throughout the periodic table, and with much more diverse geometrical and electronic structures, must have a richer intermolecular chemistry.

The conventional concepts and computational methodologies for nonbonded and intermolecular interactions are essentially atom based [48], and therefore are conceptually and operationally suitable for molecules comprising the gamut of the elements, and where functional groups are often less relevant. Atom-based evaluation of intermolecular interaction energies has not yet been challenged by an alternative, and will underlie the remainder of this article.

3.4 Non bonded Interaction Energies: The Intermolecular Force Field

Just as intramolecular energies can be evaluated with a force field for bonds, defined according to chemically distinguishable atom types, calculations of intermolecular energies use an intermolecular force field, usually requiring fewer atom types. The intermolecular force field has three components: the attractive van der Waals (dispersion) energy, the repulsive energy which keeps nonbonded atoms apart and the coulombic energies resulting from partial charges on atoms in molecules. The components of the intermolecular interaction are evaluated for atom pairs and summed over all relevant atom pairs.

The resultant of the attractive and repulsive van der Waals (vdW) energies between two atoms has the general shape shown in Figure 7. In the relevant distance regime this curve can be represented by several algebraic expressions: the discussion here is in terms of the Lennard–Jones expression (equation 1) for atoms i and j separated by d_{ij}

$$E_{ij}^{vdW} = e_{ij}^a[(d_{ij}^a/d_{ij})^{12} - 2(d_{ij}^a/d_{ij})^6]$$ (1)

in which d_{ij}^a is the distance between atoms i and j at which the interaction energy is most negative, with the magnitude e_{ij}^a [49, 50]. The first term in equation (1) represents the repulsive energy, the second the attractive energy.

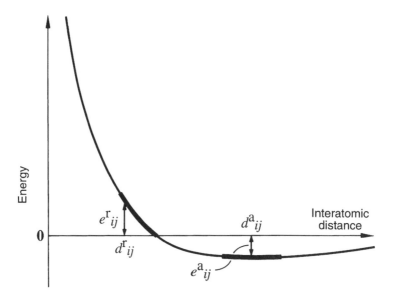

Figure 7 The variation of non-bonded van der Waals energy with distance for the atom pair ij. The significant attractive and repulsive regions are accentuated, and the attractive (superscript a) and repulsive (superscript r) distance and energy parameters are defined

Relationships between equation (1) and alternative expressions in which the repulsion is expressed exponentially are presented in the extensive literature on van der Waals interactions.

The two parameters e^a_{ij} and d^a_{ij} in equation (1) for any pair of atom types are usually obtained by combination of atom parameters e^a_i and r^a_i

$$d^a_{ij} = r^a_i + r^a_j \tag{2}$$

$$e^a_{ij} = (e^a_i \, e^a_j)^{0.5} \tag{3}$$

In this way the vdW energies are derived from parameters compiled for atom types in molecules.

In Figure 7 a second region of the interaction energy is emphasized, namely that where the interaction energy is between zero and a relatively small positive value e_{ij}^r [49]. This represents the closest that atoms i and j can approach without incurring repulsions which would overbalance the attractions. This is the connection between the interaction energy curve and the conventional concept for non-bonded contacts, involving the impenetrable surface of the molecule. This is the fundamental concept of Bondi's classical analysis [51] of van der Waals volumes and radii, and is the concept which underlies the innumerable discussions of van der Waals contacts in the literature of molecular crystal structures. The shortest observed distances are slightly repulsive distances: the assumption is that in molecular association some atom pairs will be at slightly repulsive distances in order that the bulk of the interatomic non-bonded distances are in the attractive distance domain, and the lattice energy is minimal. The distances d_{ij}^r are the distances represented by sums of conventional van der Waals radii r_i^r. Clearly the r_i^r are distinct from and less than the r_i^a.

Some confusion has arisen in the literature where ionic radii of metal ions and of anions have been used as non-bonded radii r_i^a in force fields. This usage is inappropriate, because the charges on metal atoms in domain $\mathbb{M}$ and the charges on electronegative donor atoms in domain $\mathbb{L}$ are not as extreme as the formal oxidation states or the charges used for lattice energy calculations of non-molecular ionic solids. The non-bonded radii of metals are substantially larger than the formal cation radii. For the purposes of estimating the sizes of metal atoms and electronegative atoms it is better to consider them to be relatively neutral, with sizes scaled according to covalent radii rather than ionic radii.

How then are the radius and energy parameters for the van der Waals interactions obtained? It is necessary to determine the atom types for which differentiable parameters can be justified. For the standard light elements of organic compounds the atom types (functional groups) are established, as are their van der Waals parameters through the successful modelling of many structural and thermochemical data [52–54]. For the remaining hetereogeneous elements of inorganic compounds, force field development is relatively unsophisticated. As far as van der Waals parameters are concerned there is often just one atom type per element, although with very widely different oxidation states this could be an oversimplification. Certainly the coulombic contributions to intermolecular energies, discussed in detail below, will vary with oxidation state.

The valuable determination of r^r values by Bondi [51] relied on the closest non-bonded contacts in key selected compounds. With the volume of crystal structure data now available for interrogation through the Cambridge Structural Database [55], it is possible to use statistical analysis to extract nonbonded contact distances [56, 57]. Thus, to ascertain r^r for atom type Y, it is possible to choose a probe atom Z and then extract all of the reliable crystal structures containing both Y and Z not mutually bonded, and then generate a histogram of the $Y \cdots Z$ intermolecular separations. The shortest $Y \cdots Z$ non bonded contacts are thus reliably evaluated, and r_Y^r follows from knowledge of r_Z^r. In deployment of this method for metals and other heavy atoms [58], hydrogen is a good choice for the probe atom Z because it occurs frequently on the periphery of organic ligands, its partial charge is low and so coulombic contributions are minimized, and r_H^r is well known. It is of course necessary to be careful to avoid compounds where the assumption of non-bonding contact is invalid, such as in agostic H–M bonding interactions, but the statistical examination of voluminous data allows exclusion of anomalously short contacts. This method of evaluation of r^r is general and straightforward.

Evaluations of e^a and r^a parameters are less straightforward. The relevant experimental data to which these parameters can be fitted are crystal structure and lattice energy (heat of sublimation); however, this method of estimation has not been undertaken with inorganic molecules to the extent that it has with small organic molecules. A tacit assumption is that the e^a and r^a parameters follow the general periodicities of atomic properties, with e^a following electric polarizabilities and r^a tracking both bonding radii and r^r. Values of r_i^r for an atom type can be used to estimate r_i^a and e_i^a in the following manner, using a known probe atom type j. Knowledge of the form of the van der Waals curve (such as equation 1) and the coordinates of the point (d_{ij}^r, e_{ij}^r) (Figure 7) for an atom pair allows the curve and thus e_{ij}^a and r_{ij}^a (and thence e_i^a and r_i^a) to be determined. The largest uncertainty in this calculation is the magnitude of the small repulsive energy e^r at the point of closest observed contact between i and j. For the well-understood $H \cdots H$ interaction e^r is estimated to be $+1.4 \, \text{kJ mol}^{-1}$ [58], while Gavezzotti [59] provides evidence that $C\text{-}H \cdots O$ repulsions amount to $+(0.8\text{-}2.5) \, \text{kJ mol}^{-1}$. For $M \cdots H$ interactions a value of $e^r = +1.6 \, \text{kJ mol}^{-1}$ has been used [58]. In evaluating parameters of the inter-molecular force field for H, C, N, O and F, Williams and Houpt [60a] noted that the point d_{ij}^r for Pauling radii r_i^r represented a repulsive force between atoms of $10^{-10} \, \text{N}$, which corresponds to larger e^r values in the range $10\text{-}20 \, \text{kJ mol}^{-1}$; Spackman [60b] discussed repulsive interactions, and used $4 \, \text{kJ mol}^{-1}$ for e^r.

For first-row transition metal atoms, Mingos and Rohl [61] scaled van der Waals radii according to the metal covalent radii r^c: $r^r = 2.0 r_{\text{metal}}^c / r_{\text{Fe}}^c$, assuming $r_{\text{Fe}}^c = 2.0 \, \text{Å}$. This scaling approach (but with less bias by one element) will be more widely useful in assessing r^r and r^a.

All resulting e_i^a and r_i^a parameters should be checked through their consistency with inorganic periodic principles. A universal force field (UFF) has been presented for atom types throughout the entire periodic table [62], and includes values for $x_I (\equiv 2 r_i^a)$ and $D (\equiv e_i^a)$. The UFF values were derived from atomic properties, observed and calculated, with considerable assumption and uncertainty. Nevertheless there is general agreement between the UFF values for uncharged elements and other literature values, and these UFF values demonstrate the expected periodicities. However, UFF values for metal cations are quite different, and sometimes dubious.

3.5 Coulombic Influences

A principal difference between supramolecular inorganic chemistry and the longer-known supramolecularity of organic and biomacromolecular systems is the presence of significant, if not dominant, coulombic interactions. Coulombic effects can have influences even for hydrocarbons. The edge-to-face geometry preferred for adjacent benzene molecules and systems with phenyl substituents is a consequence of the H atoms with small positive charge being attracted to the π-density central to the aromatic ring, even though most of the attractive energy between the rings is London dispersion [63] (see Section 5.1). Polar bonds and appreciable partial charges on atoms occur frequently in inorganic compounds. Metal atoms are characteristically electropositive, over a wide range, while atoms in the L domain are frequently electronegative.

The evaluation of coulombic energies commonly uses the approximation of atom-based charges. That is, atoms in the structure are assigned partial charges q (in electron units), and the coulombic energy is evaluated as the sum of terms

$$E = q_i q_j / \varepsilon d_{ij} \qquad (4)$$

for all relevant atom pairs separated at distance d_{ij} by a medium of dielectric constant ε. Only nonbonded atom pairs are summed. Coulombic energies between bonded atoms are otherwise included in force fields, and are not relevant for supramolecular considerations. The location of excess charge at the nuclear coordinates is clearly an approximation, but there is simplicity of concept and ease of evaluation associated with atom partial charges. For relatively well-defined smaller molecules and functional groups, coulombic interactions have been evaluated using molecular dipoles and quadrupoles [64].

A key factor is the dielectric constant (or relative permittivity) ε to be used in this calculation, in the present context for solids, not solutions. In the 'vacuum' between contiguous charged atoms $\varepsilon = 1$, while dielectric constants in the range 2–5 are generally regarded as accounting for the polarizability of molecules with immobile polar bonds. Therefore it is common practice to describe the intercharge medium with a distance-dependent dielectric constant $\varepsilon = r_{ij}$ which will provide a good approximation of the dielectric properties over the most significant interaction ranges. This could overestimate the dielectric constant (and underestimate coulombic energies) for situations where charged sections of a supramolecular assembly are separated by material such as hydrocarbon with low dielectric. Such situations do occur in supracoordination chemistry, e.g. metallomesogens. Computational procedures to provide improved representations of the variations of the dielectric and electrostatic potential

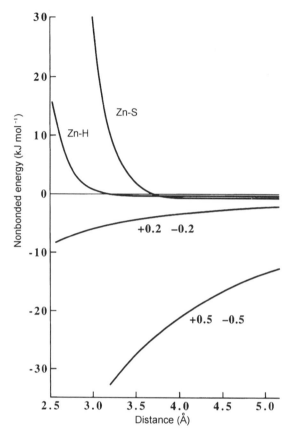

Figure 8 Plots of non-bonded energy (kJ mol^{-1}) versus distance (Å). The curves labelled Zn–H and Zn–S are the van der Waals energies from equation (1) using e^a (kJ mol^{-1})/r^a (Å) values of 0.63/2.06, 0.85/2.11 and 0.20/1.50 for Zn, S and H, respectively. The curves labelled '+0.2 −0.2' and '+0.5 −0.5' are the coulombic energies for oppositely signed charges of these values, calculated according to equation (4) with a screened dielectric constant $\varepsilon = r_{ij}$

in macromolecular systems, based on numerical solution of the Poisson–Boltzmann equation, are available [65].

The significance and consequences of coulombic interactions are demonstrated by the energy curves in Figure 8, where the coulombic energies (using $\varepsilon = r_{ij}$) for partial charges of 0.2 and 0.5 are compared with the van der Waals energies for two representative atom pairs, Zn–H and Zn–S, over typical non-bonded distance ranges.

Two conclusions are rampant. Both the magnitude and the extent of the coulombic energies outstrip those of the dispersion energies. At the distance of closest contact of atoms in domain ℙ (ca. 3–4.5 Å), the coulombic energies, even for small charges of ca. 0.1–0.2, are greater by about an order of magnitude than the dispersion forces. As will be seen, it is not uncommon to have partial atomic charges of ± 0.5 in coordination compounds, in which case the coulombic energies can be almost two orders of magnitude greater than the dispersion forces. The radial extent of the coulombic energies is also special: for the dielectric conditions described in Figure 8, the energy for two charges of 0.5 is 3.4 kJ mol^{-1} at 10 Å and 0.84 kJ mol^{-1} at 20 Å. Note that if the dielectric constant ε in equation (4) is fixed at 5, the coulombic energies for charges of 0.5 are 13.5 kJ mol^{-1}, 6.7 kJ mol^{-1} and 3.4 kJ mol^{-1} at 5 Å, 10 Å, and 20 Å, respectively. There may be uncertainty about the extent to which the coulombic interactions are ameliorated by intervening atoms, but clearly individual coulombic interactions will be dominant over long distances in inorganic supramolecular assemblies.

The dispersion energies are always negative at long distances, and thus cumulative, while oppositely signed coulombic energies mean that the net coulombic energy for a supramolecular assembly may not be large. However, the coulombic interactions provide the mechanism for directionality in the supramolecular assembly through geometrical arrangement to maximize attractions. Even for a pair of benzene molecules the coulombic contribution is almost 20% of the total energy (see below), and therefore the scope for coulombic influence in assemblies of more polarized inorganic molecules is apparent.

The long-range coulombic energies, extending well into the domain $\mathbb{E}$, have implications for the computation of non-bonded energies. Termination of the summation of equation (4) via cut-off spheres beyond which the coulombic energies are not computed causes truncation difficulties, because the energy terms for interactions either side of the boundary are numerous and not close to zero. For periodic systems, in which domain $\mathbb{E}$ is the remainder of the crystal lattice, the Ewald summation [66], partly in reciprocal space, is used [67]. In other cases a spline function can be used to grade the coulombic interactions to zero and keep the computations manageable. Supramolecular chemistry in practice is more concerned with the local effects of coulombic energies, particularly where they are anisotropic, and therefore splined computations which discount the longer-range (and more isotropic) interactions are unlikely to lead to erroneous conclusions.

What are the partial charges, and how can they be assigned? They are not formal charges or oxidation states [68]. In very few instances there are experimental data on dipole or quadrupole moments, from which atom charges can be derived. The more common procedure is to estimate atom charges from the calculated electronic structure of the molecule. This can be based on a population analysis of the molecular orbitals, usually a Mulliken analysis. Alternatively, atomic charges can be obtained by evaluating the electric potential surrounding the molecule from the computed electronic structure, and then fitting the atom-based charges to reproduce the electric potential [60, 69]. There are various procedures and accuracies associated with the calculated electronic structure of the molecule. Extended Hückel theory (EHT) is easily applied, but in not requiring self-consistency can lead to excessive charges. Of the *ab initio* methods, Hartree–Fock self-consistent field (HFSCF) methods, improved by configurational interaction (CI) to account for electron correlation, can be used for medium-size molecules. In recent times the established density functional theory (DFT) of solid-state physics is being increasingly applied to large molecules and molecules with large atoms, i.e. inorganic molecules [21]. Partial charges derived from *ab initio* calculations of electronic structure can be sensitive to the quality of the basis set [70]. Partial atomic charges for organic and biochemical molecules and functional groups have evolved with the benefit of considerable experience in molecular modelling.

Rappé and Goddard [71] provide a review of methods, and have presented a general method (called QEq, charge equilibration) for prediction of atomic partial charges in molecules, based on available atomic ionization energies, electron affinities and radii. The model uses atomic chemical potentials which depend on the geometry of all charges in the molecular assembly, and adjusts atomic charges to equalize these atomic potentials. The atomic charges generated by this method are dependent on molecular geometry and are evaluated iteratively. This method is implemented in the program Cerius^2 [72].

3.6 Hydrogen Bonding and Other Interactions

Hydrogen bonding involving electronegative atoms in domain $\mathbb{L}$ or $\mathbb{P}$ of coordination compounds and organometallic compounds is expected to be generally similar in properties to that well known in non metal compounds [73, 74]. It has been demonstrated that hydrogen bonding to thiolate and sulfide sulfur atoms in domain $\mathbb{L}$ can influence properties, particularly of biomolecules [75]. Similarly, the polarity of halogen–metal bonds could induce such halogen atoms in domain $\mathbb{P}$ to be more receptive to hydrogen bonds than are organohalogen atoms.

Intermolecular hydrogen bonds with an electron-rich metal as acceptor, and NH as donor, have been recognized [76a]. The better-known agostic C–H···M interactions [76b–76d] are predominantly intramolecular.

Another type of intermolecular interaction is that between halogen atoms in domain $\mathbb{P}$. This has been elucidated for organic systems [77a, 77b] but not yet for inorganic compounds. There is a considerable amount of literature on the intermolecular interactions between halogen atoms in domain $\mathbb{P}$ of metal clusters and metal atoms exposed at the surface of the cluster [77c, 77d]. Attention has been drawn to weakly attractive intermolecular interactions between sulfur (as S_8) and metal-bound halide [78].

4. QUESTIONS AND OBJECTIVES

An important objective is to identify the principles of association for inorganic molecules. These principles encompass shape factors, the influences of localized and delocalized charges, and any particular interactions specific to inorganic functional groups and ligand types. This goal includes an understanding of the magnitudes and the rankings of the energies of interactions, and an ability to sum all of the interactions.

In one sense there is a fundamental problem, because most of the definite and detailed data about supramolecular interactions come from crystals, and one of the questions is the extent to which the molecular structure is distorted in the crystal. In practice this predicament is not serious, because there are numerous rigid molecules unlikely to be distorted by the crystal field.

Crystal structures are complicated, and it is not easy to appreciate fully all of the nuances of crystal packing. High-level interactive graphics, preferably with space-filling representation and full use of colour, are essential tools. It is even more difficult to communicate those interactions in static print. Further, in searching for supramolecular features it is not feasible to examine individually all of the extant crystal structures. However, hypotheses about supramolecular principles in crystals can be tested efficiently through designed interrogation of the crystal structure databases and with statistical analyses.

Crystal polymorphism and phase changes are valuable phenomena because they provide access to additional points on the supramolecular topology–energy surface [73]. It is unfortunate that interest in crystal and molecular structure by chemists bent on synthesis usually ceases after the first crystal structure is determined, and that other crystallizations are not investigated. Braga and Grepioni [79] comment on the limitations of a 'first come (crystallized) first served (put on a diffractometer)' style of research.

Another objective is to uncover and understand compounds where the crystal field has modified the molecular structure. In inorganic systems the degree of modification can vary from conformational change, as occurs in the $[M(SPh)_4]$ system already mentioned, through modification of the coordination stereochemistry of atoms in domains M and L to more drastic changes in the coordination numbers and the connectivity pattern. These last two types of changes are well known to occur in the chemistry of Cu, Ag, Au and Hg with ligands such as halides and pseudohalides [80], but are probably much more prevalent than has been recognized. Thus one goal is to uncover further compounds where there are different structures in domains M and L as a consequence of interactions between domains P and E. In the case of charged molecules these effects can be recognized through instances of different molecular structure consequent on changing 'inert' counterions.

5. π-INTERACTIONS, ARYL INTERACTIONS AND THE PHENYL FACTOR

Specific intermolecular interactions are known to exist between aromatic groups: the high melting point and the distinctive herringbone crystal structure of benzene are common evidence of these interactions. Inorganic compounds frequently contain aromatic functionalities, heterocyclic ligands and/or phenyl substituents in domain $\mathbb{P}$, and therefore similar interactions are expected to be prevalent and significant in inorganic supramolecular chemistry.

5.1 Background and Fundamentals

Aryl–aryl interactions and related nonbonded π-interactions have now been investigated in considerable detail [52c, 57a, 73, 81–83]. The interactions are attractive and geometrically oriented. The principal interaction geometries are illustrated for toluene in Figure 9. The conformations in Figures 9(a) and 9(b) have one or two hydrogen atoms of one ring directed towards the π-cloud of the other ring. Both are named the edge-to-face (**ef**) conformation. The conformation in Figure 9(c) has the rings parallel but displaced such that hydrogen atoms of one ring overlay carbon atoms of the other, and is named the offset face-to-face (**off**) conformation. Lesser stabilization is provided by the conformation in Figure 9(d), with parallel rings, co-linear normals and greater separation of the ring planes than in Figure 9(c). The conformation in Figure 9(d) occurs where enforced, as in the stacked base pairs of nucleic acid polymers.

The geometry of the interaction of a pair of phenyl rings can be viewed more generally in terms of the angle between the ring planes and the offset of the centroid of one ring from the normal to the other [83b]. Scatter plots of the experimental values of these two attributes (from crystal structures) reveal the more and the less attractive geometries [84]. The occurrence, characteristics and importance of aryl–aryl interactions have been assessed for proteins [84a, 84c, 85, 86], nucleic acids [83b] and inclusion compounds [87].

The **ef** conformation is responsible for the herringbone motif in crystalline benzene and for the crystal structures of other aromatic hydrocarbons [88]. In the gas phase the benzene dimer, $(C_6H_6)_2$, also has the **ef** structure [89]. The benzene dimer in various solvent environments has been successfully modelled [82, 90].

Theoretical understanding of these interactions is well advanced. The geometrical orientation is a consequence of coulombic interactions super-imposed on the dispersion attractions [52c, 81]. The coulombic interactions can be modelled in terms of atomic partial charges (carbon $\delta-$, hydrogen $\delta+$) [81, 82] or in terms of the negative π-density and a positive σ-framework [83]. The

(a)   (b)

(c)   (d)

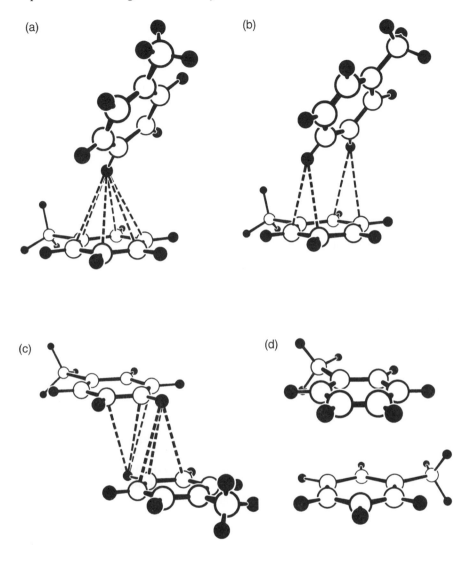

Figure 9 Principal geometries for the non-bonded interactions of aromatic rings, illustrated for toluene: (a, b) edge-to-face (**ef**) geometries in which one or two hydrogen atoms of one ring are directed towards carbon atoms of the other ring; (c) offset face-to-face (**off**) geometry with parallel ring planes offset such that hydrogen atoms of one ring overlap carbon atoms of the other; and (d) parallel overlapping rings

ef conformation is a consequence of the attraction of the partially positive H atoms of one ring to the partially negative C atoms of the other, while in the **off** conformation the offset allows some H atoms of each ring to overlap C atoms of the other. The atomic partial charges that have been used for benzene are $+0.153$ (H), -0.153 (C) [81] and $+0.115$ (H), -0.115 (C) [82]. The theoretical modelling has been applied to the benzene dimer in solution [82, 90] and to clusters of benzene molecules in relation to the crystal structure of benzene [81].

The nonbonded interaction energy for the benzene dimer is estimated experimentally to be $-10.0\pm1.7\,\mathrm{kJ\,mol^{-1}}$ [91] and the theoretically computed energies for dimers and oligomers of benzene with **ef** interaction geometries are also within this range, namely $-9.6\,\mathrm{kJ\,mol^{-1}}$ for the dimer and ca. $-5.5\,\mathrm{kJ\,mol^{-1}}$ per ring for oligomers [81, 82]. The theory indicates that the $9.6\,\mathrm{kJ\,mol^{-1}}$ net attraction between two benzene rings in the **ef** conformation is composed of $15.1\,\mathrm{kJ\,mol^{-1}}$ attractive dispersion, $1.9\,\mathrm{kJ\,mol^{-1}}$ coulombic attraction and $7.4\,\mathrm{kJ\,mol^{-1}}$ repulsion [82].

Interactions of the $H^{\delta+}$ atom of water with the negative density of benzene are also recognized [92]. A nonbonded interaction of methyl groups with negative density on phenyl groups is postulated to influence the structures and chirality of metal complexes with chelating ligands [93].

With this background, the occurrences and significance of aryl–aryl interactions in inorganic compounds can be investigated.

5.2   The Prevalence of the Phenyl Factor

Consider the following ligands which occur frequently in inorganic compounds: Ph_3P; $Ph_2PCH_2CH_2PPh_2$ (dppe); PhS^- and PhO^-; $PhCH_2$; $PhCO_2^-$; pyridine; imidazole and derivatives; 2,2′-bipyridyl (bipy); 1,10-phenanthroline (phen); $\eta^5\text{-}C_5H_5$; $\eta^6\text{-}C_6H_6$; $\eta^1\text{-}C_6H_5$; porphyrin and tetraphenylporphyrin; phthalocyanine; and salicylaldimines and derivatives. All of these ligands are able to populate domain $\mathbb{P}$ with Ph groups or exposed aromatic heterocycles. In metal compounds where there is a preponderance of these ligands, the domain $\mathbb{P}$ is mainly phenyl in character. Charged metal compounds crystallized with the commonly used cations Ph_4P^+, $Ph_3PNPPh_3^+$ and the anion Ph_4B^- will similarly have a phenyl-rich domain $\mathbb{P}$.

The widespread occurrence of these ligands is not fortuitous, but reflects the folklore of laboratories undertaking synthesis, where it is appreciated that compounds with phenyl substituents and aromatic peripheries often have lesser solubilities in synthetic reaction mixtures, and therefore crystallize. It is also common but undocumented knowledge that compounds with phenyl substituents frequently form high-quality crystals. The prevalence of phenylated ligands and aromatic peripheries and the influences of specific phenyl···phenyl interactions in supramolecular chemistry are dubbed the phenyl factor.

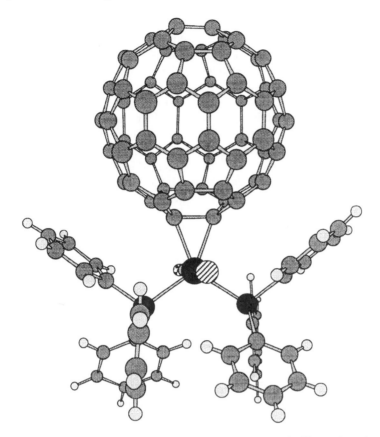

Figure 10 The molecular structure [(Ph₃P)₂IrCl(CO)C₆₀], illustrating the wrapping chelation of the C_{60} surface by two phenyl groups (Ir black, P black, Cl striped). Note also that there are **ef** phenyl–phenyl interactions between the parts of the Ph₃P ligands which are not involved in chelation of C_{60}

The phenyl factor appears in some more exotic systems, such as the addition of the complex [(Ph₃P)₂IrCl(CO)] to the fullerenes C_{60} and C_{70}. Figure 10 shows the molecular structure of [(Ph₃P)₂IrCl(CO)C₆₀] where the association of the phenyl substituents with the surface of the fullerene is a stabilizing factor [94]; the same occurs in the C_{70} adduct [95]. Indeed, the phenyl groups which are peripheral for the metal complex function as chelating agents for the fullerene, although there is a size mismatch relative to the conventional concept of chelation because the chelated object is larger than the chelator. By substitution of a flexible string of aromatic functions in place of one Ph in the phosphine ligand it is possible to engineer more extensive chelation of the fullerene [96]. Further illustration of the ubiquity of the phenyl factor comes on perusal of the structural diagrams in the recent review [7b] on metal complexes of the ligands $Ph_2PCH_2P(Ph)CH_2PPh_2$, $Ph_2PCH_2As(Ph)CH_2PPh_2$, $RC(PPh_2)_3$ and 2,6-bis(diphenylphosphino)pyridine. A compound such as $[Au(PPh_3)_3]$ BPh_4 [97] obviously has a crystal dominated by phenyl groups.

There is a corollary to the influence of the phenyl factor in lowering solubilities and therefore influencing the products which crystallize from reaction mixtures. Where the problem in synthesis is one of low solubility, as often arises with extensive heterocyclic ligand systems, the solution is to alkylate the edges of the planar aromatic sections, thereby interfering with the specific energetic interactions, decreasing lattice energies and increasing solubility. Illustrations of this occur in the increased solubilities of complexes of η^5-pentamethylcyclopentadienyl (Me_5C_5) relative to η^5-cyclopentadienyl homologues [98], and in the use of 3,5-di-*t*-butylphenyl-substituted meso-porphyrins rather than phenyl-substituted mesoporphyrins to increase the solubilities of large molecules containing these porphyrins and their metal complexes [28a, 99]. The inorganic literature now contains numerous compounds in which phenyl groups on the donor atoms are decorated with bulky substituents (Pr^i, Bu^t) in the 2, 6-positions in order to increase the bulk of the ligands and reduce the coordination number of the metal [100]. In all of these compounds the specific phenyl–phenyl interactions of Figure 9 are blocked, and the intramolecular and intermolecular interactions energies are less directional (although not necessarily less attractive) [58]. An extreme case of control of metal coordination occurs in $[Fe\{SC_6H_3\text{-}2,6\text{-}(Mes)_2\}_2]$ in which the 2,6-substituents, mesityl, are themselves 2,4,6-substituted phenyl, resulting in the iron thiolate coordination being twofold (but with secondary mesityl coordination) [100a]. These compounds are generally quite soluble; their supramolecularity is yet to be investigated.

Examination of the lattice packing in crystals of compounds with a preponderance of Ph in $\mathbb{P}$ gives an impression that these crystals are very similar to those of crystalline aromatic hydrocarbons, except for a small proportion of interstitial atoms from domains $\mathbb{M}$ and $\mathbb{L}$. The aromatic hydrocarbon analogy is not useful, however, because the configurations and conformations of the Ph groups are restricted by their connections to the heteroatoms in domain $\mathbb{L}$.

In analysing the effects and details of the phenyl factor, it is important to differentiate compounds according to the degree of rigidity of the phenyl groups in domain $\mathbb{P}$. As illustration, terminal M–E–Ph ligation is usually conformationally fluxional around both the M–E bond and the E–C_{ipso} bond (E = O, S, Se, Te), while bridging M–E(Ph)–M has fewer degrees of freedom, terminal M–PPh$_3$ has restricted rotation around the P–C_{ipso} (and possibly M–P) bonds and there is no conformational flexibility for octahedral $[M(phen)_3]$.

6. INFLUENCE OF THE CRYSTAL FIELD

A phenyl group in domain $\mathbb{P}$ of a globular molecule in a crystal must have more intermolecular than intramolecular interactions when interactions at all distances are included. It is instructive to contrast the intramolecular nonbonded Ph–Ph energies in compounds such as [Si(SPh)$_4$] [101], (Me$_4$N)$_2$[Cd(SPh)$_4$] [102] and (Ph$_4$P)$_2$[Fe(SPh)$_4$] [37] with the intermolecular Ph–Ph energies for their crystals (see Table 1). These three representative compounds have different crystal structures and three different molecular conformations for [M(SPh)$_4$] as shown in Figure 3(b) (Fe) and Figure 11 (Si, Cd).

Significant specific points emanating from the energies in Table 1 are as follows.

(1) The energies in the first row are relatively small, are dependent on the conformation of the [M(SPh)$_4$] molecule and are very much less than the maximum that could be considered for the most favourable Ph–Ph orientations, as for instance in a cluster of four C$_6$H$_6$ molecules [81].
(2) The energies in the second row represent the sums of all interactions between Ph groups of one [M(SPh)$_4$] molecule and other Ph groups in the

Table 1 Contributions to the intramolecular and intermolecular non-bonded energies for the crystalline compounds [Si(SPh)$_4$], (Ph$_4$P)$_2$[Fe(SPh)$_4$], (Me$_4$N)$_2$[Cd(SPh)$_4$] (undifferentiated totals are the sums of the vdW and coulombic energies[a]; each interaction is counted once)

Energy contribution (kJ mol^{-1})	[Si(SPh)$_4$]	(Ph$_4$P)$_2$[Fe(SPh)$_4$]	(Me$_4$N)$_2$[Cd(SPh)$_4$]
Total intramolecular Ph–Ph energy within [M(SPh)$_4$]	−6.0	−7.1	−5.7
Total intermolecular Ph–Ph energy for Ph$_4$ of one [M(SPh)$_4$] molecule to all other Ph	−167	−201	−66
One Ph to remainder of [M(SPh)$_4$] molecule (average value)	−9.2	−13.4	−13.4
One Ph to total lattice (average value)	−51	−61	−48
[M(SPh)$_4$] molecule to total lattice	−257 (vdW −244, coulombic −13)	−530 (vdW −272, coulombic −258)	−581 (vdW −214, coulombic −367)

[a]Partial charges: Si + 0.6; Fe, Cd + 0.4; P + 1.0; S − 0.6; N − 0.2; C$_{ipsoPh}$ − 0.015; C$_{Ph}$ − 0.1; H + 0.1; C$_{Me}$ + 0.141; H$_{Me}$ 0.053.

(a)

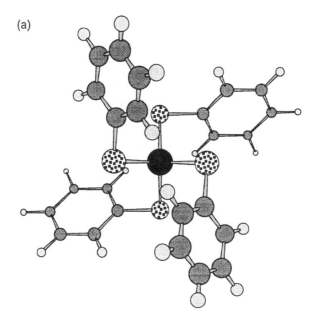

(b)

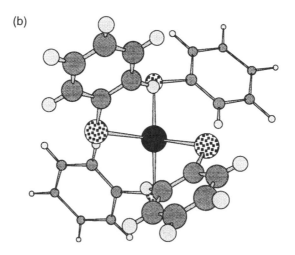

Figure 11 Conformations of the $[M(SPh)_4]^z$ molecules in the crystal structures of (a) $[Si(SPh)_4]$ (molecular symmetry C_2) and (b) $(Me_4N)_2[Cd(SPh)_4]$. See also Figure 3(b). Note that while $[Si(SPh)_4]$ and $[Cd(SPh)_4]^{2-}$ possess similar S–M–S–C torsional angles, the M–S–C–C torsional angles are different and the S–Cd–S angles are distorted

crystal lattice. The key result is that this intermolecular Ph–Ph attractive energy is at least an order of magnitude larger than the total intramolecular Ph–Ph attractive energy in row one. In $(Me_4N)_2[Cd(SPh)_4]$ the Me_4N^+ cations partially insulate the $[Cd(SPh)_4]^{2-}$ phenyl groups from those of the lattice, and the intermolecular Ph–Ph attractive energy is less than in the other two compounds.

(3) In rows three and four of Table 1 the focus is on the average influences on any one phenyl group, with the non-bonded influences of the M and S atoms included. The energy (in row three) of one Ph group to the remainder of the $[M(SPh)_4]$ molecule shows the attractions of the M and S atoms. The significant result is that the energetic influence of the lattice (in row four) is five to six times larger than the intramolecular influence (row three) for $[Si(SPh)_4]$ and $(Ph_4P)_2[Fe(SPh)_4]$ where the crystals are heavily populated by phenyl groups, and is about four times larger for $(Me_4N)_2[Cd(SPh)_4]$. These are key indices of the influence of the crystal field for conformationally flexible molecules.

(4) The last row gives the total non-bonded interaction energy between a target $[M(SPh)_4]$ molecule and the lattice. The two compounds with ions in the lattice have substantially larger coulombic contributions. Again the significant result is the magnitude of the crystal field bearing on one of these molecules.

The paradigm is clear: in molecules subject to the phenyl factor, the crystal field energy is large and dominant. The classic organic example where the crystal field dominates molecular structure is biphenyl, where the isolated molecule has its rings twisted by 42° but the crystal lattice imposes coplanarity; it is estimated that the intramolecular strain in the coplanar molecule is $8.3\,kJ\,mol^{-1}$, while the crystal field is $19.4\,kJ\,mol^{-1}$ [52c].

6.1 SbPh₅

A convincing illustration of the influence of the crystal field on molecular structure through the phenyl factor is $SbPh_5$. Homologous compounds EPh_5 (E = P, As) are trigonal bipyramidal, but in crystalline $SbPh_5$ the molecules are close to square pyramidal [103]. From cyclohexane $SbPh_5(C_6H_{12})_{0.5}$ crystallizes, and contains trigonal bipyramidal $SbPh_5$ and disordered cyclohexane [104]. In both crystals there are multiple phenyl–phenyl interactions between molecules, as illustrated in Figure 12. In the crystal containing the square pyramidal molecule, two square pyramids are packed base to base across a centre of symmetry, with an Sb–Sb separation of 6.56 Å and four **ef** interactions. The apical Ph of each molecule is involved in an **off** interaction with another apical Ph across a centre of symmetry, while on the other side of each apical Ph there is an **ef** interaction with a basal Ph of another

(a)

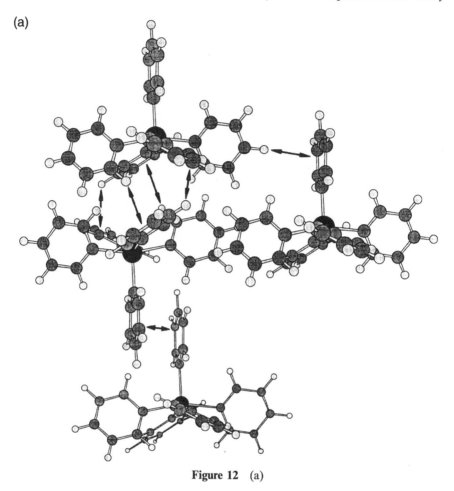

Figure 12   (a)

nearby molecule. In the crystal structure containing the trigonal bipyramidal molecule, there are multiple intermolecular attractive interactions between Ph groups, **ef** and **off**, some of which are shown in Figure 12(b) for part of the crystal structure.

The intramolecular energies and the crystal field energies for $SbAr_5$ can be estimated. Measurements of the ^1H and ^{13}C NMR spectra of these $SbAr_5$ compounds in solution at subambient temperatures demonstrate their stereochemical non-rigidity, and reveal that the barrier to interconversion between different stereochemistries at Sb is $6-6.5\,kJ\,mol^{-1}$ [105]. The crystal field energy for $SbPh_5$ is calculated to be $-257\,kJ\,mol^{-1}$ [106]. Thus the

(b)

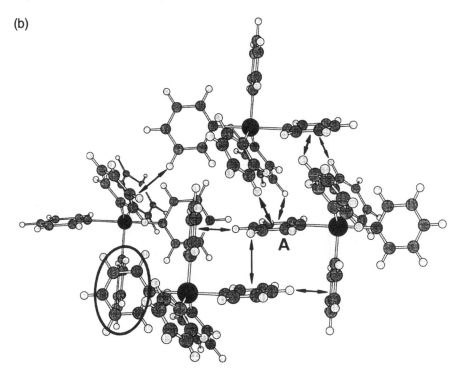

Figure 12 (*continued*)   (a) Part of the crystal structure of SbPh$_5$, showing four adjacent molecules and significant Ph···Ph attractive interactions between them. Centres of symmetry relate the molecules which are base to base and apex to apex. (b) Four neighbouring SbPh$_5$ molecules in the crystal structure of SbPh$_5$(C$_6$H$_{12}$)$_{0.5}$, with significant Ph···Ph attractive interactions between them. The upper-right centrosymmetric pair of molecules (Sb···Sb 7.8 Å) has four **ef** interactions involving two equatorial rings and one apical ring on each molecule, and the lower-left centrosymmetric pair (Sb···Sb 8.0 Å) has two **ef** interactions. Ring A is involved in four attractive Ph···Ph interactions even with this incomplete set of neighbours

intramolecular conformational energies are again at least an order of magnitude less than the supramolecular influence of the crystal field, and indeed are less than the attractive interaction energy for one of the intermolecular **ef** phenyl–phenyl interactions. The discrepancy between crystal field energy and intramolecular conformational energy in SbPh$_5$ is much greater than in biphenyl.

The distorting influence of the crystal field in these compounds is further illustrated by the crystal structure of penta-*p*-tolylantimony [107] which has a different molecular structure, close to trigonal bipyramidal but still distorted by **ef** intermolecular interactions between aryl rings.

6.2   Injunction: The Crystal Field and Molecular Structure

These examples of the decisive influence of the crystal field have widespread ramifications for the common practice of using a crystal to determine molecular structure by diffraction methods. Common practice is to ignore the crystal field, or to approach the crystal packing of a molecule by documenting the shortest (and thus repulsive) intermolecular contacts, or to assume that the crystal field is spherically symmetrical. The deficiency in examination of the shortest intermolecular contacts is that the crystal field is determined by the longer and more numerous attractive interactions, which override the fewer shorter repulsive contacts.

It is understandable that the crystal field and its influences on the molecular structure are usually ignored, or discounted, because there is no routine procedure to assess the net crystal field. The production and publication of a lattice-packing diagram are helpful, but the diagram rarely raises questions or resolves any ambiguities. The caveat expressed here needs to be accompanied by a readily implementable response. In general terms, a qualitative but valuable method is to explore the crystal lattice for the specific and directed intermolecular interactions, such as those identified in this article. Simple but useful qualitative analysis can be undertaken using standard software which generates a molecule and some of its surrounding molecules and presents their space-filling (van der Waals) surfaces: examination of various assemblages from many angles usually enables a good understanding of the intermolecular interactions.

The difficulty with extraction of reliable molecular structure data from a crystal structure is that there is usually only one crystal structure available. The question is what are the (unknown) alternatives? Clearly, multiple crystal structures for the one molecule would provide a much less ambiguous image of the structure of the molecule. It is fortunate that in some crystals there is more than one molecule per asymmetric unit. Similarly, it is desirable to seek and characterize polymorphs — different crystalline forms of the one compound — to remove ambiguities about the influence of the crystal field. However, few investigators continue experiments on crystallization once a first crystal structure is obtained for a new compound. The occurrence of solvent molecules trapped in the lattice is good intimation that alternative crystals can be formed by change of solvent. In the case of charged molecules, there is a straightforward approach to alternative crystallization by change of the counterion.

6.3 Two Examples of Uncertainties about Crystal Field Effects

To illustrate typical questions about the relative influences of intramolecular forces and the crystal field, two examples, from 1983 and 1994, are mentioned here. In a paper [108] on the structures of $[Cu(1)_2](BPh_4)_2$ and $[Cu_2(2)_4](ClO_4)_2$, the molecular distortions in pseudotetrahedral $[Cu(1)_2]^{2+}$ were considered to be 'an intrinsic property of the molecule associated with metal-to-ligand $d_\pi-p_\pi{}^*$ charge transfer rather than a consequence of lattice packing effects'. Further, the metal coordination in $[Cu_2(2)_4]^{2+}$ is very different from that in $[Cu(1)_2]^{2+}$, despite the similarity of the ligands, but because no unusually short intermolecular contacts could be identified 'it seems unlikely that packing effects could account for the different coordination spheres in the two structures'. However, in retrospect it is seen to be necessary to investigate at least the possible occurrence of the phenyl factor in the first compound.

(1) (2)

A recent instance of the question of the relative influences of molecular structure and crystal field occurred in the structures of bis(imidazole) complexes of an FeN_4 macrocycle with Fe^{II} and Fe^{III} oxidation states [109]. These compounds are important models for cytochromes. The orientations of the imidazole rings relative to the macrocyclic coordinate system change with the oxidation state, as does the intermolecular network of hydrogen bonds. However, the introduction of a counterion to balance the change in oxidation state also influences the crystal packing, introducing some ambiguity into the interpretation of the influences of the intramolecular bonding.

7. CRYSTALS CONTAINING Ph_4P^+

The cation Ph_4P^+ and its relatives such as Ph_4As^+ and $Ph_3PNPPh_3{}^+$, are popular with synthetic chemists preparing and crystallizing anionic metallocompounds. This is demonstrated by the presence in the Cambridge Structural Database of 794 crystals (out of 115 000 in 1993) containing at least one Ph_4P^+ cation, while 350 crystals contain at least one Ph_4As^+. One reason

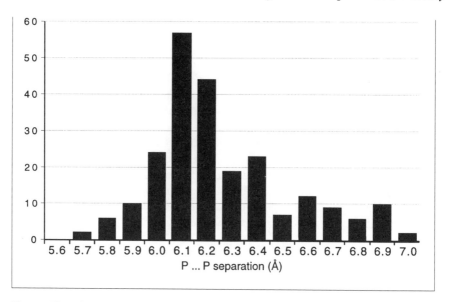

Figure 13 Histogram of the shorter P$\cdots$P separations in crystals containing Ph$_4$P$^+$, compiled using data from the Cambridge Structural Database

for this is that these cations often yield less soluble compounds and are valuable with combinations of anions and preparative solvents where there is otherwise high solubility. The combination of dimethylformamide (DMF) as preparative solvent, Ph$_4$P$^+$ as crystallizing cation, and diethyl ether as precipitant is a common example of this. Further, in practice the use of Ph$_4$P$^+$ (as opposed to cations such as Et$_4$N$^+$, Bu$_4$N$^+$) enhances the quality of the crystals and of the subsequent diffraction analysis.

There are supramolecular reasons for this. Figure 14 is a representative histogram of P$\cdots$P distances in crystals containing Ph$_4$P$^+$. The significant feature is the majority of distances in the range 6.0–6.4 Å, very much less than 13.5 Å which is twice the full van der Waals radius of Ph$_4$P$^+$ (P$\cdots$H vdW surface). This intimates that there is substantial interpenetration of Ph$_4$P$^+$ cations in these crystals, which is not surprising since the periphery is phenylated and open. Ph$_4$P$^+$ cations have restricted conformational freedom, as has been analysed by Kitaigorodskii [9].

The most frequently occurring supramolecular motif for Ph$_4$P$^+$ cations is the sextuple embrace involving interaction of three phenyl rings on each of two cations, as illustrated in Figure 14 (as it occurs in the crystal structure of (Ph$_4$P)$_2$[Pb(Se$_4$)$_2$] [110]). The three phenyl groups on each cation are canted relative to the P$\cdots$P vector such that there are multiple **ef** interactions between

(a)

(b)

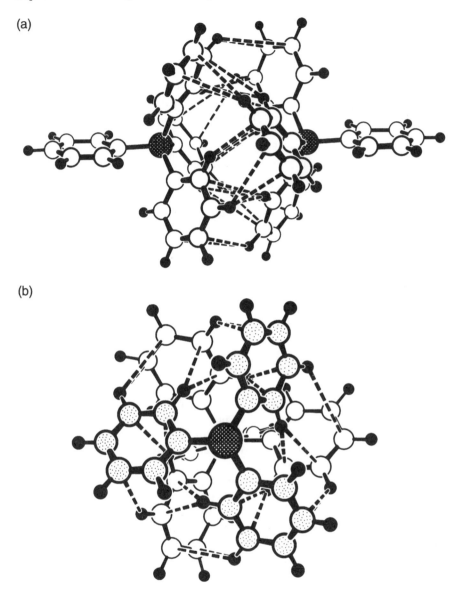

Figure 14 The sextuple embrace of two Ph_4P^+ cations: (a) view normal to the P···P vector (6.09 Å); and (b) view along the P···P axis, with the nearest phenyl group removed to reveal the pseudo-S_6 symmetry of the six interacting phenyl groups. The C atoms of the upper three phenyl groups are lightly stippled. The dashed lines are the numerous **ef** H···C contacts in the range 2.75–3.4 Å which contribute to the attraction and determine the geometry of the sextuple embrace

the six phenyl groups. The sextuply embracing pair is usually centrosymmetric, and often (as in Figure 14) the six phenyl groups involved have approximate S_6 local symmetry. In this case there are four H⋯C ef interactions between each phenyl group and each of its two partners, and all of these contacts (shown in Figure 14) are in the distance range already outlined above for ef phenyl interactions.

The ubiquity of the sextuple embrace is readily determined by analysis of the Cambridge Structural Database. There are at least 100 crystals containing Ph_4P^+ demonstrating over 200 instances of the sextuple embrace. The revealing characteristics are the P⋯P separation (generally 6.0–6.4 Å, full range 5.7–6.9 Å), often a centre of symmetry, the six separations of the centroids of the phenyl rings (ca. 4.9 Å) and the interplanar angles of the interacting phenyl planes [111].

A single Ph_4P^+ cation can participate in two sextuple embraces, and extended networks can exist. This is illustrated in Figure 15, showing sections of the crystal structure of $(Ph_4P)_4[In_2(S_4)_2(S_6)_2(S_7)]$ [112]. There exist layers which contain only Ph_4P^+ cations, with the relatively thin, elongated indium polysulfide anions located between the layers. The layers contain extended chains of sextuply embracing Ph_4P^+, each pair with a centre of symmetry and a P⋯P separation of 6.30 Å (actually 6.297 Å, 6.302 Å, not crystallographically related). Between these chains there are other centrosymmetric pairs of cations, inclined slightly to the layer plane, with P⋯P separations of 6.81 Å and with the sextuple-embrace conformation. Other intermolecular interactions between phenyl groups exist in this layer between sextuply embracing chains.

This highly structured arrangement of Ph_4P^+ cations is not unique to $(Ph_4P)_4[In_2(S_4)_2(S_6)_2(S_7)]$. We have identified it and related arrangements in a considerable number of crystals, particularly of metal polychalcogenides and related metal complex anions [111]. In addition to the widespread sextuple embrace, interpenetrating Ph_4P^+ cations can involve two phenyl groups on each cation as a quadruple embrace.

The Ph_4P^+ cation is thus seen to be far from innocent or isotropic in crystals. The conventional notion of crystal chemistry that cations are mutually repulsive, and the lattice is an array of interspersed cations and anions, may be far from the truth in many compounds. The ef conformation of the interacting phenyl rings in the sextuple embrace is a consequence of coulombic attractions (see above), and the sextuple embrace and other interaction motifs are due to attractive coulombic interaction. Therefore it is erroneous to assume (as have many authors) that Ph_4P^+ cations in crystals are mutually repulsive.

The Ph_4P^+ cation has clearly evident supramolecular influence, and its presence is very likely to impose on the molecular structure of the anion in the crystal. Quantification of the vectorial crystal field imposed by the structured array of Ph_4P^+ cations on the anions, and particularly on fluxional metal complex anions, is yet to occur. Also, it is clear that phenylated cations of this type will be important components of further crystal engineering.

8. METAL POLYCHALCOGENIDES

Metal polychalcogenides are metal complexes with E_n^{2-} ligands, where E is S, Se or Te [113, 114]. Since the mid-1980s metal polychalcogenide chemistry has developed from the original few analytical polysulfide solutions to a plethora of new and unprecedented compounds in the general class $[M_x(E_n)_y]$. This progress was possible through the recognition of the role of aprotic solvents in activating the anionic ligands involved, and with the use of creative reaction pathways. Most of these $[M_x(E_n)_y]^z$ complexes are anions, and the choice of cation for crystallization is a variable of the synthesis. The number of homoleptic metal polychalcogenide compounds characterized by crystal structure determination has risen from about four in the early 1980s to something like 300 in 1994. In addition there are numerous organometallic polychalcogenide compounds.

The prevalence of Ph_4P^+ as countercation is evident in the set of polyselenide and polytelluride compounds in a recent review [114]. Of 95 crystalline compounds (excluding those with organic ligands or heteroligands) 43 (45%) contain Ph_4P^+; the numbers of other cations are seven PNP^+, three Me_4N^+, 12 Et_4N^+, three Pr_4N^+, one $Bu^n_4N^+$, 12 alkali metal and 12 alkali crown cations.

In parallel with this development of the crystallography, the use of NMR (^{113}Cd [115, 116], ^{119}Sn [116, 117], ^{199}Hg [116, 118], ^{77}Se [113a, 114, 116, 119, 120], ^{125}Te [114, 116]) has allowed precise characterization of some of these complex systems in solution.

Four important characteristics of this chemistry will be illustrated in the following paragraphs: they all demonstrate the special significance of supramolecular influences for metal polychalcogenides.

(1) Metal polychalcogenide complexes in solution are not necessarily the same as those in the crystals which grow from those solutions:
  (a) solutions of $[Cd(S_6)_2]^{2-}$ salts in DMF contain also $[Cd(S_5)(S_6)]^{2-}$ and $[Cd(S_6)(S_7)]^{2-}$ [115];
  (b) solutions of mercury polysulfides contain $[Hg(S_x)(S_y)]^{2-}$ where x and $y = 4$ and 4, 4 and 5, 5 and 5 or 5 and 6, but only the $x=4$, $y=4$ and $x=6$, $y=6$ combinations have been crystallized [118];
  (c) the three complexes $[In_2(Se_4)_4(Se_5)]^{4-}$, $[In_3Se_3(Se_4)_3]^{3-}$ and $[In_2Se_2 (Se_4)_2]^{2-}$ when dissolved all give the same ^{77}Se NMR spectrum which is thought to be that of $[In_3Se_3(Se_4)_3]^{3-}$ [114, 121]; and
  (d) the complex $[Tl_2(S_4)_2]^{2-}$ dissociates in DMF [122].

(2) The compositions and structures of anionic complexes which crystallize from preparative solutions are quite dependent on the identity of the 'inert'

(a)

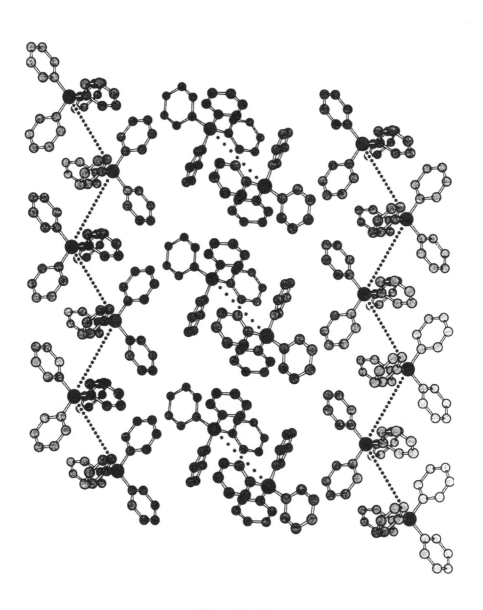

Figure 15 (a)

(b)

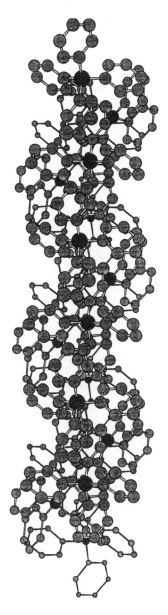

Figure 15 (*continued*)   The layered network of interacting Ph_4P^+ cations in crystalline $(Ph_4P)_4[In_2(S_4)_2(S_6)_2(S_7)]$. The views are parallel (a) and perpendicular (b) to the layer, with hydrogen atoms omitted for clarity. The chains of Ph_4P^+ cations connected by close dots are separated by 6.30 Å and involve extended sextuple embraces, while the sparse dots identify single sextuple embraces (6.81 Å separation)

cation. This effect is additional to the known incorporation of Na^+ into metal chalcogenide and polychalcogenide molecules. Some examples are as follows:

(a) an ethylenediamine solution of $K_2Hg_2Te_3$ crystallizes $[Hg_4Te_{12}]^{4-}$ with Bu_4N^+ but $_\infty[\{Hg_2Te_5\}^{2-}]$ with Ph_4P^+ [123];

(b) in the Ag^I–Se_x system, different cations yield four different structures [124], namely one-dimensionally nonmolecular $_\infty[\{AgSe_4\}^-]$ with Ph_4P^+, one-dimensionally nonmolecular $_\infty[\{AgSe_5\}^-]$ with Me_4N^+, molecular $[Ag_4(Se_4)_4]^{4-}$ with Et_4N^+ and molecular $[Ag_4(Se_4)_3]^{2-}$ with Pr_4N^+; and

(c) in the Pd^{II}–Se_x system, a simple monometallic complex $[Pd(Se_4)_2]^{2-}$ crystallizes with Ph_4P^+ [125], but with K^+ a complex nonmolecular structure has been found [114, 126].

(3) Coordinated polychalcogenide chains in crystalline complexes are sometimes disordered in length, connectivity and/or conformation. Examples are as follows:

(a) the compounds $(Ph_4P)_2[Cd(S_6)_2]$ and $(Ph_4P_2[(S_6)Cd(S_6)_{0.44}(S_7)_{0.56}]$ demonstrate variability of cadmapolysulfane ring sizes and ring conformations in virtually the same crystal lattice [115];

(b) the four different ions $[W_3Se_9]^{2-}$, $[W_2Se_4(Se_2)(Se_3)]^{2-}$, $[W_2Se_4(Se_2)(Se_4)]^{2-}$ and $[W_2Se_4(Se_3)_2]^{2-}$ have similar shapes and sizes, and their Ph_4P^+ salts cocrystallize [119];

(c) $(Ph_4P)_2[Cu_4(Se_4)_{2.4}(Se_5)_{0.6}]$ and $(Ph_4P)_2[Cu_4(Se_4)_{2.1}(Se_5)_{0.9}]$ [127];

(d) the anions $[In_2S(S_5)(S_4)_2]^{2-}$ and $[In_2S(S_5)(S_4)(S_6)]^{2-}$ equally populate the anion site in the crystal with Ph_4P^+ [112];

(e) $[Sn(S_4)_2(S_6)_{0.6}(S_4)_{0.4}]^{2-}$ [128]; and

(f) $[Mo_2S_6]^{2-}$ or $[Mo_2S_7]^{2-}$ occupies the same site in crystalline $(Ph_4P)_2[Mo_2S_{6.63}]$ [129].

(4) Polymorphism occurs frequently:

(a) there are at least four lattices for the straightforward molecular compounds $(Ph_4P)_2[M(Se_4)_2]$, two for $M = Fe$, Zn and another two for $M = Hg$ [114];

(b) $(Ph_4P)_2[Se(Se_5)_2]$ crystallizes in two different lattices [116, 130]; and

(c) $(Ph_4P)_2[Tl_2(S_4)_2]$ crystallizes in three different lattices, one of which also appears to afford different-coloured crystals [122].

The clear conclusions from these observations are that (1) metal polychalcogenides are unusually adaptable coordination systems, which is not surprising, and (2) the crystal fields in many of the solids are demonstrably distorting both the compositions and the stereochemistries. The structures of the crystals do not faithfully represent the most stable coordination chemistry.

Properties 1 and 2 are analogous to the phenomenon of polymorphism in organic crystals [73], which is another indicator of the influence of the crystal field. The distorting influence of the crystal field on the anionic metal polychalcogenide complexes is understandable in the many cases with Ph_4P^+, given the strongly vectorized nature of $Ph_4P^+ \cdots Ph_4P^+$ interactions as described above.

These conclusions are reinforced by examination of the crystal structures of metal polychalcogenides, which are generally rich sources of insight into supramolecular inorganic chemistry [111]. Figure 16 shows the crystal structures of $(Ph_4P)_2[Pb(Se_4)_2]$, $(Ph_4P)_2[Hg(Se_4)_2]$ and $(Ph_4P)_2[Ni(Se_4)_2]$ which crystallize with different unit cells but demonstrate similar crystal structures. There is a similar segregation of cations in each of the structures, with the anions also segregated along low-volume channels between the cations. In each structure the Ph_4P^+ cations interact tightly, with sextuple embraces and other phenyl–phenyl interactions. The networks of interactions between Ph_4P^+ cations are not the same in these three structures, but they are similar and pronounced. Each of the $[M(Se_4)_2]^{2-}$ ions has a different four-coordinate stereochemistry at M, namely pyramidal for Pb, tetrahedral for Hg and planar for Ni, and Figure 16 (Plate I) shows that the cavities for the metal complexes have slightly different shapes. These lattices are dominated by the cations, in terms of both relative volume and lattice-packing energies, but the cation connection networks are sufficiently variable to accommodate different shapes of anions.

Part of the crystal structure of $(Ph_4P)_4[In_2(Se_4)_4(Se_5)]$ is pictured in Figure 17 (Plate II). The polyselenide anion is clearly buried in a matrix of cations, with multiple attractive interactions between the cations dominating the lattice packing. The stereochemical requirements of the In atoms, of the Se_4 chelate rings and of the Se_5 chain which connects the two In atoms as it threads through a narrow channel would appear to be subservient (energetically and geometrically) to the phenyl factor in the cation matrix. This is consistent with the evidence from ^{77}Se NMR that this anion does not exist as such in solution [114, 121]. Crystal structures similar to that shown in Figure 17 occur in many crystalline metal polychalcogenides.

9. COMPOUNDS [M(SPh)₄]

About 10 of these compounds are known, with charges 0, $1-$, $2-$ and tetrahedral metal coordination, for metals ranging from Mn^{II} to Si^{IV}. Four of the many molecular conformations possible for these compounds are illustrated in Figure 18: it is clear that these molecules have open, penetrable peripheries. Two conformations, idealized to D_{2d} (Figure 18a) or S_4 (Figure 11a) molecular symmetry, are commonly observed in crystal structures. For

188

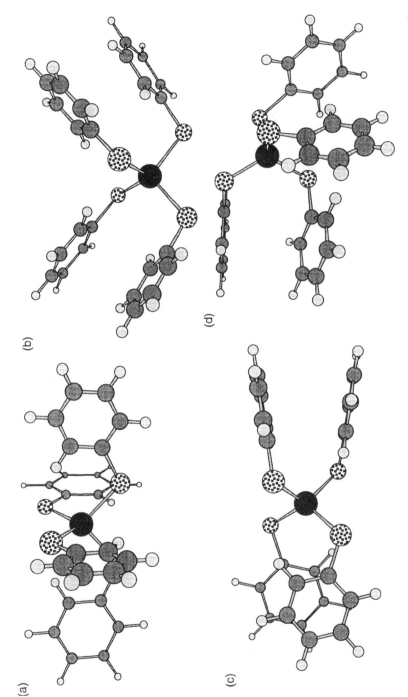

Figure 18 Conformations of [M(SPh)₄]: (a) idealized D_{2d} conformation; (b, c) two different energy minima for an isolated molecule demonstrating **off** intramolecular ring interactions; and (d) an energy minimum for an isolated molecule demonstrating two **ef** interactions and one **off** interaction

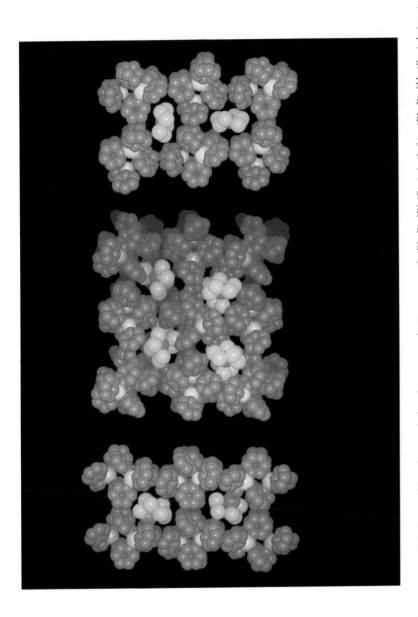

PLATE 1 Figure 16 Comparable views of the three crystal structures of (Ph₄P)₂[Pb(Se₄)₂] (left), (Ph₄P)₂[Hg(Se₄)₂] (centre) and (Ph₄P)₂[Ni(Se₄)₂] (right) showing the similar lattices of tightly interacting Ph₄P⁺ cations (hydrogen atoms omitted) that allow different shapes of cavities for the different shapes of [M(Se₄)₂]²⁻ complexes (M blue, Se yellow, P purple, C green)

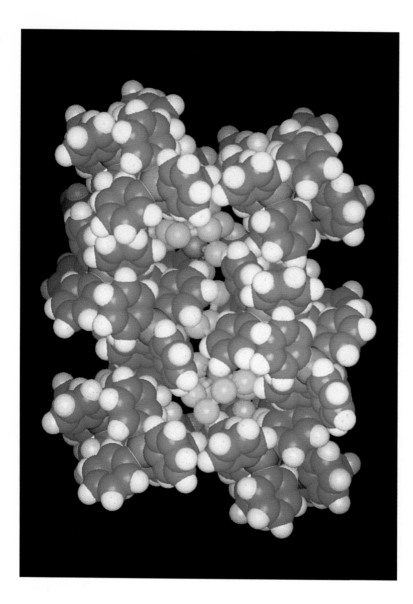

PLATE II Figure 17 Part of the crystal structure of $(Ph_4P)_4[In_2(Se_4)_4(Se_5)]$ (In brown, Se yellow, P purple)

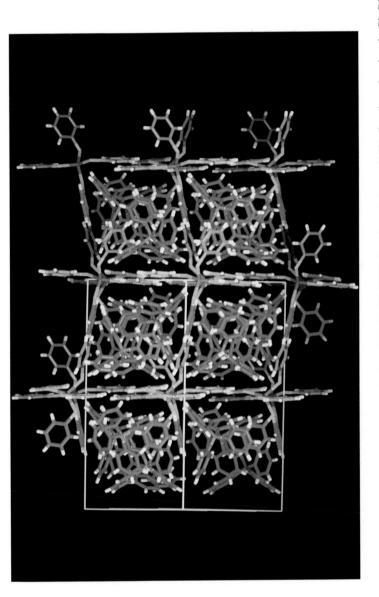

PLATE III Figure 19 Representation of the crystal lattice of $(Ph_4P)_2[Fe(SPh)_4]$ as viewed along a [110] direction, showing the Ph_4P^+ cations in segregated columns and the planes containing most of the SPh ligands of the $[Fe(SPh)_4]^{2-}$ ions. Fe (purple) is largely obscured at the intersection of the SPh planes; S is yellow; the four independent Ph rings of $[Fe(SPh)_4]^{2-}$ are coloured mauve, pink, blue and brown. P is purple, and the Ph rings of the Ph_4P^+ are green and white. The key feature is the alignment of the anion Ph rings in the {001} and {110} planes, enclosing the Ph_4P^+ domains. Some of the blue and brown Ph rings of $[Fe(SPh)_4]^{2-}$ project into the Ph_4P^+ domain

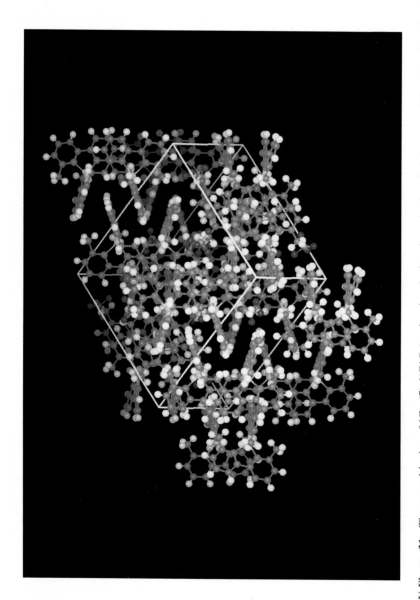

PLATE IV Figure 31 The crystal lattice of $[Cu_2(7)_2](ClO_4)_2$, showing the interleaved splayed chelates parallel to the view direction. The chelate ligands which are mutually parallel in each molecule are all aligned perpendicular to the view direction

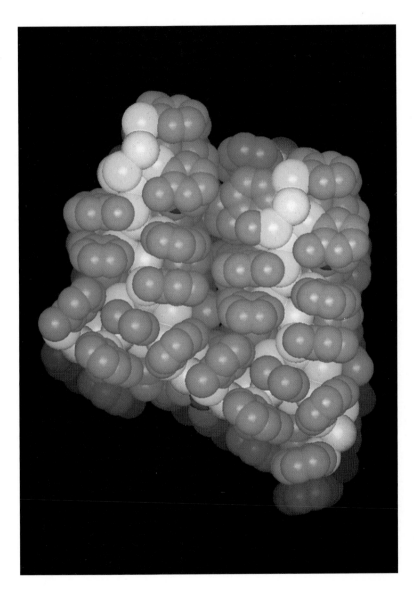

PLATE V Figure 37 Part of the crystal structure of $_\infty[Cd_7(SC_6H_4\text{-}2\text{-}Me)_{14}(DMF)_2]$, with hydrogen atoms omitted, showing the dominance of aryl-aryl interactions and the insignificant contribution of the Cd and S atoms to the volume. The O (red) and N (blue) atoms of the DMF molecules are just visible, and it is evident that the DMF molecules play a space-filling role

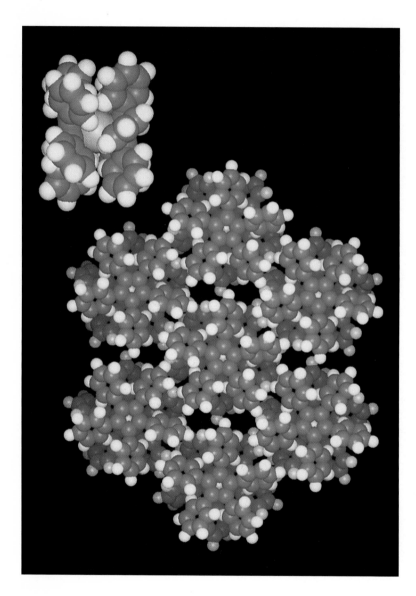

PLATE VI Figure 38 The molecular structure of $[Sn(C_5Ph_5)_2]$, and part of the crystal structure showing the pseudohexagonal array of fivefold molecules in a layer

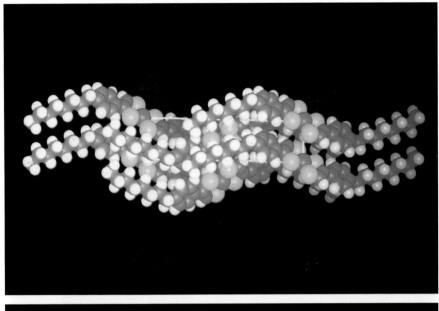

(a)

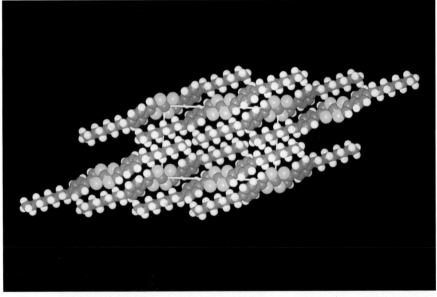

(b)

PLATE VII Figure 40 Two views of the crystal structure (a=13.33 Å, b=5.38 Å, c=22.16 Å, β=90.9°, $P2_1/c$) of the nickel dithiolene complex (**10**) (Ni blue, S yellow). The c-axis is horizontal in both views. (a) view along the a-axis, illustrating the seahorse conformations of the molecules and the parallel contiguous chelates along the short b-axis (b) view along the b-axis, showing the c-glide operation which generates **ef** interactions between the phenyl spacers of contiguous molecules along c. Note also that the NiS$_4$ central region of each molecule is enclosed by the alkyl chains

PLATE VIII Figure 41 Part of the crystal structure of (**13a**) (Pd purple, S yellow, O red)

isolated molecules, force field methods indicate that there are many other local energy minima (such as the structures in Figures 18b–18d) which use the advantages of the **off** and **ef** conformations of phenyl rings [58].

Particular attention has focused on the group of compounds $(Ph_4P)_2[M(SPh)_4]$ (M = Mn, Fe, Co, Ni, Zn, Cd) which are isomorphous. There are distortions from perfect MS_4 tetrahedral stereochemistry in these compounds, originally attributed to electronic effects at the metal [131] and subsequently to intramolecular repulsions between *ortho*-hydrogen atoms of the SPh rings and other S atoms of the MS_4 coordination [37]. The influences of crystal packing were discounted in the anaylsis of intramolecular geometry [37]. However, the one crystal structure adopted by all of the $(Ph_4P)_2[M(SPh)_4]$ compounds is quite remarkable.

Figure 19 (Plate III) shows the $(Ph_4P)_2[Fe(SPh)_4]$ crystal lattice as projected along a [110] direction, while Figure 3(a) shows the conformation of the $[Fe(SPh)_4]^{2-}$ in this lattice. The anions are aligned mainly in an approximately orthogonal array of planes which largely separate the volumes of the cell occupied by the Ph_4P^+ cations into stacks. As is apparent, two of the Ph rings of the anion are directed into the stacks of cations. Sextuply embracing pairs of Ph_4P^+ occur within the stacks of cations. In this lattice there are multiple phenyl–phenyl interactions between cation and cation and between cation and anion, with most of the interactions being of the **ef** type.

Figure 20 shows the details of some of the phenyl–phenyl H$\cdots$C interactions between the rings of the $[Fe(SPh)_4]^{2-}$ anion and the rings of just two of the Ph_4P^+ cations. It is difficult to escape the conclusion that the total interaction energy of the four phenyl rings of the $[Fe(SPh)_4]^{2-}$ anion with the surrounding cations will be substantial, anisotropic at any ring, and a principal force determining the molecular conformation of the anion in the crystal [132]. Another group of isomorphous $[M(SAr)_4]^{2-}$ compounds is formed with the 2-phenylbenzenethiolate ligand, crystallized with Et_4N^+, for M = Mn, Fe, Co, Ni, Zn, Cd and Hg [133]. Full analyses of the intermolecular and intramolecular influences in all of these compounds are required.

10. METALLAMOLECULES WITH Ph$_3$P AND RELATED LIGANDS

In the Cambridge Structural Database (1993) more than 3000 crystals contain at least one Ph_3PM moiety. Clearly the three phenyl groups will be directed outwards from the periphery of the molecule, and thus the phenyl factor is expected to play an important role in intermolecular interactions. The $MPPh_3$ group is more constrained than the Ph_4P^+ group discussed above, owing to the other ligands bonded to M, and as a consequence the intermolecular interactions between $MPPh_3$ groups are not as tight as those of Ph_4P^+. The

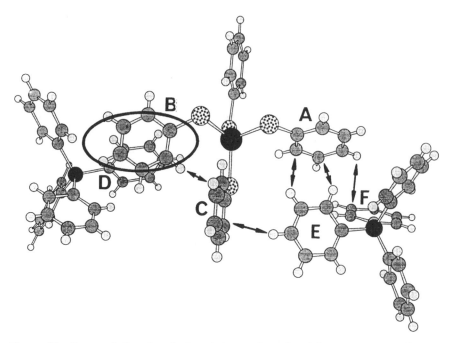

Figure 20 Some of the phenyl–phenyl interactions involving an $[Fe(SPh)_4]^{2-}$ ion (centre, rings A, B, and C) and two Ph_4P^+ cations (rings D, E, and F) in the crystal structure of $(Ph_4P)_2[Fe(SPh)_4]$. P atoms are black, Fe is dark grey. Anion ring A is multiply contacted (**ef**) by two of the hydrogen atoms of cation ring E; one hydrogen atom on anion ring A interacts (**off**) with cation ring F; one hydrogen atom of cation ring E contacts (**ef**) three carbon atoms of anion ring C; two hydrogen atoms of cation ring D make four connections (**ef**) with carbon atoms of anion ring B (enclosed section). All of these attractive contacts are less than 3.2 Å

histogram of all intermolecular P–P distances has its maximum close to 6.9 Å, with a very small proportion of distances less than 6.5 Å, in contrast to Figure 13. The majority of P–P distances are less than twice the van der Waals radius of isotropic PPh_3, and the intermolecular interactions between hemispherical $MPPh_3$ groups are structured by the phenyl factor. The sextuple embrace with six relatively short (ca. 5 Å) distances between the centroids of the interacting phenyl groups still occurs frequently. There are variations on the sextuple-embrace geometry as described above, and there are also other types of ordered phenyl–phenyl interactions between $MPPh_3$ functions, including the **off** geometry. Details of the operation of the phenyl factor in compounds with PPh_3 and related ligands will be described separately [111]. In the following sections attention is drawn to the molecular structures of some of the metal clusters where these ligands dominate domain $\mathbb{P}$.

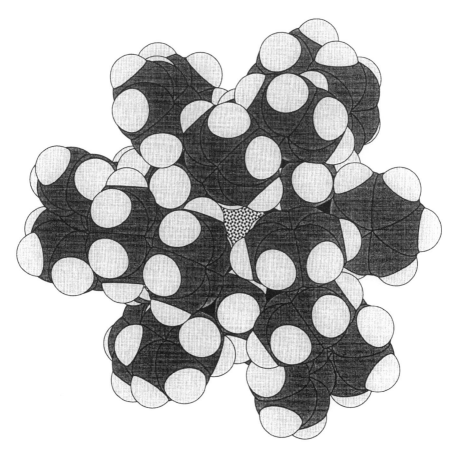

Figure 21 The van der Waals surface of [Au$_{18}$S$_8$(Ph$_2$PCH$_2$CH$_2$PPh$_2$)$_6$] viewed along its pseudo-threefold axis (P atoms black, S speckled)

The cluster [Au$_{18}$S$_8$(Ph$_2$PCH$_2$CH$_2$PPh$_2$)$_6$] has been prepared and characterized crystallographically [134]. The molecule has pseudo-threefold symmetry, with the eight S atoms comprising a hexagonal bipyramid. The 24 Ph groups do not fully envelope the molecule, but engage in a considerable number of **ef** and slightly canted **off** interactions, as shown in Figure 21.

A cluster with 36 Ph groups closely packed over its surface is [ClAg$_{14}$(SPh)$_{12}$(Ph$_3$P)$_8$]Cl [135]. The core of the cluster is approximately cubic, with a face-centred cube of Ag atoms, Cl at the body centre, μ_3-SPh ligands displaced slightly from the midpoints of the edges of the cube and terminal PPh$_3$ ligands on the cube vertices. The atomic skeleton is shown in Figure 22(a) and a view of the van der Waals surface is shown in Figure 22(b).

(a)

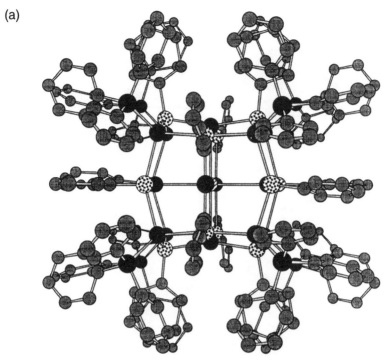

Figure 22   (a)

Another heavily phenylated cluster (30 Ph groups) is $[Cu_{14}(\mu\text{-S})(\mu\text{-SPh})_{12}(PPh_3)_6]$ [136].

There is a large number of gold clusters with terminal phosphine ligands. An intriguing subset of these, synthesized by Schmidbaur *et al.*, contains $AuPPh_3$ groups radiating from a central atom, like a porcupine [137]. Examples of these compounds are $[S(AuPPh_3)_4](CF_3SO_3)_2(CH_2Cl_2)_2$ [138], $[N(AuPPh_3)_4]F$ [139], $[C(AuPPh_3)_5]BF_4(CH_2Cl_2)_3$ [140], $[N(AuPPh_3)_5]$ $(BF_4)_2(THF)_2$ [141] and $[C(AuPPh_3)_6](MeOBF_3)_2$ [137]. Clearly the peripheral domains of these compounds are populated exclusively by phenyl groups, and therefore the role of the phenyl factor is to be questioned. Figure 23 shows side views of $[C(AuPPh_3)_6]^{2+}$, and illustrates that the phenyl groups are in contact over the surface but certainly do not fully envelope the CAu_6 core. The phenyl packing density is lower in the clusters $[X(AuPPh_3)_5]$ and $[X(AuPPh_3)_4]$. The porcupine geometry diminishes the intramolecular phenyl packing density relative to molecules such as $[Pt(PPh_3)_4]$, where the PPh_3 ligands are all bonded to the one atom. Intramolecular **ef** and **off** packing conformations are evident over the surfaces of the gold clusters.

(b)

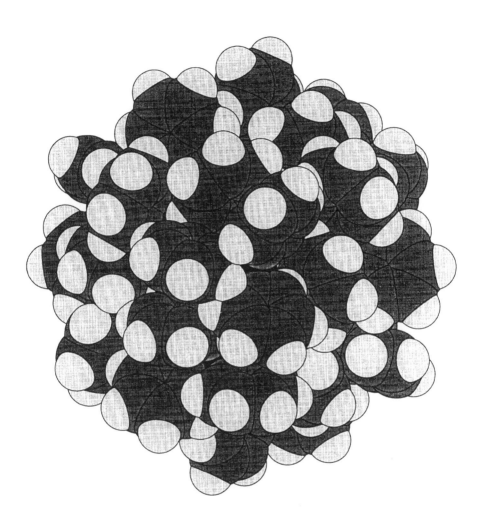

Figure 22 (*continued*) Two views of the molecular structure of $[ClAg_{14}(SPh)_{12}(Ph_3P)_8]^+$. (a) View of the skeleton (without H atoms) along an approximate twofold axis: Ag atoms (black) comprise the central face-centred cube, S atoms are speckled, and the eight P atoms (black) constitute the outer cube. The SPh ligands are in approximately coplanar sets, and the Ph planes of the terminal PPh_3 ligands form roughly parallel sets such that the molecular symmetry is approximately T_h. (b) Van der Waals surface from a view direction along a pseudo-threefold axis of the core. The close packing of the surface Ph groups is evident

(a)

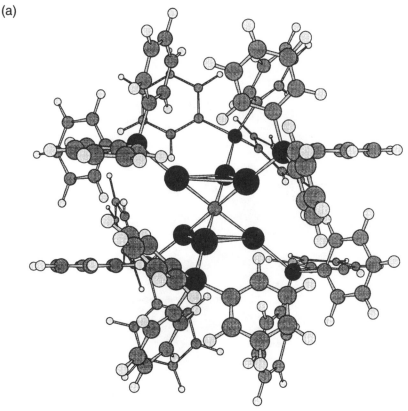

Figure 23   (a)

While the crystals of these compounds necessarily have numerous intermolecular phenyl···phenyl interactions, they do not in general demonstrate close sextuple embraces: a porcupine embrace is aloof? This is probably because the gold clusters are relatively globular, and not intramolecularly flexible. There are interesting supramolecular questions to be raised for these compounds. What is determining the molecular conformations? It is evident that core compositions $[X(AuP)_n]$ with $n = 4$, 5, and 6 are sterically feasible. However, rotation about X–Au–P bonds is related to the intramolecular phenyl–phenyl nonbonded interactions, and do these override any X–Au–P bonding influences? And what is the relative influence of the intermolecular phenyl–phenyl interactions? These questions are still unanswered.

Only two aspects of the supramolecularity of these compounds in their crystals will be illustrated here. In the simpler compound with quinuclidine, $[\{HC(CH_2CH_2)_3N\}AuPPh_3]BF_4$ [142], there is a close centrosymmetric interaction like a sextuple embrace, but this embrace involves two Ph groups and the $AuNR_3$ with the third Ph distal along the interaction axis, as shown in

(b)

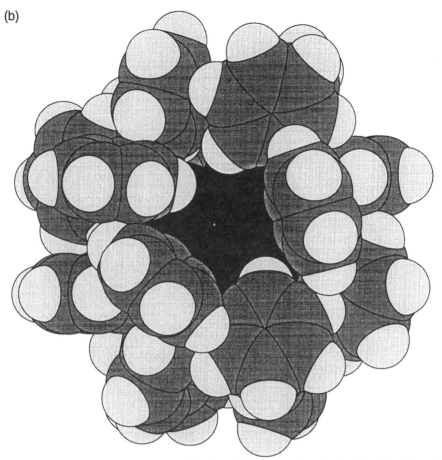

Figure 23 (*continued*) Side views of [C(AuPPh$_3$)$_6$]$^{2+}$ showing the density of packing of phenyl groups over the molecular surface (P atoms black, Au very dark grey)

Figure 24. As well as the **ef** Ph$\cdots$Ph interactions for the partial embrace, there is an attractive interaction between a quinuclidine H atom and the face of a Ph group. The crystal lattice involves additional **ef** Ph$\cdots$Ph interactions and is constructed such that there are strata containing only Ph groups and strata containing the quinuclidine sections and the BF$_4^-$ anions.

The lattice of the compound [N(AuPPh$_3$)$_4$]F has threefold symmetry (space group $P\bar{3}$c1) with the tetrahedral molecules located on and translated along threefold axes. The off-axis (AuPPh$_3$)$_3$ array of the tetrahedral cluster is relatively open (and maintained by **ef** interactions), exposing the NAu$_3$ core and providing a cavity which loosely accommodates the apical PPh$_3$ of the next molecule along the axis. This is shown in Figure 25.

(3)

(4)

(5)

(6)

(7)

11. HELICAL METAL COMPLEXES WITH OLIGOBIPYRIDYL AND RELATED LIGANDS

The coordination chemistry of oligopyridyl ligands such as 2,2'-bipyridine (bipy), 1,10-phenanthroline (phen) and, to a lesser extent, 2,2',2''-terpyridine (terpy) has been known for a long time. It has been shown that coupling of these and related chelating ligands to generate a string of chelating functions along a strand can lead to the formation of oligostranded helical complexes [143, 144]. Some of the ligands involved are sexipyridine (3), linked bipyridines

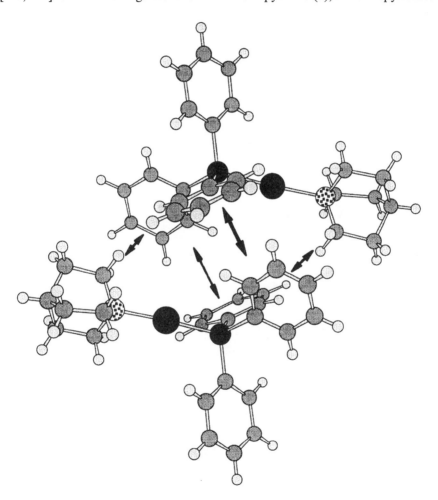

Figure 24 A pair of cations in the crystal structure of [{HC(CH$_2$CH$_2$)$_3$N}AuPPh$_3$]BF$_4$, showing the pseudo-sextuple embrace (N atoms speckled, P atoms black, Au very dark grey). The arrows indicate the attractive interactions

(a)

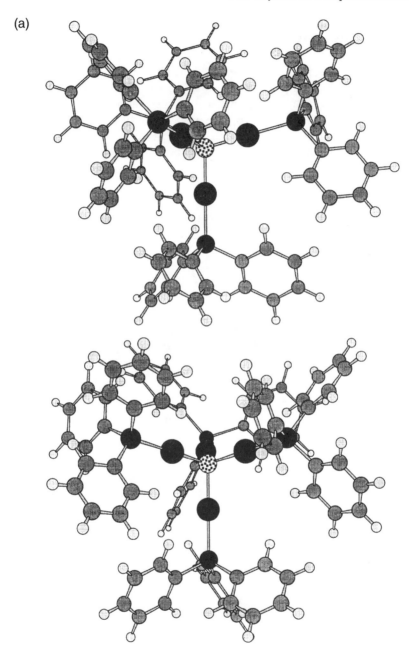

Figure 25 (a) Two adjacent cations in crystalline [N(AuPPh$_3$)$_4$]F, showing how the PPh$_3$ of one molecule faces the (AuPPh$_3$)$_3$ array of the next molecule along the (vertical) threefold axis (atom shadings as for Figure 24)

(b)

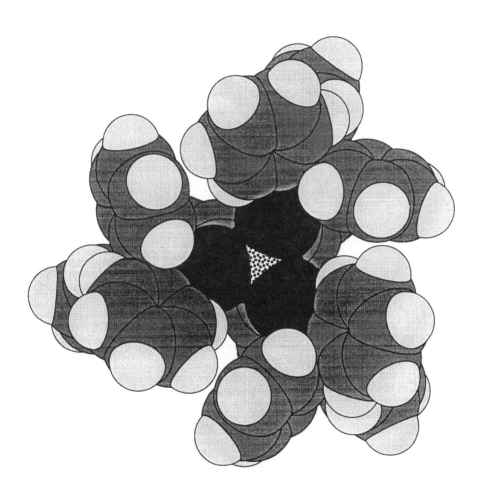

Figure 25 (*continued*)   (b) View into the (AuPPh₃)₃ array showing the phenyl–phenyl interactions

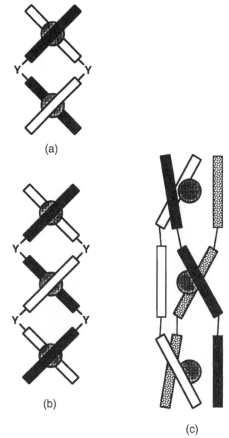

(a)

(b)

(c)

Figure 26 Diagrammatic representations of helical metal complexes. The circles are metal atoms and each long block represents a rigid chelate unit with connectors Y or a single bond; distinct ligands are shaded differently: (a) the dihelicate where each chelate can be bidentate (tetrahedral metal coordination) or tridentate (octahedral metal a coordination); (b) the triple dihelicate; and (c) the triple trihelicate, in which each chelate is bidentate and the metal coordination is octahedral

such as (4) and the linked pyridylbenzimidazoles of (5). To illustrate this class of compounds, the ligand (6), which is a pair of bipyridyl ligands linked at the 6,6'-positions (Y may be a direct bond (null Y) or CH_2CH_2 or CH_2OCH_2) coordinates with tetrahedral metal ions such as Cu^I or Ag^I to give the double helix shown diagrammatically in Figure 26(a) [145], The molecular structure of one such dihelicate using the ligand (7) and tetrahedral Cu^I coordination is shown in Figure 27 [145]. The phenanthroline equivalent of this structure is also known [146]. With extension of the ligand to a tris(bipyridine), a dihelicate with three tetrahedral metals is formed (Figure 26b).

(a) (b)

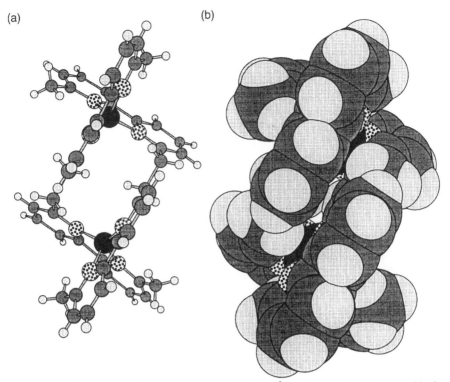

Figure 27 The helical molecular structure of $[Cu_2(7)_2]^{2+}$ in its crystals (Cu atoms black, N speckled)

The sexipyridine (**3**) when coordinated with metals engaging in octahedral hexacoordination forms a dihelicate of the type shown in Figure 26(a), in which each of the blocks represents a planar terpyridyl ligand unit. The molecular structure of $[Cd_2(\text{sexipyridine})_2]^{4+}$ is shown in Figure 28 [147] and involves a 57° rotation about the central linkage of the two terpyridyl ligands.

In these structures the oligopyridyl chelating units are planar. π-Stacking of parallel contiguous chelates along the axis of the helicate is often invoked as a characteristic contributor in these structures, and is apparent in Figure 28 and the front section of Figure 27. However, the rear chelates in Figure 27 are not parallel, a feature which is interpreted below.

While suitably spaced tridentate ligands around an octahedral metal yield a dihelicate, stereochemically suitable bidentate ligands and octahedral metals should allow organization of a three-stranded trihelicate. This was established with the introduction of the designed ligand (**5**), which has four donor atoms but is not able to function as a tetradentate ligand [148]. While this ligand forms a dihelicate with tetrahedral Cu^+, it forms the trihelicate with Co^{2+}, i.e. $[Co_2(5)_3]^{4+}$, with the structure shown in Figure 29. While the contiguous ligands in this structure are parallel, they are separated by 4.2–4.6 Å, and the π-stacking interactions are considered to be weak.

The incorporation of oligobipyridine ligands in trihelicates is enhanced by connecting the spacers at the 5,5′-positions, as in (8). When $Y = CH_2CH_2$, the trihelicate with three metal atoms, e.g. $[Ni_3(8)_3]^{6+}$, is formed with a structure as shown in Figure 26(c) [144a]. It is interesting that crystallization of this compound (with ClO_4^-) involves spontaneous resolution. It has been shown that chiral oligobipyridine ligands with the chiral centre located in the bridge Y cause high induction of chirality in the assembled helicates [149]. It has also been demonstrated that elaboration of the periphery of these metallohelicates is possible with groups such as aminodeoxynucleosides [150].

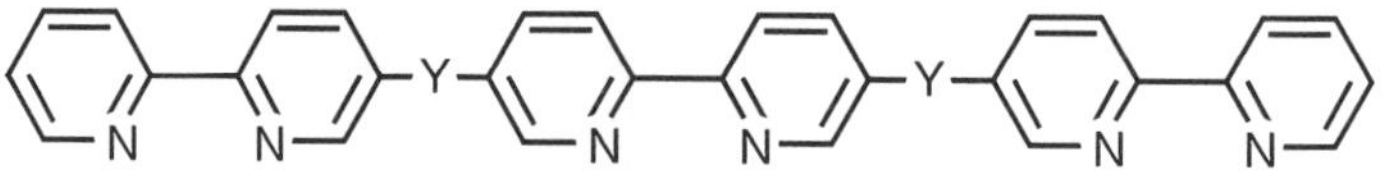

(8)

In a different type of metal-assisted self-organization of a helicate structure, pyridyl or bipyridyl ligands were connected to a strand of amphiphilic protein. Coordination of the ligands around a metal ion provides an association centre from which the proteins assemble as three-helix bundles or four-helix bundles [151]. The metal (Ru^{II}) must undergo very slow subsequent exchange.

11.1 Natural Philosophy

There are significant principles of molecular self-organization evident in these helicates [144, 145, 149]. There must be suitable matching of (1) the chelation geometry, (2) the metal coordination number and stereochemistry, and (3) the spacing of the chelating ligands along the oligochelate chains which wrap around the metal atoms along the helix axis. The number of linked chelating units in one ligand affects the length of the helicate. Alternative intrastrand chelation of one metal by the oligochelate would not generate a helicate. The self-assembly of helicates is elegantly described by Lehn in terms of molecular programming, in which the instructions are embodied in the stereochemistry of the ligand, while the metal coordination stereochemistry determines the organizational algorithm [144]. These concepts have been taken further by Lehn through demonstration of self-recognition and selection in the processes of helicate assembly. Thus, mixtures of oligobipyridine ligands of different lengths with Cu^+ yielded only helicates which were homogeneous in ligand length; unsubstituted and substituted oliobipyridine ligands assembled mainly with their like in the formation of metal helicates; and mixtures of

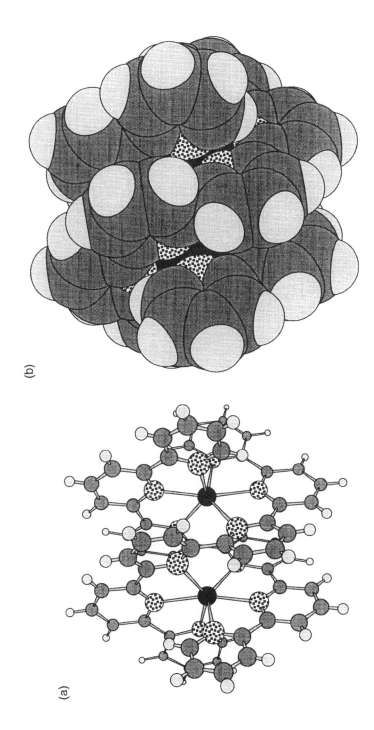

Figure 28 The molecular structure of the dihelicate $[Cd_2(sexipyridine)_2]^{4+}$ (Cd atoms black, N speckled)

(a)

(b)

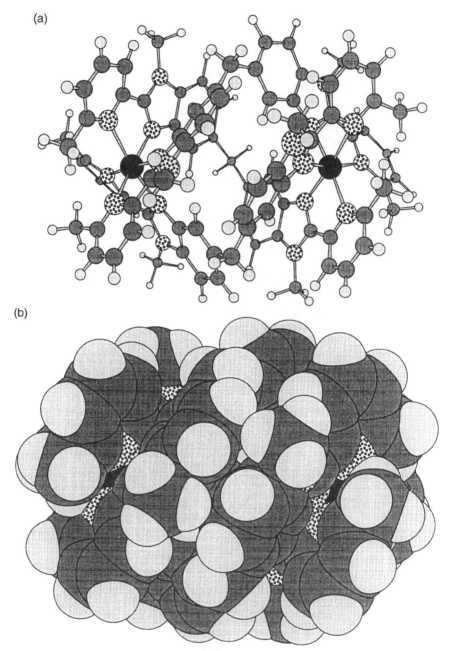

Figure 29 The molecular structure of $[Co_2(5)_3]^{4+}$ viewed along a virtual twofold axis (Co atoms black, N speckled). Two of the three ligands are in front of the helix axis, the other is behind

oligobipyridines with Cu^+ and Ni^{2+} yielded only homogeneous double or triple helicates, respectively [144b]. Lehn expands these results with the natural philosophy of instructed mixtures in synthetic chemistry, consequent upon supramolecular effects. Again this is the replication of molecular biology. There is a notion of chemical soups for synthesis, instructed by components using supramolecular algorithms. The use of chemical soups is already familiar in metallochemistry and crystal chemistry where mixtures of multiple cations have been used in the crystallization of difficult anions according to the supramolecular algorithms of crystal packing.

The research challenge now is to improve our rather poor understanding of these supramolecular algorithms for crystals: this is the *raison d'être* of this volume.

11.2 Lattice Supramolecularity for Molecular Helicates

While the self-organization of the helicates involves supramolecular influences, the resulting helicates are strictly supermolecules, entirely bonded. However, the presence of heterocyclic ligands suggests the possibilities of directed non-bonded interactions between helicate molecules and a higher dimension of supramolecularity in condensed phases. In some cases where the helicate bears a substantial positive charge $(4+, 6+)$, the counterions (and often solvent) in the lattice separate the positively charged helicates, and specific interhelicate interactions are not apparent. However, the dihelicate $[Cu_2(7)_2]^{2+}$, as the perchlorate [145], demonstrates significant crystal supramolecularity. This can be understood by first appreciating the details of the molecular structure, as shown in Figures 27 and 30(a). Two of the bipyridine sections are approximately parallel, closer to the virtual twofold axis, and separated by ca. 3.7 Å, while the other two bipyridine sections are more widely separated and splayed by about 30°: these are designated parallel and splayed chelate rings in the following. Figure 30(b) shows three contiguous molecules in the crystal. The two centrosymmetrically related molecules on the left-hand side of Figure 30(b) have their splayed chelates interleaved closely, with one chelate ring of each molecule interleaved between the two of the other. The parallel bipyridine chelates, A and A', are separated by only 3.4 Å and have an ideal **off** relationship as shown in Figure 30(c). The canted chelate ligands A and B also interact very closely, as shown in detail in Figure 30(d): the hydrogen atoms at positions 3, 4, 3' and 4' of the bipyridine ring A approach closely to the nitrogen and 2,2'-carbon atoms of the bipyridine of ring B. Figure 31 (Plate IV), a larger section of the crystal lattice, clearly shows that the interleaved splayed chelate ligands are all aligned in one direction in the crystal. The chelate ligands which are mutually parallel in each molecule are all aligned parallel to each other in the crystal. There are further significant non-bonded contacts in this crystal: hydrogen atoms at positions 4 and 5 at the edge of a

(a)

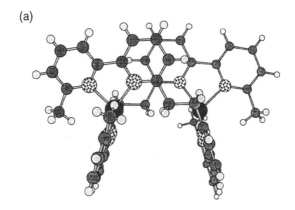

(b)

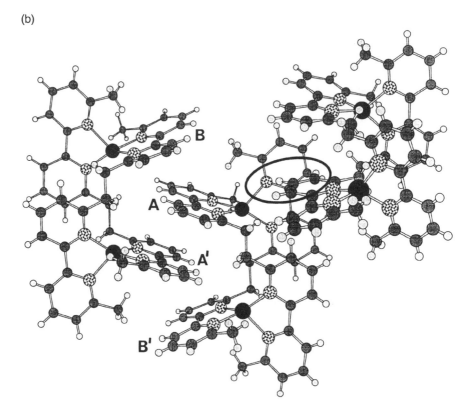

Figure 30 Molecular and crystal structures of the dihelicate $[Cu_2(7)_2](ClO_4)_2$ (Cu atoms black, N speckled): (a) view of the molecule from a direction orthogonal to that of Figure 27, illustrating the splayed bipyridine sections of each ligand; (b) three contiguous helicate molecules of the crystal lattice

(c)

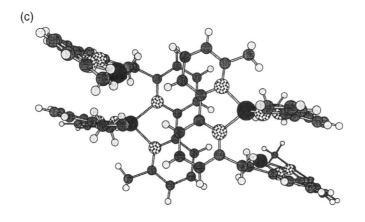

(d)

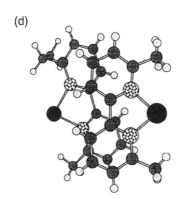

Figure 30 (*continued*)   (c) details of the offset of bipyridine chelates A and A'; (d) detail of the interaction of chelate rings B and A (underneath)

'splayed' bipyridine chelate approach closely the face of a 'parallel' bipyridine chelate (H$\cdots$N 3.2 Å, H$\cdots$C 2.7 Å): the circled region of Figure 30(b) illustrates this interaction.

This crystal structure of $[Cu_2(\mathbf{7})_2](ClO_4)_2$ suggests further possibilities in the crystal engineering of redox-active components. The analogous double helicate of Cu^I with the phenanthroline ligand 1,2-bis(9-methyl-1,10-phenanthroline-2-yl)ethane [146] crystallizes in a more open lattice containing CH_2Cl_2 without the lattice supramolecularity of $[Cu_2(\mathbf{7})_2](ClO_4)_2$.

Another significant compound is formed by the ligand (**9**) and Cu^+ with the composition $[Cu_4(\mathbf{9})_4]^{4+}$ [152]. The molecular structure is shown in Figure 32. The ligands form two parallel pairs with an interplanar separation of 3.47 Å, yielding pseudotetrahedral coordination at each of the copper atoms. There are informative subtleties in this molecular structure. π-Stacking of the ligands is obviously present, but the idealized structure with tetrahedral stereochemistry at copper would require that the ligands be superimposed, with molecular symmetry D_{2d}. The molecule has exact C_2 symmetry, approximate D_2 symmetry, but is clearly (see Figure 32b) distorted from D_{2d} symmetry with the ligand planes angled about 25° from orthogonal. This allows the contiguous parallel ligands to be offset and decreases slightly the interplanar separation, both of which are clear preferences for the interligand stereochemistry. However, a consequence of this is distortion of the Cu^IN_4 stereochemistry from tetrahedral towards planar, and this is partially resisted in the actual structure by distortion of the CuC_2N_2 chelate rings from planarity. The advantages of offset π-stacking of heterocyclic aromatic ligands are clearly demonstrated in this molecular structure.

(9)

NMR data indicate that this molecule has higher symmetry in solution, and there is the possibility that the molecular structure is fluxional between two D_2 conformations by concerted sliding of parallel ring planes with average symmetry D_{2d}.

What of the supramolecular chemistry of this interesting molecule? Intermolecular **off** or **ef** arrangements might be expected. In the crystalline compound the molecules are aligned such that all ligands are parallel, i.e. the C_2 axes of all molecules are parallel (space group $P2/n$). However, the triflate counterions and methanol separate the molecules. This suggests that slight modification of the ligand such that it is able to bear one negative charge and form an uncharged molecule $[Cu_4L_4]$ could yield an interesting supramolecular material. The demonstrated electron reservoir properties of $[Cu_4(\mathbf{9})_4]^{4+}$ enhance this interest [152].

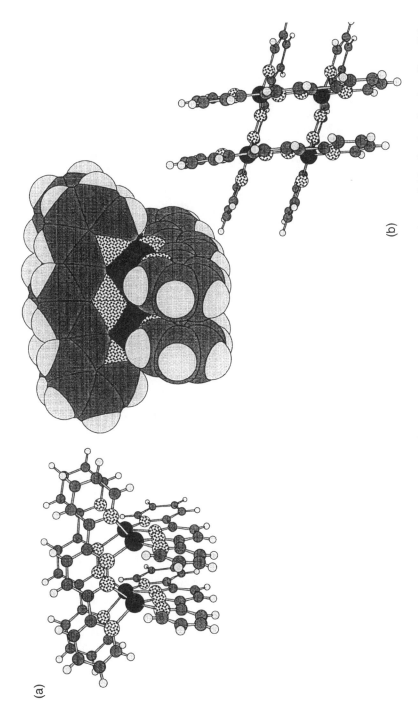

Figure 32 (a) The molecular structure of $[Cu_4(\mathbf{9})_4]^{4+}$. (b) projection view close to the molecular twofold axis, showing the offset of the parallel ligands and the related distortion of the chelate ligands from planarity and the copper coordination from tetrahedral (Cu

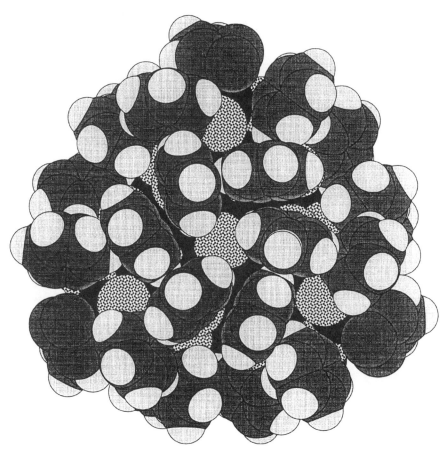

Figure 33 View along the threefold axis of the supertetrahedral cluster $[S_{14}Cd_{32}(\mu\text{-}SPh)_{36}(DMF)_4]$, with the four terminal DMF ligands on the threefold axes represented instead by NMe_3

12. CADMIUM ARENETHIOLATES

There exists an expanding class of large molecular clusters containing Cd^{2+}, S^{2-} and ArS^-, where Ar is usually Ph, and of non-molecular compounds $Cd(SAr)_2$ in which the supramolecularity of Ar groups in domain $\mathbb{P}$ is significant [153]. These compounds include $[Cd_4(SPh)_{10}]^{2-}$, $[SCd_8(SAr)_{16}]^{2-}$ [154] $[SCd_8(SAr)_{12}X_4]^{2-}$ (X = halide) [155], $[S_4Cd_{10}(SPh)_{16}]^{4-}$ [155, 156] (see Figure 4d), $[S_4Cd_{10}(SPh)_{12}X_4]^{4-}$ [155], $[S_4Cd_{17}(SPh)_{28}]^{2-}$ [157], $[S_{13}Cd_{20}(SPh)_{19}]^{5-}$ [158], $[S_{14}Cd_{32}(SPh)_{36}(DMF)_4]$ [159] and other derivatives with substituted phenyl.

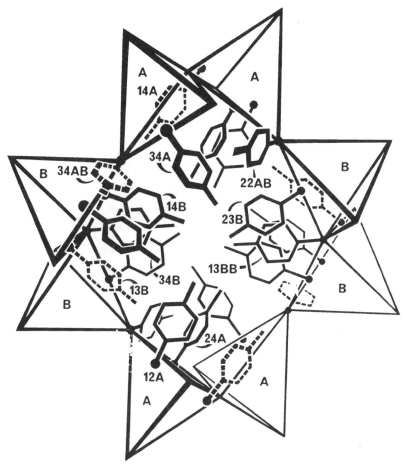

Figure 34 Part of the crystal structure of $_\infty$[Cd(SC$_6$H$_4$-4-Me)$_2$], showing the protrusion of 4-methylbenzenethiolate ligands into an internal cavity surrounded by adamantanoid [Cd$_4$(μ-SAr)$_6$(μ-SAr)$_{4/2}$] cages, which are symbolized by the large tetrahedra

These molecular cluster structures have PhS$^-$ on the periphery in doubly bridging and terminal coordination positions. The configurational isomerism at doubly bridging SPh and the conformational variability of the terminal SPh ligands allow many arrangements of the Ph groups over the surfaces of the clusters: there are four configurational isomers for [M$_4$(SPh)$_{10}$]$^{2-}$ and 186 configurational isomers possible for [S$_4$Cd$_{10}$(SPh)$_{16}$]$^{4-}$ [156]. The CdS$_4$ coordination is invariably tetrahedral, and therefore the separation of nearest S atoms is about 4 Å, which allows some rotation about the S–C$_{ipso}$ bonds and variability of the Ph conformations. Figure 33 illustrates the surface

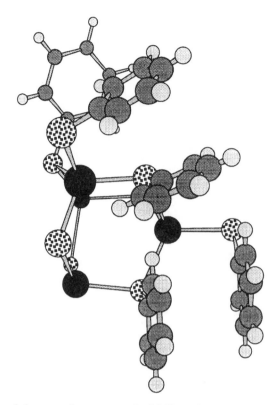

Figure 35 Part of the crystal structure of $_\infty$[Cd(μ-SPh)$_2$], showing an adamantanoid cage and five of the eight doubly bridging thiolate ligands in the asymmetric unit (Cd atoms black, S atoms speckled). The two parallel rings at the bottom right are mutually **off**, and each is **ef** to the ring above in the centre: H–C = 2.69 Å, 2.97 Å, 3.09 Å and 3.13 Å. The two rings in the upper front are mutually **off**. The rear H atom of the top front ring is directed to the centre of the ring behind in **ef** conformation with H–C = 2.87 Å, 2.87 Å, 2.95 Å, 2.96 Å, 3.03 Å and 3.04 Å

packing of Ph groups for the largest cluster, super-tetrahedral [S$_{14}$Cd$_{32}$(μ-SPh)$_{36}$(DMF)$_4$] [159], with the common **ef** and **off** arrangements evident. The crystal packing of very large phenylated clusters (such as this uncharged instance with no other components in the lattice, and the above-listed anions associated with smaller cations) reveals aspects of crystal supramolecularity which may be useful in crystal design [160].

In addition to the above-listed large molecules there are signficant non-molecular structures, including $_\infty$[Cd(SPh)$_2$] [161], $_\infty$[Cd(SC$_6$H$_4$-4-Me)$_2$] [161b], $_\infty$[Cd$_8$(SC$_6$H$_4$-4-Br)$_{16}$(DMF)$_3$] [162] and $_\infty$[Cd$_7$(SC$_6$H$_4$-2-Me)$_{14}$(DMF)$_2$] [163]. These non-molecular structures have open filigreed Cd–S networks, commonly with linked adamantanoid cages. A general characteristic of these structures is that the ArS ligands are directed towards internal cavities where non-bonded interactions between Ar groups are present. There are regions of these crystals where the local molecular surface, domain $\mathbb{P}$, is concave rather than convex, as illustrated in Figure 34 for part of the structure of $_\infty[Cd(SC_6H_4-4-Me)_2]$ [161b].

Detailed analyses of the molecular and non-molecular structures will be presented separately, but here I illustrate the strong influence of the phenyl factor in the crystal structures of $_\infty[Cd(\mu-SPh)_2]$ and $_\infty[Cd_7(SC_6H_4-2-Me)_{14}(DMF)_2]$. Figure 35 shows part of the adamantanoid cage forming the asymmetric unit of $_\infty[Cd(\mu-SPh)_2]$, in which the adamantanoid cages are linked with the same topology as SiO_2 (α-cristobalite) [161]. The figure clearly shows the **off** and **ef** relationships between five of the phenyl groups of bridging SPh ligands. In fact these phenyl–phenyl relationships (six relationships involving five rings) are close to ideal in geometry, and an unusual **off(ef)$_2$** interaction occurs amongst the rings in the bottom right of the figure. For $_\infty[Cd_7(SC_6H_4-2-Me)_{14}(DMF)_2]$, Figure 36 shows the Cd and S atoms of part of the three-dimensionally nonmolecular Cd–S filigree [163]. Each Cd has tetrahedral coordination, each SAr is doubly bridging and there are local cycles of $Cd_2(\mu-SAr)_2$ and $Cd_3(\mu-SAr)_3$ (the DMF is weakly connected to three Cd atoms in this cycle) which are connected in macrocycles as shown in Figure 36: the minimum traverse around the macrocycle involves 24 Cd and 24 S atoms. In the elongated tetragonal cell these very large cycles are linked in a more complex network. However, this description in terms of the Cd–S network misses the point that the bulk of the volume of this lattice is occupied by the substituents. This is illustrated in Figure 37 (Plate V), which shows part of the lattice and the dominance of the Ar–Ar interactions, which are both **ef** and **off**. The less significant role of the Cd and S atoms and the DMF is illustrated by their minor appearance in Figure 37. Thus the Cd–S network is not a determinant of the crystal structure, but is a consequence of the supramolecularity of this crystal. The DMF appears to fill a void, weakly oriented for coordination of Cd, but not required for coordination. In our original paper we fell into the trap of neglecting supramolecular effects in the interpretation of this structure.

13. DECAPHENYLSTANNOCENE

Decaphenylstannocene, $[Sn(C_5Ph_5)_2]$, is a molecule with a totally phenylated periphery but with severely restricted conformational flexibility. In this sense it is the antithesis of molecules such as $[M(SPh)_4]$ discussed above. Figure 38 (Plate VI) shows the molecular structure and part of the crystal structure [164]. Within the molecule the phenyl groups substituting the two parallel C_5 rings are canted like gears, or a 'double-opposed paddlewheel' [164], and are largely locked into place by **ef** interactions between Ph groups alternating from one

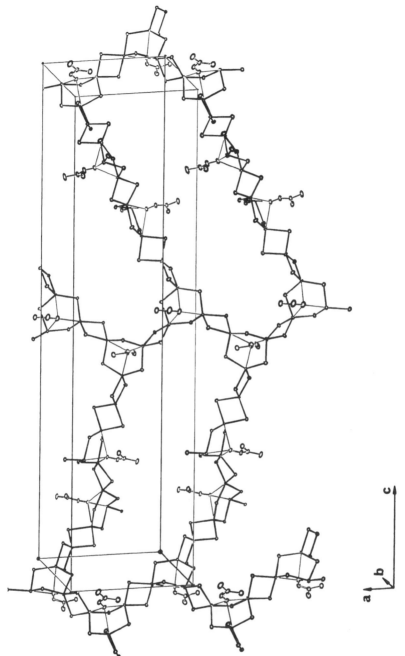

Figure 36 The filigree of Cd–S bonds and the linked large cycles in the crystal structure of $_\infty$[Cd$_7$(SC$_6$H$_4$-2-Me)$_{14}$(DMF)$_2$]. Reproduced with permission from *Inorg. Chem.*, **29**, 1571 (1990)

ligand to the other. The molecular structure is distinctly different from other less substituted stannocenes where the angle between the two cyclopentadienyl ligand planes is ca. 35°. The $5s^2$ electrons have lost their normal stereochemical activity. In the crystal the pentagonal molecules are arranged in pseudohexagonal layers (see Figure 38), but the association between the molecules within the layers is not efficient, and there are only a few phenyl–phenyl interactions at the distances found in crystals of more flexible molecules. Some variations in the canting of the phenyl rings are evident in the crystal structure, dependent on the local intermolecular contacts.

14. METALLOMESOGENS

Mesogens are compounds which generate mesophases, intermediate between the crystal phase and the liquid or solution phase. Mesophases possess some of the order of crystals, but are fluid like liquids. There are different classes of mesophases, principally thermotropic where temperature is the principal variable in phase changes and lyotropic where there is additional solvent and concentration is an important variable. The different classes of thermotropic phases relate to the different types of partial ordering of the molecules: in nematic phases there is orientational order of the molecules, and in smectic phases there is translational order.

There are many metal complexes, metallomesogens, which deomonstrate liquid crystal or mesomorphic properties, and the organization of metal complexes in fluid phases is definitely supramolecular behaviour. These compounds were well reviewed in 1991 [20a], 1992 [20b] and 1993 [20c]: the strength of this research in Europe and Japan is evident.

There are two principal molecular shapes for metallomesogens: calamitic (rod-like) and discotic (disc-like). Many of the calamitic metallomesogenic molecules possess the characteristics illustrated diagrammatically in Figure 39. The metal and its ligand frame (called the promesogenic ligand by Espinet *et al.* [20b]) are typically four-coordinate dithiolene, dithiocarboxylate, salicylal-dimine, cyclometallated arylimine or β-diketonate. Aryl- or arylnitrile–metal coordination is also possible. Connected close to this central frame is a

Figure 39   Generalized scheme of the components of calamitic metallomesogens, with a central metal bound to its ligand frame (dotted), connected to rigid sections (bubbles), connected to extended flexible chains

relatively rigid group, the proximal rigid spacer, which is often an aryl group or an aryl extension of the chelate frame. A flexible chain, usually alkyl or alkoxy, extends from the rigid spacer.

These characteristics are illustrated by the calamitic nickel dithiolenes such as (10). Another common type comprises the cyclometallated diarylazines and arylimines, exemplified by (11). While many calamitic metallomesogens are uncharged, the silver stilbazole compounds such as (12) are positively charged, and allow large design variability though modification of the anion, including mesogenic anions.

$$CH_3(CH_2)_7 \quad\quad\quad\quad\quad\quad\quad (CH_2)_7CH_3$$

Even though the presence of the extended flexible tails is an obvious characteristic of metallomesogens, aryl–aryl interactions are significant for their supramolecular structures. The proximal rigid spacer of Figure 39 is often involved in **off** and/or **ef** interactions [165]. These can be illustrated in the crystal structure of the nickel dithiolene structure (10) [166]. The centrosymmetric molecules adopt a seahorse conformation in a thin ($b = 5.4$ Å) unit cell, as shown in Figure 40(a) (Plate VII). Within each molecule the atoms of the two chelate rings and the C_6H_4 proximal spacers are all close to coplanar, and the molecules are approximately parallel. As can be seen in Figure 40(b) (Plate VII), contiguous molecules along the long c-axis (22.16 Å) are related by a c-glide symmetry element, but the central planar region of the molecule is canted relative to the reflection plane such that the central planes of molecules separated by $c/2$ are approximately perpendicular. This allows each C_6H_4 group to engage an **ef** interaction with the $C_6H_4C_2H$ section of its neighbouring molecule. Details of this are shown in Figure 40(b). Along the short b-direction of the lattice, the canting of the molecular plane means that contiguous molecules displaced by 5.38 Å along b in fact have **off** aryl groups (Figure 40a): H atoms of one aryl group are 3.23 Å from two C atoms of the neighbouring group. Thus there are significant but standard aryl–aryl interactions along the b- and c-directions of the lattice. In the a-direction,

(11)

(12)

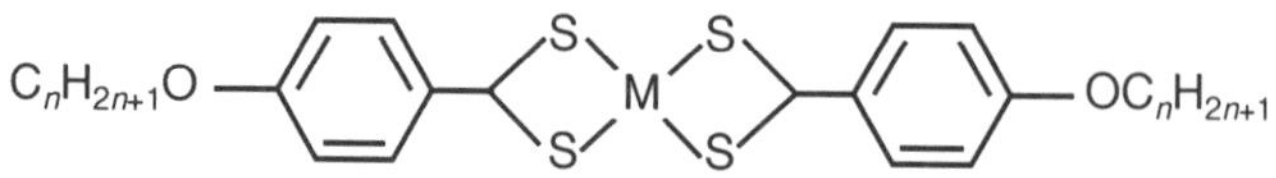

(13)

(13a) M=Pd, n=8

(13b) M=Zn, n=4

each $NiS_4C_4H_2$ coordination plane is flanked by two extended octyl chains which are almost coplanar with the coordination plane.

In the similar crystal structures of the two dithiobenzoate chelate compounds (13a) and (13b) the two aryl proximal spacers are essentially coplanar with the metal chelate region, and the planes of all molecules are essentially parallel [167]. The molecular stacking is then such as to allow off interactions between contiguous molecules. Figure 41 (Plate VIII) pictures the Pd structure looking almost parallel to the planes; the Zn structure (13b) is similar in crystal packing, but the coordination around the Zn atom rearranges for a pair of contiguous molecules such that the four primary Zn–S bonds are approximately tetrahedral and involve three dithiocarboxylate ligands. It is interesting to note that there is virtually no crystallographic symmetry in these lattices (both space groups are $P\bar{1}$), and thus the ordered liquid phases could have structures similar to the crystal structure shown in Figure 41. It is also noticeable that the long alkyl or alkoxy chains in these structures are arranged in the crystals to flank the metal coordination region, as in (10).

In addition to these aryl–aryl interactions, prevously underrecognized, there are some generalizations that have been made about the structure and properties of metallomesogens and their relationships [20].

(1) In many cases the uncomplexed ligand also demonstrates mesogenic behaviour, but incorporation of a metal increases the phase transition temperatures over those of the mesogenic ligands.

(2) The larger polarizabilities of metal atoms and often of their donor atoms (domains $\mathbb{M}$ and $\mathbb{L}$) contribute generally to the increased intermolecular interaction energies and higher transition temperatures, while in some cases there may be contributions from secondary coordination interactions.

(3) Whereas in non-metal mesogens there is often a requirement to increase intermolecular interaction energies in order to cause mesomorphism, in metallomesogens the object is often to reduce the intermolecular interaction energies in order to reduce the transition temperatures to below the regime of thermal decomposition.

(4) It is natural for coordination chemistry to focus on the metal atom and its electron configuration and stereochemistry. However, metal coordination stereochemistry is not a major determining factor for metallomeso-morphs, and the properties of the complete metallomesogen need to be evaluated.

Silver thiolate compounds (AgSR) adopt a bilayer structure in the crystalline state [168] with a central slab of μ_3-SR and trigonal planar Ag. On melting, these compounds form lamellar, cubic, micellar and hexagonal columnar mesophases, depending on the length of R [169].

Even though metallomesogen research began only in the 1980s, it is now well advanced at the phenomenological level. There has been generated a large amount of information, structural and thermal, still to be fully interpreted. The occurrence of the liquid crystalline state and the nature of the semiordering provide valuable information about intermolecular interactions in relation to molecular shape. The combined knowledge of the crystal structure of the fully ordered phase of a thermotropic compound and of the structure type for the liquid crystalline phase yields a considerable amount of information about intermolecular forces. Use of this information will aid the development of intermolecular force fields for systems containing metals, while the deployment of those force fields in the molecular mechanics and molecular dynamics simulations of metallomesophases is expected to accelerate their development and application. Espinet *et al.* [20b] have drawn attention to the numerous opportunities for advancing the science of supramolecular interactions and structure in metallomesogens, and to the prospects for 'specifically ordered arrangements in the solid state, either as materials or as precursors'.

15. CRYSTAL STRUCTURES OF METAL CARBONYLS AND ORGANOMETALLIC COMPOUNDS

Braga and Grepioni have undertaken a series of analyses of the crystal structures of metal carbonyl clusters, π-arene complexes, mixed carbonyl π-arene complexes and some substituted derivatives [170]. Their analyses are based on examination of the packing details and motifs in the crystal structures, recognition of specific interactions, calculation of the packing potential energy for the crystal and consideration of contributions to the lattice volume. Their objective, to investigate the interplay between molecular structure and crystal structure, is very similar to the perspective in this article.

Carbonyl ligands undergo intermolecular interpenetration of domains $\mathbb{P}$: there is an interlocking of CO ligands, like a 'Velcro' interaction [79, 171]. In π-arene molecules such as [M(arene)$_2$], [(arene)M(CO)$_3$] and [(arene)-Ru$_6$C(CO)$_{11}$], both the **off** and **ef** arene–arene conformations occur and are influential, but are thwarted in alkylated arene ligands such as $\eta^6\text{-}C_6Me_6$ [172].

Some metal carbonyl clusters demonstrate a variable geometrical relationship between the metal framework in domain M and the array of CO ligands in the periphery P. When viewed at the molecular level this is interpreted as the CO ligands migrating over the cluster core with little energy cost by interchange between terminal M–CO and bridging M(μ-CO)M bonding. However, the intermolecular interlocking of CO ligands in crystals leads to the alternative view of the geometrical flexibility, in which the CO ligand framework is fixed and the metal framework moves within it. Metal carbonyl clusters and arene-substituted derivatives provide further demonstration that the crystal field can influence molecular structure, and that interpretation of molecular structure from a crystal structure determination can be misleading. The polymorphism of crystal structures of metal carbonyl clusters is termed crystal isomerism [170].

The dynamics of the reorientations of arene rings in these crystalline organometallic compounds are similar to those in the pure crystalline arenes [173].

There is a considerable number of metal carbonyl cluster anions whose intermolecular interactions in crystals can be ameliorated by the presence of countercations. These crystal structures have also been analysed [79, 174]. The relative volume of the countercation is the significant factor in these crystal structures, with large cations such as PPN$^+$ and Ph$_4$P$^+$ resulting in mixed crystal packing according to the principles which govern that of uncharged compounds. Specific cation–anion pairing is not a characteristic of these compounds, and the coulombic contributions are considered to be so diluted and isotropic as to have no specific influences on the supramolecularity in the crystals. Specific cation–cation interactions can be recognized [174]. Smaller cations such as Me$_4$N$^+$ and alkali metal ions (often with associated water) naturally allow aggregation of the anions into one- and two-dimensional patterns which can be related to the volume ratio V_{anion}/V_{cation} [174].

16. MOLECULAR COMPOUNDS CONTAINING IONS

Mingos and Rohl [61, 175, 176] have analysed the packing of inorganic molecular ions ranging in size from monatomic ions (volume ca. 10 Å^3) to $[Au(PPh_2Me)_4]^+$ (volume 795 Å^3) and $[Os_6(CO)_{18}]^{2-}$ (volume 467 Å^3). They have used and evaluated a range of concepts and measures of the sizes and shapes of ions. These measures (some of which derive from the earlier analyses of organic structures by Gavezzotti [177]) are the volume (V); the surface area (S); the shape, calculated from moments of inertia as indices of spherical, cylindrical and discoidal shape; the effective radius (R_{eff}) derived from the volume; the maximum radius (R_{max}) as the largest distance between the centroid of the ion and its outer surface; the exposure ratio (E_r) as the ratio of volume to surface area; $I = SV^{-0.67}$ as a dimensionless indicator of non-spherical shape; and interpenetration indices for pairs of molecular ions (same or opposite charge) expressed as distances between molecular ions minus the distance expected for impenetrable spheres. They have also calculated standard packing coefficients (molecular volume as a percentage of lattice volume) and the lattice energies (with $\varepsilon = 1$).

This analysis was presented in the context of the 'folklore' of the 'art' of crystallization. Mingos and Rohl considered the crystallization generalization of Basolo [178] that 'solid salts separate from aqueous solution easiest for combinations of either small cation plus small anion or large cation plus large anion, preferably with systems having the same but opposite charges on the counterions', and also the expectations of McDaniel [179] that there is only a lower critical limit to the size of the cation and no upper limit. While these generalizations were limited in purview (water as solvent), and every synthetic chemist knows that there are innumerable exceptions, they were for a long time the only guiding generalizations for crystallizations of salts. Mingos and Rohl [175] looked particularly at cations needed to stabilize unusual anionic coordination complexes, and concluded that favourable crystallization occurs when the cation is not smaller than the anion. There are, however, many well-defined crystals of the small cation Me_4N^+ associated with large anionic clusters.

The analyses of Rohl and Mingos focus on the lattice array of cations and anions, as a function of the radius ratio, in comparison with the idealized arrays of monatomic cations and anions which are based on radius ratios. The deviations from the monatomic behaviour are interpreted in terms of the shapes of the polyatomic ions and the degree of interpenetration. These analyses have been published for PF_6^- salts of cations ranging from Me_4N^+ to $[Au(PPh_2Me)_4]^+$ [176], and for 'soft salts' composed of large polarizable metal clusters as both cation and anion [180]. In addition to these analyses of the ion-packing arrangements, the relative contributions of the van der Waals and coulombic energies to the lattice energy were calculated. In general, as the size of the ions increases the van der Waals contribution increases and the coulombic contribution decreases, with the total lattice energy approximately constant. The coulombic contributions to the lattice energy were obtained by summation of atom–atom contributions, but were probably overestimated because the partial atomic charges were overestimated by extended Hückel methods [176]. Alternatively, the coulombic contributions to the lattice energies in these compounds can be approximated by assuming unit charges at the centres of the ions.

Specific and local structure-determining interactions were neglected in this analysis of the salts as close-packed ions. Whereas the coulombic contributions as evaluated by Rohl and Mingos will favour the symmetrical arrays, as occur for monatomic ions, local factors such as phenyl–phenyl interactions, other H to phenyl interactions and hydrogen bonding are in general likely to influence crystal structures. These specific interactions also have a coulombic component which causes them to be structure directing. Thus there is raised a general question about the relative influences of general coulombic effects (yielding close packing) and local coulombic interactions (possibly causing deviations from close packing) in the structures of crystals containing ions. As the sizes of the ions increase the overall coulombic energy which favours regular structure diminishes in influence, while each local effect does not. Therefore it would be expected that salts containing smaller ions would be more likely to be regular and determined by ionic size and shape, while salts containing larger ions would be more susceptible to local interactions and peculiarities.

Some calibration of this qualitative assertion is found in the structures of salts of Ph_4P^+, which are dominated by the phenyl factor as described above. The Ph_4P^+ cation is not particularly large (317 Å^3), and the partial charges which direct the phenyl–phenyl interactions are small (± 0.12). Given the common occurrence in inorganic compounds of molecular ions larger than Ph_4P^+, and the occurrence of partial charges larger than ± 0.1, it could be expected that localized structural features due to coulombic influences occur widely.

The influences of cation properties on the crystal structures of halocuprates and -argentates have been documented [80].

17. SUPRAMOLECULAR BUT NONMOLECULAR INORGANIC LATTICES

A supramolecular but nonmolecular lattice might be considered to be an oxymoron. However, as mentioned earlier, there is a considerable number of crystals which contain interpenetrating but unconnected infinite lattices. In the context of supramolecularity, the special property of these crystals is that there are two or more sets of atoms which are connected within the sets but not between the sets (just as in molecular structures), and there are extended non-bonding boundaries (as in molecular lattices) but there are no finite molecules. The crystals are maintained by the interactions across non-bonding boundaries.

The most common lattice type is diamondoid. The isomorphous compounds $Zn(CN)_2$ and $Cd(CN)_2$ possess two interpenetrating diamondoid lattices in a cubic lattice [23], as does Cu_2O. Other coordination compounds with interpenetrating lattices include $_\infty[SCd_8(SBu^i)_{12}(CN)_{4/2}]$ [181] (two interpenetrating diamondoid lattices), $_\infty[Cu(4,4'\text{-bipy})_2(PF_6)]$ [182] (four interpenetrating diamondoid lattices), $_\infty[Cu(1,4\text{-dicyanobenzene})_2(BF_4)]$ [183] (five interpenetrating diamondoid lattices), $_\infty[Cu\{NC(CH_2)_4CN\}_2(NO_3)]$ [184] (six interpenetrating diamondoid lattices), $_\infty[Cu_2(pyrazine)_3(SiF_6)]$ [182] (two interpenetrating honeycomb lattices), $_\infty[Cd(4,4'-bipy)_2\{Ag(CN)_2\}_2]$ [185] (two interpenetrating rectilinear lattices), $_\infty[Cd(pyrazine)\{Ag_2(CN)_3\}\{Ag(CN_2)\}]$ [185] (three interpenetrating rectilinear lattices) and $_\infty[Ag_2(O_2CCH_2CH_2CO_2)]$ [186] (three interpenetrating diamondoid lattices of Ag_4 squares). Crystallization of $Cd(CN)_2$ from polar and non-polar solvents yields diamondoid lattices with large volumes of included solvent, often mobile, as unique supramolecular systems [23, 33, 187], while incorporation of rigid non-chelating ligand systems also generates open non-molecular lattices with supramolecular properties [23, 188]. The relationships between these systems and the gas-phase helicates of Zn and Cd with composition $[M_n(CN)_{2n+1}]^-$ [22], mentioned earlier raise questions about supramolecularity.

The compound $_\infty[K_2PdSe_{10}]$ [126] provides an intriguing extra facet of this supramolecular chemistry. There are two interpenetrating but different diamondoid lattices, one with $[Pd(Se_4)_{4/2}]^{2-}$ units and the other with $[Pd(Se_6)_{4/2}]^{2-}$ units. The local coordination at Pd is square planar, even though the Pd atoms are located at the tetrahedral connection sites of the diamondoid lattice. The conversion between tetrahedral and square planar geometries is achieved by means of the flexible Se_x chains. Note the discussion in Section 8 about the cation influences on the crystallization of palladium polyselenide compounds.

18. TWO-DIMENSIONAL METAL-BASED STRUCTURES AND SURFACE ENGINEERING

An increasingly significant field of supramolecular coordination chemistry involves ordered monolayers of organic molecules on a metal base. A well-studied instance is the formation of self-assembled monolayers of amphiphilic thiolates and derivatives on the surface of gold or silver. Two-dimensional ordering is a consequence of the intermolecular interactions between substituents. For leading references see elsewhere [189, 190]. Related to these are the two-dimensionally non-molecular structures of silver (and mercury) thiolates [39, 168]. There are many other inorganic layered structures where supramolecularity in one dimension is significant [32, 191, 192].

19. CONCLUDING AND SUMMARIZING REMARKS

A sound and prescient outlook on the supramolecularity of inorganic compounds should include at least the following perspectives.

(1) Concepts of molecularity need to be more pliable than in organic compounds, because interatomic interactions range continuously from strong primary bonding through weak secondary bonding to weak dispersion at long distance. A continuum between bonding and non-bonding interactions is possible, and consequently molecular boundaries are sometimes poorly defined. This requires more specific definition of the internal and peripheral domains of the assembly, and expands the scope of supramolecularity.

(2) The intermolecular interactions between inorganic molecules can have substantial magnitudes, with both the van der Waals attractions and the coulombic interactions being larger than for organic molecules.

(3) Atomic partial charges in inorganic molecules are often around ± 0.5, and the coulombic components of intermolecular energies frequently are the largest components for inorganic assemblies. Ordered distributions of charge in molecules can introduce directionality into intermolecular interactions, and can be supramolecular structure-determining.

(4) Supramolecular interactions are attractive. The concepts of 'steric bulk' and implicit repulsions between groups still pervade much of the discussion of molecular chemistry, and while it is true that a few peripheral atoms are sometimes forced together to the extent that their interaction is repulsive, the majority of interactions in supramolecular assemblies are attractive. Ligands such as tertiary phosphines have been analysed in terms of their volume, often expressed as a cone angle at the metal or phosphorus atoms, with the underlying notion of exclusion and intramolecular repulsion. However, specific attractions between ligands (especially Ph_3P ligands), intramolecular and intermolecular, paint a more realistic supramolecular landscape.

(5) Specifically directional attractive interactions commonly occur between phenyl groups and other unsubstituted aromatic (and heterocyclic) groups. These interactions, dubbed the phenyl factor, are ubiquitous and significant in coordination compounds and organometallic molecules.

(6) Paradoxically, polyatomic cations such as Ph_4P^+ are not mutually repulsive, but can have attractive, structure-determining supramolecular interactions.

(7) The energy of the crystal field (or the supramolecular field in non-crystalline assemblies) is substantial, and can distort the structures of plastic molecules. This is more likely to occur in inorganic systems than in organic compounds, because the intermolecular energies are greater and the intramolecular conformational energies smaller in inorganic systems. The caveat is that the interpretation of molecular structure from crystal structure data can be subject to subtle errors.

Much further research on inorganic supramolecularity is needed. There is a need to search for additional supramolecular motifs, particularly for molecules with atoms other than C and H in domain $\mathbb{P}$, using statistical analyses of crystal structure data. The crystallography of inorganic molecular compounds provides many opportunities for investigation of supramolecularity. Crystals with wider ranges of mixed components need to be explored: while some binary crystals are known, as inorganic host–guest inclusion complexes, ternary and quaternary molecular associations in crystals are still to be developed.

Crystal polymorphism for any one compound is a valuable and informative phenomenon [73]. In contrast, crystallographically isomorphous series of compounds can be misinterpreted as reflecting favourable molecular structure, when what is demonstrated is favourable crystal structure. Crystal polymorphism is almost certainly more widespread than recognized because crystallization experiments usually cease after the first crystal structure of a compound is determined.

Looking forward to the design of supramolecular assemblies of inorganic molecules and their crystal engineering, what are the prospects? As for organic molecules [73], it is important to recognize possibilities for distortions of plastic molecules: the engineering of crystals based on the design of molecules is feasible only where the molecule cannot be deformed by the crystal field. Accepting this, a key design tool is molecular shape, as illustrated in the work on metal mesotetraphenylporphyrin (TPP) compounds: it has already been demonstrated that it will be possible to elaborate the TPP, or coordination at the metal, or included molecules, with a fair degree of certainty about the overall structure of the crystal, and applications have already been suggested in the papers of Strouse [193]. Other design tools include the phenyl factor motifs and the sextuple embrace in compounds containing Ph_3P and/or Ph_4P^+ or similar peripheries.

Despite the progress so far, supramolecular inorganic chemistry is in the early stages of growth and development.

20. ACKNOWLEDGEMENTS

I gratefully acknowledge financial support from the Australian Research Council and the University of New South Wales. Discussions with Professor Gautam Desiraju clarified my early thinking on intermolecular interactions and crystal energies, and I also thank Dr Marcia Scudder and Dr Dawit Gizachew for discussions and valuable local assistance. I am grateful to Professors Mercuri Kanatzidis, Alan Balch, Doug Rees, Dieter Fenske, Hubert Schmidbaur and Kaluo Tang who kindly provided crystallographic details, and to Dr Andrew Rohl who provided unpublished information about Ph_4B^- salts.

NOTES AND REFERENCES

1. J.-M. Lehn, *Angew. Chem., Int. Ed. Engl.*, **29**, 1304 (1990).
2. The negative name inorganic is entirely anachronistic for this field of chemistry. Inorganic chemistry does encompass the chemistry of living organisms, and this facet is demonstrated in inorganic supramolecularity.
3. (a) V. Balzani and F. Scandola, *Supramolecular Photochemistry*, Ellis Horwood, Chichester, 1991; (b) P. Gutlich and A. Hauser, *Coord. Chem. Rev.*, **97**, 1 (1990); (c) S. M. Baxter, W. E. Jones, E. Danielson, L. Worl, G. Strouse, J. Younathan and T. J. Meyer, *Coord. Chem. Rev.*, **111**, 47 (1991); (d) K. Kalyanasundaram and M. Grätzel, *Photosensitization and Photocatalysis Using Inorganic and Organometallic Compounds*, Kluwer, 1993.
4. (a) J. M. Williams, H. H. Wang, T. J. Emge, U. Geiser, M. A. Beno, P. C. W. Leung, K. D. Carlson, D. L. Thorn, R. J. Thorn and A. J. Schultz, *Prog. Inorg. Chem.*, **35**, 51 (1987); (b) P. Zanello, *Coord. Chem. Rev.*, **83**, 199 (1988).
5. (a) E. König, *Prog. Inorg. Chem.*, **35**, 527 (1987); (b) M. Bacci, *Coord. Chem. Rev.*, **86**, 245 (1988); (c) J. S. Miller and A. J. Epstein, *Angew. Chem., Int. Ed. Engl.*, **33**, 385 (1994); (d) D. Gatteschi, A. Caneschi, L. Pardi and R. Sessoli, *Science*, **265**, 1054 (1994).
6. (a) O. Kahn and J. P. Launay, *Chemtronics*, **3**, 140 (1988); (b) J. Zarembowitch and O. Kahn, *New J. Chem.*, **15**, 181 (1991).
7. (a) C. Mealli, C. A. Ghilardi and A. Orlandini, *Coord. Chem. Rev.*, **120**, 361 (1992); (b) A. L. Balch, *Prog. Inorg. Chem.*, **41**, 239 (1994).
8. (a) H. Schumann, A. Lentz, R. Weiman and J. Pickardt, *Angew. Chem., Int. Ed. Engl.*, **33**, 1731 (1994); (b) A. F. Masters, L. D. Field and T. W. Hambley, personal communication.
9. A. I. Kitaigorodskii, *Molecular Crystals and Molecules*, Academic Press, New York, 1973, p. 186.
10. B. K. Burgess, *Chem. Rev.*, **90**, 1377 (1990).
11. (a) J. Kim and D. C. Rees, *Nature (London)*, **360**, 553 (1992); (b) M. K. Chan, J. Kim and D. C. Rees, *Science*, **260**, 792 (1993); (c) J. T. Bolin, A. E. Ronco, T. V. Morgan, L. E. Mortenson and N. H. Xuong, *Proc. Natl. Acad. Sci. USA*, **90**, 1078 (1993); (d) J. T. Bolin, N. Campobasso, T. V. Morgan, S. W. Muchmore and L. E. Mortenson, *ACS Symp. Ser.*, **535**, 186 (1993).
12. I. G. Dance, *Aust. J. Chem.*, **47**, 979 (1994).
13. S. Mann, J. Webb and R. J. P. Williams, *Biomineralization: Chemical and Biochemical Perspectives*, VCH, Weinheim, 1989.
14. (a) P. M. Harrison, G. C. Ford, D. W. Rice, J. M. A. Smith, A. Treffrey and J. L. White, in *Frontiers in Bioinorganic Chemistry*, (ed. A. V. Xavier), VCH, Weinheim, 1986, p. 268; (b) C. T. Dameron, R. N. Reese, R. K. Mehra, A. R. Kortan, P. J. Carroll, M. L. Steigerwald, L. E. Brus and D. R. Winge, *Nature (London)*, **338**, 596 (1989); (c) C. T. Dameron, B. R. Smith and D. R. Winge, *J. Biol. Chem.*, **264**, 17 355 (1989); (d) C. T. Dameron and D. R. Winge, *Inorg. Chem.*, **29**, 1343 (1990).
15. (a) A. M. Pyle and J. K. Barton, *Prog. Inorg. Chem.*, **38**, 413 (1990); (b) J. K. Barton, in *Bioinorganic Chemistry*, (eds I. Bertini, H. B. Gray, S. J. Lippard and J. S. Valentine), University Science Books, Mill Valley, CA, 1994, Chapter 8.
16. (a) J. M. Berg, *Prog. Inorg. Chem.*, **37**, 143 (1989); (b) M. S. Lee, G. P. Gippert, K. V. Soman, D. A. Case and P. E. Wright, *Science*, **245**, 635 (1989); (c) N. P. Pavletich and C. O. Pabo, *Science*, **252**, 809 (1991); (d) D. L. Merkle, M. H. Schmidt and J. M. Berg, *J. Am. Chem. Soc.*, **113**, 5450 (1991).
17. V. C. Culotta, T. Hsu, S. Hu, P. Fürst and D. Hamer, *Proc. Natl. Acad. Sci. USA*,

86, 8377 (1989); (b) J. G. Wright, H.-T. Tsang, J. Penner-Hahn and T. V. O'Halloran, *J. Am. Chem. Soc.*, 112, 2434 (1990); (c) S. A. Raybuck, M. D. Distefano, B. K. Teo and C. T. Walsh, *J. Am. Chem. Soc.*, 112, 1983 (1990).

18. S. J. Lippard, in *Bioinorganic Chemistry* (eds I. Bertini, H. B. Gray, S. J. Lippard and J. S. Valentine), University Science Books, Mill Valley, CA, 1994, Chapter 9.

19. A. H. Krotz, B. P. Hudson and J. K. Barton, *J. Am. Chem. Soc.*, 115, 12 577 (1993).

20. (a) A.-M. Giroud-Godquin and P. M. Maitlis, *Angew. Chem., Int. Ed. Engl.*, 30, 375 (1991); (b) P. Espinet, M. A. Esteruelas, L. A. Oro, J. L. Serrano and E. Sola, *Coord. Chem. Rev.*, 117, 215 (1992); (c) S. A. Hudson and P. M. Maitlis, *Chem. Rev.*, 93, 861 (1993).

21. I. G. Dance, *Chem. Aust.*, 61, 241 (1994) and references therein.

22. I. G. Dance, P. A. W. Dean and K. J. Fisher, *Angew. Chem., Int. Ed. Engl.*, 34, 314 (1995).

23. B. F. Hoskins and R. Robson, *J. Am. Chem. Soc.*, 112, 1546 (1990).

24. E. König, G. Ritter and S. K. Kulshreshtha, *Chem. Rev.*, 85, 219 (1985).

25. J.-A. Real, B. Gallois, T. Granier, F. Suez-Panama and J. Zarembowitch, *Inorg. Chem.*, 31, 4972 (1992).

26. F. C. J. M. van Veggel, W. Verboom and D. N. Reinhoudt, *Chem. Rev.*, 94, 279 (1994).

27. (a) A. Juris, V. Balzani, F. Barigelletti, S. Campagna, P. Belser and A. von Zelewsky, *Coord. Chem. Rev.*, 84, 85 (1988); (b) G. Denti, S. Serroni, S. Campagna, A. Juris, M. Ciano and V. Balzani, in *Perspectives in Coordination Chemistry* (eds A. F. Williams, C. Floriani and A. E. Merbach), VCH, Weinheim, 1992, p. 153; (c) S. Campagna, G. Denti, S. Serroni, M. Ciano, A. Juris and V. Balzani, *Inorg. Chem.*, 31, 2982 (1992); (d) S. Serroni, G. Denti, S. Campagna, A. Juris, M. Ciano and V. Balzani, *Angew. Chem., Int. Ed. Engl.*, 31, 1493 (1992).

28. (a) J.-C. Chambron, A. Harriman, V. Heitz and J.-P. Sauvage, *J. Am. Chem. Soc.*, 115, 6109 (1993); (b) J.-C. Chambron, A. Harriman, V. Heitz and J.-P. Sauvage, *J. Am. Chem. Soc.*, 115, 7419 (1993).

29. (a) A.-C. Chambron, C. Dietrich-Buchecker and J.-P. Sauvage, *Top. Curr. Chem.*, 165, 131 (1993); (b) A.-C. Chambron, C. O. Dietrich-Buchecker, J.-F. Nierengarten and J.-P. Sauvage, *J. Chem. Soc., Chem. Commun.*, 801 (1993); (c) C. O. Dietrich-Buchecker, J.-F. Nierengarten, J.-P. Sauvage, N. Armaroli, V. Balzani and L. De Cola, *J. Am. Chem. Soc.*, 115, 11 237 (1993).

30. H. L. Anderson, A. Bashall, K. Henrick, M. McPartlin and J. K. M. Sanders, *Angew. Chem., Int. Ed. Engl.*, 33, 429 (1994).

31. H. Reuter, *Angew. Chem., Int. Ed. Engl.*, 31, 1185 (1992).

32. (a) M. S. Whittingham and R. A. Jacobson, *Intercalation Chemistry*, Academic, 1982; (b) R. Setton, *Chemical Reactions in Organic and Inorganic Constrained Systems*, D. Reidel, 1986, p. 165; (c) J. Rouxel and R. Brec, *Annu. Rev. Mater. Sci.*, 16, 137 (1986); (d) R. Schollhorn, *Angew. Chem., Int. Ed. Engl.*, 27, 1392 (1988); (e) D. O'Hare, *Chem. Soc. Rev.*, 121 (1992); (f) L. Herman, J. Morales, L. Sanchez, J. L. Tirado and A. R. Gonzalez-Elipe, *J. Chem. Soc., Chem. Commun.*, 1081 (1994); (g) R. L. Bedard, S. T. Wilson, L. D. Vail, J. M. Bennett and E. M. Flanigen, in *Zeolites: Facts and Figures*, (eds P. A. Jacobs and R. A. van Santen), Elsevier, Amsterdam, 1989, p. 375.

33. B. F. Abrahams, B. F. Hoskins, J. Liu and R. Robson, *J. Am. Chem. Soc.*, 113, 3045 (1991); (b) B. F. Abrahams, M. J. Hardie, B. F. Hoskins, R. Robson and G. A. Williams, *J. Am. Chem. Soc.*, 114, 10 641 (1992); (c) B. F. Abrahams, M. J. Hardie, B. F. Hoskins, R. Robson and E. E. Sutherland, *J. Chem. Soc., Chem. Commun.*, 1049 (1994).

34. (a) G. D. Stucky, *Prog. Inorg. Chem.*, **40**, 99 (1992); (b) P. Behrens and G. D. Stucky, *Angew. Chem., Int. Ed. Engl.*, **32**, 696 (1993); (c) D. D. Archibald and S. Mann, *Nature (London)*, **364**, 430 (1993) and references therein; (d) C. J. Drummond and B. W. Ninham, *Chem. Aust.*, **59**, 529 (1992) and references therein.
35. E. Haeckel, *Art Forms in Nature*, Dover, New York, 1974.
36. D. Braga and F. Grepioni, *Organometallics*, **11**, 711 (1992).
37. D. Coucouvanis, D. Swenson, N. C. Baenziger, C. Murphy, N. Sfarnas, A. Simopoulos, A. Kostikas and D. G. Holah, *J. Am. Chem. Soc.*, **103**, 3350 (1981).
38. (a) D. Altermatt and I. D. Brown, *Acta Crystallogr., Part B*, **41**, 240 (1985); (b) I. D. Brown and D. Altermatt, *Acta Crystallogr., Part B*, **41**, 244 (1985); (c) H. H. Thorp, *Inorg. Chem.*, **31**, 1585 (1992); (d) W. Liu and H. H. Thorp, *Inorg. Chem.*, **32**, 4102 (1993); (e) E. Shustorovich, *Adv. Catal.*, **37**, 101 (1990).
39. K. A. Fraser, K. J. Fisher, M. L. Scudder, D. C. Craig and I. G. Dance, unpublished results.
40. D. C. Bradley and N. R. Kunchur, *J. Chem. Phys.*, **40**, 2258 (1964).
41. G. A. Bowmaker, I. G. Dance, B. C. Dobson and D. A. Rogers, *Aust. J. Chem.*, **37**, 1607 (1984).
42. I. G. Dance, L. J. Fitzpatrick, A. D. Rae and M. L. Scudder, *Inorg. Chem.*, **22**, 3785 (1983).
43. (a) H. Schmidbaur, *Gold Bull.*, **23**, 11 (1990); (b) H. Schmidbaur, *Interdisciplin. Sci. Rev.*, **17**, 213 (1992); (c) S. S. Pathaneni and G. R. Desiraju, *J. Chem. Soc., Dalton Trans.*, 319 (1993).
44. D. Perreault, M. Drouin, A. Michel, V. M. Miskowski, W. P. Schaefer and P. D. Harvey, *Inorg. Chem.*, **31**, 695 (1992).
45. (a) H. Schmidbaur, W. Graf and G. Müller, *Angew. Chem., Int. Ed. Engl.*, **27**, 417 (1988); (b) H. Schmidbaur, K. Dziwok, A. Grohmann and G. Müller, *Chem. Ber.*, **122**, 893 (1989); (c) K. Dziwok, J. Lachmann, D. L. Wilkinson, G. Müller and H. Schmidbaur, *Chem. Ber.*, **123**, 423 (1990).
46. (a) P. Pyykko and Y. Zhao, *Angew. Chem., Int. Ed. Engl.*, **30**, 604 (1991); (b) P. Pyykko, J. Li and N. Runeberg, *Chem. Phys. Lett.*, **218**, 133 (1994).
47. N. C. Payne, R. J. Puddephatt, R. Ravindranath and I. Treurnicht, *Can. J. Chem.*, **66**, 3176 (1988).
48. A. J. Pertsin and A. I. Kitaigorodskii, *The Atom–Atom Potential Method*, Springer, Berlin, 1987.
49. Superscripts a and r refer respectively to the attractive and repulsive contexts of the parameters.
50. The conventional symbol ε (for e_{ij}^a) is avoided because it could be confused with the dielectric constant which is also relevant in the evaluation of nonbonded energies.
51. A. Bondi, *J. Phys. Chem.*, **68**, 441 (1964).
52. (a) N. L. Allinger and U. Burkert, *Molecular Mechanics*, American Chemical Society, Washington, DC, 1982; (b) N. L. Allinger, *Adv. Phys. Org. Chem.*, **13**, 1 (1976); (c) J.-H. Lii and N. L. Allinger, *J. Am. Chem. Soc.*, **111**, 8576 (1989).
53. B. R. Brooks, R. E. Bruccoleri, B. D. Olafson, D. J. States, S. Swaminathan and M. Karplus, *J. Comput. Chem.*, **4**, 187 (1983).
54. (a) S. J. Weiner, P. A. Kollman, D. A. Case, U. C. Singh, C. Ghio, G. Alagona, S. Profeta and P. Weiner, *J. Am. Chem. Soc.*, **106**, 765 (1984); (b) S. J. Weiner, P. A. Kollman, D. T. Nguyen and D. A. Case, *J. Comput. Chem.*, **7**, 230 (1986).
55. F. H. Allen, J. E. Davies, J. J. Galloy, O. Johnson, O. Kennard, C. F. Macrae and E. M. Mitchell, *J. Chem. Inf. Comput. Sci.*, **31**, 187 (1991).
56. S. C. Nyburg and C. H. Faerman, *Acta Crystallogr., Part B*, **41**, 274 (1985).

57. (a) A. Gavezzotti, *Chem. Phys. Lett.*, **161**, 67 (1989); (b) A. Gavezzotti, *Acta Crystallogr., Part B*, **46**, 275 (1990).
58. D. Gizachew, PhD thesis, University of New South Wales, 1994; I. G. Dance and D. Gizachew, unpublished results.
59. A. Gavezzotti, *J. Am. Chem. Soc.*, **95**, 8948 (1991).
60. (a) D. E. Williams and D. J. Houpt, *Acta Crystallogr., Part B*, **42**, 286 (1986); (b) M. A. Spackman, *J. Chem. Phys.*, **85**, 6579 (1986).
61. D. M. P. Mingos and A. L. Rohl, *J. Chem. Soc., Dalton Trans.*, 3419 (1991).
62. A. K. Rappé, C. J. Casewit, K. S. Colwell, W. A. Goddard and W. M. Skiff, *J. Am. Chem. Soc.*, **114**, 10024 (1992).
63. C. Seel and F. Vogtle, *Angew. Chem., Int. Ed. Engl.*, **31**, 528 (1992).
64. Z. Berkovitch-Yellin and L. Leiserowitz, *J. Am. Chem. Soc.*, **104**, 4052 (1982).
65. *Program Delphi*, BIOSYM/Molecular Simulations, San Diego, CA and Cambridge, U.K.
66. P. P. Ewald, *Ann. Phys.*, **64**, 253 (1921).
67. (a) N. Karasawa and W. A. Goddard, *J. Phys Chem.*, **93**, 7320 (1989); (b) M. W. Deem, J. M. Newsam and S. K. Sinha, *J. Phys. Chem.*, **94**, 8356 (1990).
68. K. M. Merz, M. A. Murcko and P. A. Kollman, *J. Am. Chem. Soc.*, **113**, 4484 (1991).
69. (a) B. H. Besler, K. M. Merz Jr and P. A. Kollman, *J. Comput. Chem.*, **11**, 431 (1990); (b) C. M. Breneman and K. B. Wiberg, *J. Comput. Chem.*, **11.**, 361 (1990).
70. S. C. Hoops, K. W. Anderson and K. M. Merz Jr, *J. Am. Chem. Soc.*, **113**, 8262 (1992).
71. A. K. Rappé and W. A. Goddard, *J. Phys. Chem.*, **95**, 3358 (1991).
72. *Cerius²*, Molecular Simulations, Burlington, MA.
73. G. R. Desiraju, *Crystal Engineering: The Design of Organic Solids*, Elsevier, Amsterdam, 1989.
74. M. C. Etter, *Acc. Chem. Res.*, **23**, 120 (1990).
75. (a) E. Adman, K. D. Watenpaugh and L. H. Jensen, *Proc. Natl. Acad. Sci. USA*, **72**, 4854 (1975); (b) J. C. Dewan, M. A. Walters and W. P. Chung, *J. Am. Chem. Soc.*, **113**, 525 (1991); (c) M. A. Walters, J. C. Dewan, C. Min and S. Pinto, *Inorg. Chem.*, **30**, 2656 (1991).
76. (a) L. Brammer and D. Zhao, *Organometallics*, **13**, 1545 (1994) and references therein; (b) M. Brookhart, M. L. H. Green and L.-L. Wong, *Prog. Inorg. Chem.*, **36**, 1 (1988); (c) R. H. Crabtree and D. G. Hamilton, *Adv. Organomet. Chem.*, **28**, 299 (1988); (d) R. H. Crabtree, *Chem. Rev.*, **85**, 245 (1985).
77. (a) J. A. R. P. Sarma and G. R. Desiraju, *Acc. Chem. Res.*, **19**, 222 (1986); (b) R. Bishop, D. C. Craig, I. G. Dance, M. L. Scudder and A. T. Ung, *Mol. Cryst. Liq. Cryst.*, **240**, 113 (1994); (c) A. Simon, *Angew. Chem., Int. Ed. Engl.*, **27**, 159 (1988); (d) J. D. Corbett, in *Perspectives in Coordination Chemistry* (eds A. F. Williams, C. Floriani and A. E. Merbach), VCH, Weinheim, 1992, p. 219.
78. W. Baratta, F. Calderazzo and L. M. Daniels, *Inorg. Chem.*, **33**, 3842 (1994).
79. D. Braga and F. Grepioni, *Organometallics*, **11**, 1256 (1992).
80. S. Jagner and G. Helgesson, *Adv. Inorg. Chem.*, **37**, 1 (1991).
81. D. E. Williams, *Acta Crystallogr., Part A*, **36**, 715 (1980).
82. W. L. Jorgensen and D. L. Severance, *J. Am. Chem. Soc.*, **112**, 4768 (1990).
83. (a) C. A. Hunter and J. K. M. Sanders, *J. Am. Chem. Soc.*, **112**, 5525 (1990); (b) C. A. Hunter, *Philos. Trans. R. Soc. London, Ser. A*, **345**, 77 (1993).
84. (a) C. A. Hunter, J. Singh and J. M. Thornton, *J. Mol. Biol.*, **218**, 837 (1991); (b) J. Singh and J. M. Thornton, *FEBS Lett.*, **191**, 1 (1985); (c) J. Singh and J. M. Thornton, *J. Mol. Biol.*, **211**, 595 (1990).

85. R. O. Gould, A. M. Gray, P. Taylor and M. D. Walkinshaw, *J. Am. Chem. Soc.*, **107**, 5921 (1985).
86. (a) S. K. Burley and G. A. Petsko, *Science*, **229**, 23 (1985); (b) S. K. Burley and G. A. Petsko, *J. Am. Chem. Soc.*, **108**, 7995 (1986).
87. G. Klebe and F. Diederich, *Philos. Trans. R. Soc. London, Ser. A*, **345**, 37 (1993).
88. (a) A. Gavezzotti and G. Desiraju, *Acta Crystallogr., Part B*, **44**, 427 (1988); (b) G. R. Desiraju and A. Gavezzotti, *J. Chem. Soc., Chem. Commun.*, 621 (1989); (c) G. R. Desiraju and A. Gavezzotti, *Acta Crystallogr., Part B*, **45**, 473 (1989).
89. (a) K. C. Janda, J. C. Hemminger, J. S. Winn, S. E. Novick, S. J. Harris and W. Klemperer, *J. Chem. Phys.*, **63**, 1419 (1975); (b) J. M. Steed, T. A. Dixon and W. Klemperer, *J. Chem. Phys.*, **70**, 4940 (1979).
90. G. Karlström, P. Linse, A. Wallqvist and B. Jönsson, *J. Am. Chem. Soc.*, **105**, 3777 (1983).
91. J. R. Grover, E. A. Walters and E. T. Hui, *J. Phys. Chem.*, **91**, 3233 (1987).
92. (a) J. L. Atwood, F. Hamada, K. D. Robinson, G. W. Orr and R. L. Vincent, *Nature (London)*, **349**, 683 (1991); (b) S. Suzuki, P. G. Green, R. E. Bumgarner, S. Dasgupta, W. A. Goddard and G. A. Blake, *Science*, **257**, 942 (1992).
93. T. B. Karpishin, T. D. P. Stack and K. N. Raymond, *J. Am. Chem. Soc.*, **115**, 6115 (1993).
94. A. L. Balch, V. J. Catalano and J. W. Lee, *Inorg. Chem.*, **30**, 3980 (1991).
95. A. L. Balch, V. J. Catalano, J. W. Lee, M. M. Olmstead and S. R. Parkin, *J. Am. Chem. Soc.*, **113**, 8953 (1991).
96. A. L. Balch, V. J. Catalano, J. W. Lee and M. M. Olmstead, *J. Am. Chem. Soc.*, **114**, 5455 (1992).
97. P. G. Jones, *Acta Crystallogr., Part B*, **36**, 3105 (1980).
98. J. M. Manriquez, P. J. Fagan, L. D. Schertz and T. J. Marks, *Inorg. Snyth.*, **21**, 181 (1982).
99. (a) M. J. Crossley and P. L. Burn, *J. Chem. Soc., Chem. Commun.*, 39 (1987); (b) M. J. Crossley and P. L. Burn, *J. Chem. Soc., Chem. Commun.*, 1569 (1991); (c) J.-C. Chambron, V. Heitz and J.-P. Sauvage, *J. Chem. Soc., Chem. Commun.*, 1131 (1992).
100. (a) J. J. Ellison, K. Ruhlandt-Senge and P. P. Power, *Angew. Chem., Int. Ed. Engl.*, **33**, 1178 (1994); (b) K. Ruhlandt-Senge and P. P. Power, *Inorg. Chem.*, **32**, 4505 (1993); (c) P. P. Power and S. C. Shoner, *Angew. Chem., Int. Ed. Engl.*, **30**, 330 (1991); (d) K. Tang, M. Aslam, E. Block, J. R. Nicholson and J. A. Zubieta, *Inorg. Chem.*, **26**, 1488 (1987); (e) K. Tang, A. Li, X. Jin and Y. Tang, *J. Chem. Soc., Chem. Commun.*, 1590 (1991).
101. R. K. Shibao, N. L. Keder and H. Eckert, *Inorg. Chem.*, **29**, 4163 (1990).
102. N. Ueyama, T. Sugawara, K. Sasaki, A. Nakamura, S. Yamashita, Y. Wakatsuki, Y. Yamashita, H. Yamazaki and N. Yasuoka, *Inorg. Chem.*, **27**, 741 (1988).
103. (a) P. J. Wheatley, *J. Chem. Soc.*, 3718 (1964); (b) A. L. Beauchamp, M. J. Bennett and F. A. Cotton, *J. Am. Chem. Soc.*, **90**, 6675 (1968).
104. C. Brabant, B. Blanck and A. L. Beauchamp, *J. Organomet. Chem.*, **82**, 231 (1974).
105. G. L. Kuykendall and J. L. Mills, *J. Organomet. Chem.*, **118**, 123 (1976).
106. I. G. Dance and D. Gizachew, unpublished results: van der Waals component $-238\,kJ\,mol^{-1}$, coulombic $-19\,kJ\,mol^{-1}$ $(q_{Sb}+0.5, q_C=-0.1, q_H=+0.1)$.
107. C. Brabant, J. Hubert and A. L. Beauchamp, *Can. J. Chem.*, **51**, 2952 (1973).
108. M. G. B. Drew, T. R. Pearson, B. P. Murphy and M. Nelson, *Polyhedron*, **2**, 269 (1983).
109. F. Wiesemann, R. Wonnemann, B. Krebs, H. Keutel and E.-G. Jäger, *Angew. Chem., Int. Ed. Engl.*, **33**, 1363 (1994).

110. R. M. H. Banda, J. Cusick, M. L. Scudder, D. C. Craig and I. G. Dance, *Polyhedron*, **8**, 1995 (1989).
111. I. G. Dance and M. L. Scudder, *J. Chem. Soc., Chem. Commun.*, 1039 (1995).
112. S. Dhingra and M. G. Kanatzidis, *Inorg. Chem.*, **32**, 3300 (1993).
113. (a) T. B. Rauchfuss and M. Draganjac, *Angew. Chem., Int. Ed. Engl.*, **24**, 742 (1985); (b) A. Müller and E. Diemann, *Adv. Inorg. Chem.*, **31**, 89 (1987); (c) M. A. Ansari and J. A. Ibers, *Coord. Chem. Rev.*, **100**, 223 (1990); (d) L. C. Roof and J. W. Kolis, *J. Am. Chem. Soc.*, **93**, 1037 (1993).
114. M. G. Kanatzidis and S.-P. Huang, *Coord. Chem. Rev.*, **130**, 509 (1994).
115. R. M. H. Banda, I. G. Dance, T. D. Bailey, D. C. Craig and M. L. Scudder, *Inorg. Chem.*, **28**, 1862 (1989).
116. J. Cusick, PhD thesis, University of New South Wales, 1993.
117. S.-P. Huang, S. Dhingra and M. G. Kanatzidis, *Polyhedron*, **9**, 1389 (1990).
118. T. D. Bailey, R. M. H. Banda, D. C. Craig, I. G. Dance and I. N. L. Ma, *Inorg. Chem.*, **30**, 187 (1991).
119. (a) R. M. W. Wardle, S. Bhaduri, C.-N. Chau and J. A. Ibers, *Inorg. Chem.*, **27**, 1747 (1988); (b) R. M. W. Wardle, C.-N. Chau and J. A. Ibers, *J. Am. Chem. Soc.*, **109**, 1859 (1987).
120. J. Cusick and I. G. Dance, *Polyhedron*, **10**, 2629 (1991).
121. S. S. Dhingra and M. G. Kanatzidis, *Inorg. Chem.*, **32**, 1350 (1993).
122. S. S. Dhingra and M. G. Kanatzidis, *Inorg. Chem.*, **32**, 2298 (1993).
123. R. C. Haushalter, *Angew. Chem., Int. Ed. Engl.*, **24**, 433 (1985).
124. S.-P. Huang and M. G. Kanatzidis, *Inorg. Chem.*, **30**, 1455 (1991).
125. G. Krauter, K. Dehnicke and D. Fenske, *Chem.-Ztg.*, **114**, 7 (1990).
126. K.-W. Kim and M. G. Kanatzidis, *J. Am. Chem. Soc.*, **114**, 4878 (1992).
127. J. Cusick, M. L. Scudder, D. C. Craig and I. G. Dance, *Polyhedron*, **8**, 1139 (1989).
128. A. Müller, J. Schimanski, M. Römer, H. Bögge, F.-W. Baumann, W. Eltzner, E. Krickemeyer and U. Billerbeck, *Chimia*, **39**, 25 (1985).
129. A. I. Hadjikyriacou and D. Coucouvanis, *Inorg. Chem.*, **26**, 2400 (1987).
130. (a) S.-P. Huang, S. Dhingra and M. G. Kanatzidis, *Polyhedron*, **11**, 1869 (1992); (b) B. Krebs, E. Lührs, R. Willmer and F.-P. Ahlers, *Z. Anorg. Allg. Chem.*, **592**, 17 (1991).
131. D. Swenson, N. C. Baenziger and D. Coucouvanis, *J. Am. Chem. Soc.*, **100**, 1932 (1978).
132. I. G. Dance, D. Gizachew and M. L. Scudder, unpublished results.
133. A. Silver, S. A. Koch and M. Millar, *Inorg. Chim. Acta*, **205**, 9 (1993).
134. J. M. Waters, personal communication.
135. K. Tang, personal communication.
136. K. Tang, T. Xia, X. Jin and Y. Tang, *Polyhedron*, **12**, 2895 (1993).
137. F. Scherbaum, A. Grohmann, B. Huber, C. Kruger and H. Schmidbaur, *Angew. Chem., Int. Ed. Engl.*, **27**, 1544 (1988).
138. F. Canales, M. C. Gimeno, P. G. Jones and A. Laguna, *Angew. Chem., Int. Ed. Engl.*, **33**, 769 (1994).
139. E. Zeller, H. Beruda, A. Kolb, P. Bissinger, J. Riede and H. Schmidbaur, *Nature (London)*, **352**, 141 (1991).
140. F. Scherbaum, A. Grohmann, G. Müller and H. Schmidbaur, *Angew. Chem., Int. Ed. Engl.*, **28**, 463 (1989).
141. A. Grohmann, J. Riede and H. Schmidbaur, *Nature (London)*, **345**, 140 (1990).
142. A. Grohmann, J. Riede and H. Schmidbaur, *Z. Naturforsch., Teil B*, **47**, 1255 (1992).
143. E. C. Constable, *Angew. Chem., Int. Ed. Engl.*, **30**, 1450 (1991).

144. (a) R. Krämer, J.-M. Lehn, A. de Cian and J. Fischer, *Angew. Chem., Int. Ed. Engl.*, **32**, 703 (1993); (b) R. Krämer, J.-M. Lehn and A. Marquis-Rigault, *Proc. Natl. Acad. Sci. USA*, **90**, 5394 (1993).

145. M.-T. Youinou, R. Ziessel and J.-M. Lehn, *Inorg. Chem.*, **30**, 2144 (1991) and references therein.

146. Y. Yao, M. W. Perkovic, D. P. Rillema and C. Woods, *Inorg. Chem.*, **31**, 3956 (1992).

147. (a) E. C. Constable and M. D. Ward, *J. Am. Chem. Soc.*, **112**, 1256 (1990); (b) E. C. Constable, M. D. Ward and D. A. Tocher, *J. Chem. Soc., Dalton Trans.*, 1675 (1991).

148. (a) A. F. Williams, C. Piguet and G. Bernardinelli, *Angew. Chem., Int. Ed. Engl.*, **30**, 1490 (1991); (b) C. Piguet, G. Bernardinelli, B. Bocquet, A. Quattropani and A. F. Williams, *J. Am. Chem. Soc.*, **114**, 7440 (1992).

149. W. Zarges, J. Hall, J.-M. Lehn and C. Bolm, *Helv. Chim. Acta*, **74**, 1843 (1991).

150. U. Koert, M. M. Harding and J.-M. Lehn, *Nature (London)*, **346**, 339 (1990).

151. (a) M. R. Ghadiri, C. Soares and C. Choi, *J. Am. Chem. Soc.*, **114**, 825 (1992); (b) M. R. Ghadiri, C. Soares and C. Choi, *J. Am. Chem. Soc.*, **114**, 4000 (1992).

152. M.-T. Youinou, N. Rahmouni, J. Fischer and J. A. Osborn, *Angew. Chem., Int. Ed. Engl.*, **31**, 733 (1992).

153. I. G. Dance, in *Perspectives in Coordination Chemistry* (eds A. F. Williams, C. Floriani and A. E. Merbach), VCH, Weinheim, 1992, p. 165.

154. G. S. H. Lee, K. J. Fisher, D. C. Craig, M. L. Scudder and I. G. Dance, *J. Am. Chem. Soc.*, **112**, 6435 (1990).

155. I. G. Dance, K. J. Fisher and G. S. H. Lee, in *Metallothioneins: Synthesis, Structure and Properties of Metallothioneins, Phytochelatins and Metal Thiolate Complexes* (eds M. J. Stillman, C. F. Shaw and K. T. Suzuki), VCH, Weinheim, 1992, p. 284.

156. I. G. Dance, A. Choy and M. L. Scudder, *J. Am. Chem. Soc.*, **106**, 6285 (1984).

157. G. S. H. Lee, D. C. Craig, I. N. L. Ma, M. L. Scudder, T. D. Bailey and I. G. Dance, *J. Am. Chem. Soc.*, **110**, 4863 (1988).

158. N. Herron, A. Suna and Y. Wang, *J. Chem. Soc., Dalton Trans.*, 2329 (1992).

159. N. Herron, J. C. Calabrese, W. E. Farneth and Y. Wang, *Science*, **259**, 1426 (1993).

160. I. G. Dance, G. S. H. Lee and M. L. Scudder, unpublished results.

161. (a) D. C. Craig, I. G. Dance and R. G. Garbutt, *Angew. Chem., Int. Ed. Engl.*, **25**, 165 (1986); (b) I. G. Dance, R. G. Garbutt, D. C. Craig and M. L. Scudder, *Inorg. Chem.*, **26**, 4057 (1987); (c) I. G. Dance, R. G. Garbutt and T. D. Bailey, *Inorg. Chem.*, **29**, 603 (1990).

162. I. G. Dance, R. G. Garbutt, D. C. Craig, M. L. Scudder and T. D. Bailey, *J. Chem. Soc., Chem. Commun.*, 1164 (1987).

163. I. G. Dance, R. G. Garbutt and M. L. Scudder, *Inorg. Chem.*, **29**, 1571 (1990).

164. (a) M. J. Heeg, C. Janiak and J. J. Zuckerman, *J. Am. Chem. Soc.*, **106**, 4259 (1984); (b) M. J. Heeg, R. H. Herber, C. Janiak, J. J. Zuckerman, H. Schumann and W. F. Manders, *J. Organomet. Chem.*, **346**, 321 (1988).

165. I. G. Dance, unpublished results.

166. M. Cortrait, J. Gaultier, C. Polycarpe, A. M. Giroud and U. T. Mueller-Westerhoff, *Acta Crystallogr., Part C*, **39**, 833 (1983).

167. H. Adams, A. C. Albinez, N. A. Bailey, D. W. Bruce, A. S. Cherodian, R. Dhillon, D. A. Dunmur, P. Espinet, J. L. Feijoo, E. Lalinde, P. M. Maitlis, R. M. Richardson and G. Ungar, *J. Mater. Chem.*, **1**, 843 (1991).

168. I. G. Dance, K. J. Fisher, R. M. H. Banda and M. L. Scudder, *Inorg. Chem.*, **30**, 183 (1991).
169. M. J. Baena, P. Espinet, M. C. Lequerica and A. M. Levelut, *J. Am. Chem. Soc.*, **114**, 4182 (1992).
170. D. Braga and F. Grepioni, *Acc. Chem. Res.*, **27**, 51 (1994).
171. D. Braga and F. Grepioni, *Organometallics*, **10**, 1254 (1991).
172. (a) D. Braga and F. Grepioni, *Organometallics*, **10**, 2563 (1991); (b) D. Braga, P. J. Dyson, F. Grepioni and B. F. G. Johnson, *Chem. Rev.*, **94**, 1585 (1994).
173. (a) D. Braga, *Chem. Rev.*, **92**, 633 (1992); (b) D. Braga, F. Grepioni, B. F. G. Johnson, H. Chen and J. Lewis, *J. Chem. Soc., Dalton Trans.*, 2559 (1991); (c) D. Braga, P. J. Dyson, F. Grepioni and B. F. G. Johnson, *Chem. Rev.*, **94**, 1585 (1994).
174. D. Braga, F. Grepioni, P. Milne and E. Parisini, *J. Am. Chem. Soc.*, **115**, 5115 (1993).
175. D. M. P. Mingos and A. L. Rohl, *Inorg. Chem.*, **30**, 3769 (1991).
176. A. L. Rohl and D. M. P. Mingos, *J. Chem. Soc., Dalton Trans.*, 3541 (1992).
177. (a) A. Gavezzotti, *J. Am. Chem. Soc.*, **105**, 5220 (1983); (b) A. Gavezzotti, *J. Am. Chem. Soc.*, **107**, 962 (1985).
178. F. Basolo, *Coord. Chem. Rev.*, **3**, 213 (1968).
179. D. H. McDaniel, *Annu. Rep. Inorg. Gen. Synth.*, 293 (1972).
180. A. L. Rohl and D. M. P. Mingos, *Inorg. Chim. Acta*, **212**, 5 (1993).
181. I. G. Dance, R. Garbutt and M. L. Scudder, unpublished results.
182. L. R. MacGillivray, S. Subramanian and M. J. Zaworotko, *J. Chem. Soc., Chem. Commun.*, 1325 (1994).
183. R. Robson, B. F. Abrahams, S. R. Batten, R. W. Gable, B. F. Hoskins and J. Liu, in *Supramolecular Architecture* (ed. T. Bein), American Chemical Society, Washington, DC, 1992, p. 256.
184. Y. Kinoshita, I. Matsubara, T. Higuchi and Y. Saito, *Bull. Chem. Soc. Jpn.*, **32**, 1221 (1959).
185. T. Soma, H. Yuge and T. Iwamoto, *Angew. Chem., Int. Ed. Engl.*, **33**, 1665 (1994).
186. A. Michaelides, V. Kiritsis, S. Skoulika and A. Aubry, *Angew. Chem., Int. Ed. Engl.*, **32**, 1495 (1993).
187. J. Kim, D. Whang, J. I. Lee and K. Kim, *J. Chem. Soc., Chem. Commun.*, 1400 (1993).
188. B. F. Hoskins and R. Robson, *J. Am. Chem. Soc.*, **111**, 5962 (1989).
189. S.-C. Chang, I. Chao and Y.-T. Tao, *J. Am. Chem. Soc.*, **116**, 6792 (1994).
190. H. Sellers, A. Ulman, Y. Shnidman and J. E. Eilers, *J. Am. Chem. Soc.*, **115**, 9389 (1993).
191. G. Ouvrard and R. Brec, *Eur. J. Solid-State Inorg. Chem.*, **27**, 477 (1990).
192. (a) J. Rouxel, in *Supramolecular Architecture* (ed. T. Bein), American Chemical Society, Washington, DC, 1992, p. 88; (b) P. Palvadeau, J. Rouxel, M. Queignec and B. Bujoli, in *Supramolecular Architecture* (ed. T. Bein), American Chemical Society, Washington, DC, 1992, p. 114.
193. M. P. Byrn, C. J. Curtis, S. I. Khan, P. A. Sawin, R. Tsurumi and C. E. Strouse, *J. Am. Chem. Soc.*, **112**, 1865 (1990).

Chapter 6

The Protein as a Supermolecule: The Architecture of a $(\beta\alpha)_8$ Barrel

JENNY P. GLUSKER

The Institute for Cancer Research, The Fox Chase Cancer Center, Philadelphia, PA, USA

1. INTRODUCTION

Proteins are supermolecules that have been designed by nature to carry out certain specific tasks and to demonstrate distinctive properties. These molecules are built with a palette of 20 amino acids of which all but one, glycine, are chiral L-amino acids. The amino acids are strung together in a linear manner, like beads on a string, by way of peptide linkages, to form the backbone of the protein. The amino acid side chains stick out from this backbone. The architectural principles involved in constructing a supermolecule that can perform the required functions are described here.

The 20 different amino acids in the linear polypeptide chain are the components that are arranged in a unique manner for each type of protein. If there are 200 or more amino acids, the number of different ways of linking these 20 amino acids, and hence of forming possible three-dimensional structures, is 200^{20}. This means that this simple palette of amino acids leads to many possibilities for diversity in the protein itself, and that an almost infinite number of protein structures can be designed. I will describe in this chapter an example of how the long polypeptide chain of a protein consisting of many hundreds of amino acids folds in a totally unique manner to give a supermolecule with distinctive and well-controlled properties.

The precise linear arrangement of amino acids in a protein was first established for insulin by Sanger [1–3]. It is now found for each individual protein by sequencing either the amino acids or the nucleic acid that codes for

The Crystal as a Supramolecular Entity. Edited by G. R. Desiraju

them. This raises the problem of determining the extent to which, if the amino acid sequence is known, the manner of folding of the polypeptide backbone can be predicted. For certain proteins it is found that, when denatured, the protein will fold again into only one preferred folding pattern. This was shown by Anfinsen who demonstrated that ribonuclease A, which contains four disulfide bonds per molecule, can, after denaturation which involves breaking these bonds, refold in the correct manner giving the catalytically active enzyme [4–7]. The formation of the four disulfide bonds in the denatured enzyme was not random. This implies that information on how to form disulfide links between the correct cysteine SH groups is coded in some manner by data in the amino acid sequence, at least for the final folding pattern. The problem of how to determine whether protein folding is controlled mainly by kinetic or by thermodynamic factors, or by both, i.e. by the relative heights of potential energy barriers or by the relative energies of the initial and final (lower-energy) states, is still the subject of debate [8].

Molecules of soluble proteins generally fold in an organized manner and tend to be approximately spherical in shape. If protein molecules existed merely as long chains, like snakes, they could have an infinite variety of overall conformations, each of which could easily convert to another. Therefore it is more in line with the aim of nature (to form a protein that can carry out a required function or role) that the polypeptide should fold in a unique and specific manner. The most general folding pattern of any linear organic polymer is that of a random coil in which all portions of the molecule are exposed to solvent and rotation about single bonds occurs rapidly. The protein backbone chain, like that of a linear organic polymer, can also adopt such a folding pattern giving a helical structure (generally an α helix) or, if distant parts of the polypeptide chain line up side by side, a sheet-like structure (called a β sheet). These are the main components of the folded protein and help lead to a compact structure for the fully folded protein. If the protein has a molecular weight of approximately 40 000, the molecular volume would be about 50 000 Å^3, which means that the approximate diameter of the protein, if it is spherical, is 46 Å. The length of any individual linear helix within the spherically folded protein structure would then not be likely to exceed 35–40 Å (corresponding to about 25 amino acid residues in an α helix or 10 in one strand of a β sheet), allowing for the requirement that the backbone would have to fold back again in what is called a 'reverse turn' in order to give the overall globular shape of the protein.

Surely, one would think, nature has chosen an amino acid sequence that contains all the information needed to effect the designed overall folding of the protein. The rules that relate sequence and folded structure, however, are not fully understood, although some general methods for predicting folding in at least portions of the structure have been put forward. They are currently being tested against new crystal structure determinations as these are reported in the scientific literature [9]. The correctly folded state has a lower energy than the unfolded one, and no natural proteins are known to date that fold equally well into two or more distinctly different folding patterns. For each protein the overall folding is unique. How does this come about?

It seems unlikely that all possible protein conformations are sampled until the lowest-energy conformation is found. An estimate by Levinthal [10] suggests that the age of the universe would be too short for all the available conformations to have been tested in an average-sized protein by randomly rotating the backbone angles between atoms until the correct final conformation is found. This follows from the known rates of conformational changes in molecules. Experimental data on these rates for protein folding are available [11, 12]; ribonuclease A folds and forms its four disulfide bonds in milliseconds [13]. When ribonuclease A is synthesized, the course and result of folding *in vitro* are the same as for the naturally occurring enzyme [14]. Some other proteins, however, take longer to fold. An example is bovine pancreatic trypsin inhibitor which folds into an intermediate structure that has a non-native disulfide link that is later reorganized to give the final folding pattern [15].

A method for predicting protein folding from a known amino acid sequence is becoming an acute necessity now that the entire human genome is being sequenced. Presumably, considering how science is progressing, it can be expected that by the end of the 1990s the sequences of most proteins will be known. But how do they fold? Approximate atomic coordinates for three-dimensional structures of most types of proteins are currently available [16]. The structural information that has been obtained has not yet been analyzed with sufficient sophistication for it to be possible to predict folding with complete accuracy directly from an amino acid sequence. The basis for all such predictions is the X-ray diffraction analyses of many crystalline proteins, mainly to date the globular water-soluble ones. Nuclear magnetic resonance studies have then led to information on the three-dimensional structure in solution, and in some cases (for small proteins) have predicted the overall folding before the X-ray crystal structure analysis has been reported [17, 18].

X-ray crystal structure analyses have shown that each protein does not have an entirely different folding pattern, but that most protein structures are made up from one or more of a number of possible structural motifs, most of which have, by now, been identified. The questions that could be asked include: why does the protein have the amino acid sequence that it does? what is the significance of each amino acid? why does the molecule fold in the way that it does? and what are the functional results of such a folding? The controls imposed on protein folding include the interactions between main-chain and side-chain atoms in the folded polypeptide chain; the manner in which subunits, if any, interact with each other; and the way in which the active site, if the protein is an enzyme, is formed. These controls lead to the ultimate size and shape of the protein. The extent to which the protein folds correctly will, however, depend also upon its environment and on the need for specific cofactors to complete the folding or for molecular chaperones to guide the folding. It is not intended to dwell here on catalytic function, but rather on structure and on the individual uses that are made of amino acids and the currently identified structural motifs.

The overall architecture of a protein molecule is the arrangement of peptide groups in the polypeptide chains, its folding and the interactions between the various side-chain functional groups. Linderström-Lang [19] described protein architecture in terms of several levels which are diagramed in Figure 1. 'Primary structure' refers to the amino acid sequence, while 'secondary structure' is any regular local structure of a segment of a polypeptide chain, such as helices or hydrogen-bonded strands. 'Tertiary structure' is the overall topology of the folded polypeptide chain to give domains which typically contain 100–200 amino acid residues. 'Quaternary structure' describes the aggregation of folded polypeptides with each other by means of specific interactions [20, 21]. These secondary, tertiary and quaternary structures correspond to increasing hierarchies of supramolecular assembly. Primary and secondary structures are described in Sections 1–3, tertiary structures in Section 4 and quaternary structures in Sections 5–7 of this article. Chosen here as an example for a description of protein architecture is the enzyme D-xylose isomerase for which the crystal structure was determined by X-ray crystallographic techniques by Carrell and coworkers [22, 23]. This protein has an overall folding pattern that is like that found for several other protein structures, and therefore it serves to illustrate some of the rules for protein architecture.

From an architectural point of view the question posed is how to design a protein using only one or a few polypeptides so that the resulting supermolecule has the correct overall shape to move to the appropriate location in the body and the precise alignment of functional groups to carry out its biological role, be it, for example, specific molecular recognition or enzymic catalysis. Some studies are now underway to design proteins, such as the work of Richardson and Richardson on betabellin [24], but these are in the early stages. Therefore, I concentrate here on the dissection of an enzyme D-xylose isomerase, in order to find out how it has been put together — how it was designed from a limited selection of building materials in such a way that it would work in a biological environment.

2. THE FRAMEWORK: THE FOLDING OF THE POLYPEPTIDE BACKBONE

The polypeptide backbone of a protein, the framework that is folded to give the overall topology of this supermolecule [25–29], consists of a repeated

LEVELS OF PROTEIN STRUCTURE

PRIMARY (amino acid sequence)   SECONDARY (α helices, β sheet)

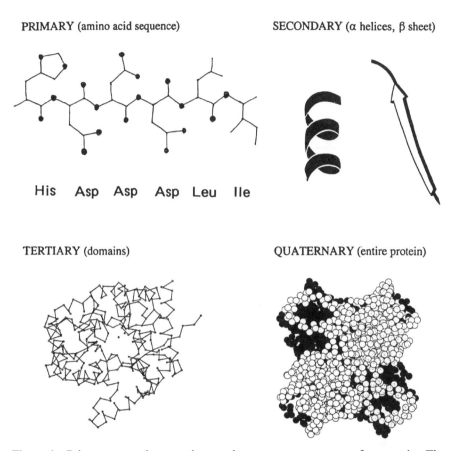

His   Asp   Asp   Asp   Leu   Ile

TERTIARY (domains)   QUATERNARY (entire protein)

Figure 1 Primary, secondary, tertiary and quaternary structure of a protein. The enzyme D-xylose isomerase is given as an example. The primary structure is the amino acid sequence, the secondary structure is the folding to α helices, β sheets and turns, the tertiary structure is the domain structure and the quaternary structure is the overall assemblage of the entire protein, often composed of several subunits

sequence of an amide nitrogen atom, an α-carbon atom (Cα) and a carbonyl carbon atom, i.e. $(-NH-C_\alpha-CO-)_n$. Schulz and Schirmer write [30] that 'the peptide bond is handy because it can be synthesized and split at comparatively low expense'. This means that it is a cheap building material, readily available and easily disposed of in a nonpolluting manner.

The linear peptide polymer has a repeat distance of approximately 3.8 Å between α-carbon atoms in a *trans*-conformation, as shown in Figure 2. Each

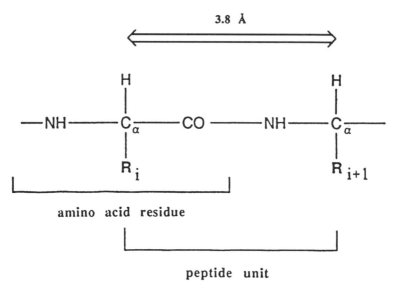

Figure 2 An amino acid residue and peptide unit in a protein, showing the repeat distance of 3.8 Å

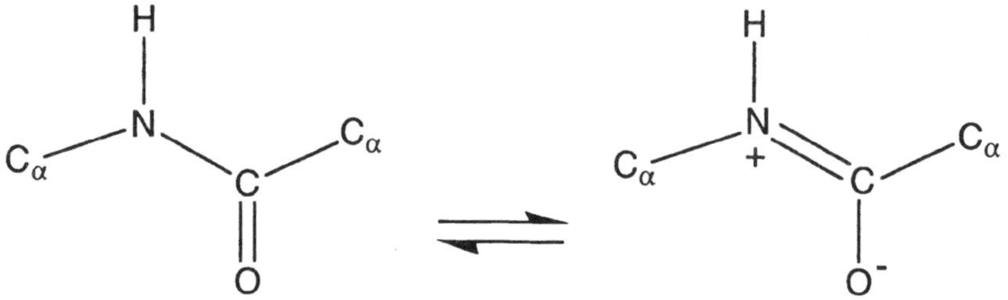

Figure 3 Resonance of possible peptide structures leads to some double-bond character in the C–N bond, causing the unit to be approximately planar. The charges on the resonance form with a carbon–nitrogen double bond give the group a dipole moment

peptide group (CαCONHCα) is approximately planar, a finding that was explained by Pauling in 1933. He showed how a partial double-bond character to the C–N bond in the peptide group would hinder rotation about the C–N bond, thereby giving a planar, rigid unit as shown in Figure 3, and as found in crystal structures of peptides and proteins [31]. This double-bond character will impart a partial negative charge on the carbonyl oxygen atom and a partial positive charge on the nitrogen atom. As a result, each such peptide unit has a permanent dipole moment of 3.5 debye units $(1.2 \times 10^{-29}\,\text{Cm})$ [8]. Thus the framework of a protein is built up of a chain of planar polar peptide units, each

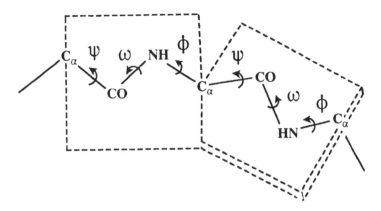

Figure 4 The conformational torsion angles in a protein and the planar peptide units that are hinged together

plane hinged to two others with various relative orientations, as indicated in Figure 4. Idealized three-dimensional coordinates for individual planar units like those shown in Figure 4 are used by the crystallographer to fit a molecular model to a protein electron density map.

Main-chain CO (carbonyl) and NH (amino) groups form hydrogen bonds to each other as the protein folds and two portions of it come close together. Baker and Hubbard [32] have listed some relevant rules for hydrogen bonding in proteins, derived from the data in reported crystal structure determinations. The most important rule is that nearly all functional groups that can form hydrogen bonds do so. The optimal angle between the donor–hydrogen atom bond direction and the acceptor atom (D–H···A) is near 180°, and the extent of interaction is reduced, they find, to near zero at 120°. Some of these hydrogen bonds are within the protein and some are to the water molecules that surround the protein. The most common hydrogen bond within proteins is between main-chain NH and CO groups, of which there are equal numbers; some 40–70% of these main-chain groups form hydrogen bonds to each other. In addition, 30–40% of them form hydrogen bonds to water molecules. Hydrogen bonds that are internal to the protein structure are generally short and strong, while those on the surface appear to be weaker. As diagramed in Figure 5(a), the NH group generally only donates a hydrogen atom in one hydrogen bond (although as shown in Figure 5b, sometimes bifurcated hydrogen bonds are formed), while the carbonyl CO group can accept two hydrogen bonds at 120° to each other (Figure 5c). Any problems caused by this discrepancy between CO and NH hydrogen-bonding capabilities are mediated by water molecules because they can donate two hydrogen bonds, but can accept either one or two.

(a)

(b)

(c)

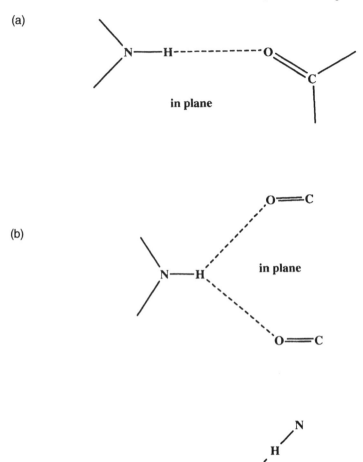

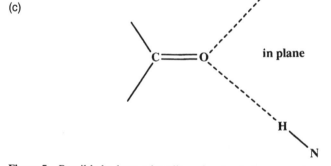

Figure 5 Possible hydrogen bonding of main-chain groups in a protein, as described by Baker and Hubbard [32]. (a) NH groups tend to form one hydrogen bond to an oxygen atom, as shown, but sometimes are found with a bifurcated hydrogen bond (b). (c) A carbonyl group tends to accept two hydrogen bonds. In each of these possibilities the system is approximately planar

Side-chain groups are covalently bound to the polypeptide at the α-carbon atoms. These side chains affect how the entire protein molecule folds. When in an unfolded conformation the protein has a large entropy value because it has so much conformational flexibility. This entropy is reduced on folding, but the loss of entropy is compensated for by energy gained by favorable interactions that are formed and unfavorable interactions that are eliminated. Important in these interactions are those resulting from the hydrophobic effect, i.e. from an energetically unfavorable disruption of the infinitely associated hydrogen-bonded structure of water by materials that do not readily form hydrogen bonds [33, 34]. When a nonpolar substance (such as an oil) is added to water, the forces between the oil and water (van der Waals forces) are weak, while those between water and water (fluctuating hydrogen bonds) are stronger. The total system (the mixture of oil and water) is more stable if the molecules in the oil can separate out and associate with other like molecules, leaving the water structure undisturbed, except at the surface of the oil. To keep the unfavorable oil–water interactions to a minimum, the oil will rise to the top of the mixture, or, if shaken with water (as for salad dressing), will temporarily form spherical droplets. The solubility of a protein in water is increased if the nonpolar side chains (which are like oil) are in the center of the molecule, and the polar side chains (which readily form hydrogen bonds and therefore interact well with water) are at or near the surface. This folded state of the protein has less of its surface exposed to solvent than does the unfolded state. In this way the necessary disruption of the water structure by the added protein is kept to a minimum. This description is, of course, a generalization, and often, as noted by Richards, 'the "grease" is by no means "buried"' [35]. Protein folding involves the interactions of hundreds of amino acid residues. Because a single hydrogen bond may release 5–10 kcal mol^{-1} in bonding energy, it might have been expected that the folding energy of proteins would be of the order of hundreds of kilocalories per mole, while in fact it is generally less than 25 kcal mol^{-1}. This relatively low folding energy is found because many of the intramolecular hydrogen bonds found in the folded protein were previously present in the unfolded protein as intermolecular hydrogen bonds with water molecules.

The efficiency of packing as the protein folds is revealed by its packing density. This is the van der Waals volume of the atoms in a protein divided by the actual space occupied. High packing densities imply that the various amino acid side chains fit snugly together, while a low packing density would imply that the recognition between amino acids is poor. Close-packed spheres have a packing density of approximately 0.74. Therefore, since protein molecules such as ribonuclease and lysozyme (without the added water of a crystal structure) have packing densities in the range of 0.72 and 0.76, it is clear that the interiors of proteins are as well packed as are crystals of small molecules [35, 36]. On the other hand, crystalline proteins, in general, contain 20–70% water in their unit cells. Matthews [37] suggested the use of V_m, the crystal volume per unit molecular weight of protein, to determine the molecular weight when the unit cell volume is known. He showed that the value of V_m has a normal range of 1.6–$3.5\,\text{Å}^3$ per unit molecular weight. The values take into account the different density of water, which consists of hydrogen-bonded molecules $2.7\,\text{Å}$ apart, giving a lower density than that found for the protein which consists of bonded atoms that are about $1.5\,\text{Å}$ apart. The calculation of V_m includes the space in the crystal filled by water as well as that filled by protein, and the values are consistent with those for a density of about $1.3\,\text{g cm}^{-3}$ for the protein itself, but less for the water surrounding it.

Nature has chosen the peptide group as the unit of a polymer in proteins. The problem is that as the protein folds and the hydrophobic side chains lie in the interior of the protein, some hydrophilic polar carbonyl and amino groups in the protein will also be buried and this is energetically unfavorable. The dilemma is overcome by reducing the hydrophilic character of these main-chain groups by hydrogen bonding. In this way hydrophobic groups can be buried within the protein without a high energy cost. Such hydrogen bonding is important in the formation of secondary structure, but its effect on the overall stability of the protein is small, because protein–water hydrogen bonds are replaced by protein–protein hydrogen bonds which have fairly similar energies [5]. All side chains internal to the protein that have hydrogen-bonding capabilities form hydrogen bonds if at all possible. Any hydrophobic patches on the surface of a protein during protein folding will tend to aggregate with other hydrophobic patches, as in subunit assembly, and as shown by the mutation of hemoglobin to sickle-cell hemoglobin. This sickle-cell mutation involves the substitution of the hydrophilic glutamic acid residue by a hydrophobic valine residue. In one subunit of the mutant protein the side chain of this valine residue fits into a hydrophobic pocket that is too small to accommodate glutamic acid in the wild-type (normal) enzyme. The result is that the mutant (sickle-cell) protein molecules adhere to each other, and sickle-cell hemoglobin fibers are formed which can have a profound physiological effect [38, 39]. Thus hydrophobic surface patches can be useful if the assembly of several proteins (or subunits) is required, but otherwise they are to be avoided in the protein folding.

There are two main ways of forming hydrogen bonds between polypeptide CO and NH groups, as diagramed in Figure 6, so that it is energetically feasible to bring the polypeptide chain into the hydrophobic center of the protein molecule. One way is to form a helical structure, as shown in Figure 6(a). Internal hydrogen bonding within the same portion of the polypeptide chain gives rise to an α helix. Alternatively, hydrogen bonds may be formed between distant strands of the polypeptide backbone that lie antiparallel or parallel to each other to give a β sheet, as shown in Figures 6(b) and 6(c). These two types of secondary structure, α helices and β sheets are the motifs most commonly found in proteins. The terms α and β were derived from X-ray diffraction studies of fibrous proteins, such as keratin, by Astbury. Keratin was found to give, under normal circumstances, a diffraction pattern referred to as the 'α-keratin X-ray pattern' with repeat distances of approximately 5.1 Å [40, 41]. On stretching, however, a different diffraction pattern, the 'β-keratin pattern', is obtained. It is possible to reconvert β-keratin (stretched) to α-keratin (folded) when the tension is relaxed. This led to the designations α and β that are still used for the α helix (folded) and the β sheet (stretched), respectively. The two types of hydrogen bonding are identified from a consideration of their conformation angles [42, 43]. The approximate values for the C–N–Cα–C and N–Cα–C–N torsion angles are, respectively, $-57°$ and $-47°$ for an α helix and $-139°$ and $135°$ for an antiparallel β sheet.

In the α helix, diagramed in Figure 6(a), the polypeptide chain is coiled in a right-handed spiral manner with 3.6 amino acid residues per turn [44, 45]. The pitch is such that one turn rises 5.4 Å parallel to the helix axis. A left-handed α helix is less favorable energetically because it would have problems with steric overcrowding The amino acid residues methionine, alanine, leucine and glutamic acid appear to favor α helix formation, while valine, isoeucine, glycine and proline are helix breakers. Hydrogen bonds in an α helix are formed between a carbonyl oxygen atom and the amide nitrogen atom four residues beyond. Since all carbonyl groups point in the same direction, nearly parallel to the helix axis, the helix has a dipole moment in which the amino terminus is positive and the carboxy terminus is negative to the same extent [46–48]. Often, however, negatively charged residues (such as aspartic acid or glutamic acid) are found near the amino end of an α helix and positively charged residues (such as lysine or histidine) near the carboxy end. In addition, negative ions, such as phosphate or sulfate ions, can often be found near the amino terminus of an α helix, while metal ions congregate near the carboxy terminus [49–53]. Caution has, however, been recommended in assigning importance to the macrodipole of an α helix in view of the many other forces, such as the presence of polar side chains, that may also account for these observations [54]. The interior of the α helix is closely packed and the van der Waals packing probably contributes more to the stability of the α helix than does the hydrogen bonding. The portion of membrane-bound proteins that spans the nonpolar membrane consists of α-helical segments with predominantly nonpolar side chains [55].

The periodicity of 3.6 residues per α-helical turn means that if the helix is viewed down its axis, the side chains stick out approximately every 100°, and the helix can be represented by a helical wheel [56], shown in Figure 7. If the α helix is on the outside of the protein, the nature of the sidechains must change from hydrophobic for the part that faces the interior of the protein to hydrophilic for the part of the protein that can interact with solvent. The helix is then described as amphipathic. This is seen in the crystal structure of a

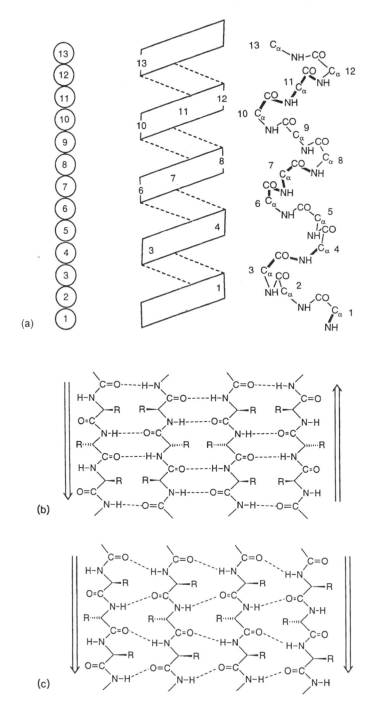

(a)

(b)

(c)

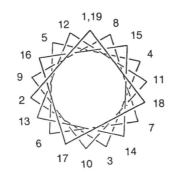

(a)

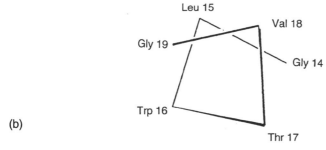

(b)

Figure 7 Views down α helices in which the side chains project every 100°. The resulting diagrams are called helical wheels and, in many cases, show that the helices have hydrophilic side chains on one side and hydrophobic on the other (amphipathic helices): (a) numbering through 19 amino acids in an α helix; and (b) an example for six amino acids

'leucine zipper' protein which contains several leucine residues that repeat every seven amino acids [57, 58]. The leucine side chains line up parallel to the α helix axis.

The first amino terminal and the last four carboxy terminal amino acid residues cannot make the types of hydrogen bonds that the amino acids internal to the helix can. Therefore hydrogen bonds must be made with either other groups in the protein or with the solvent. The first and last residues are often occupied by groups that can form such bonds. A cap is defined as the first or last residue of a helix whose α carbon atoms lie on the cylinder containing

Figure 6 (*opposite*) Hydrogen bonding between polypeptide main-chain CO and NH groups: (a) formation of an α helix; (b) formation of an antiparallel β sheet; and (c) formation of a parallel β sheet. The formation of β sheets occurs through side-by-side hydrogen bonding between β-strands

the helix backbone. Glycine, serine, threonine, aspartate and asparagine stabilize the amino cap, while glycine and asparagine stabilize the carboxy cap. Proline is destabilizing as it causes the loss of two intrahelical hydrogen bonds, while glycine is also destabilizing as it does not bury much hydrophobic surface area [59].

A convention for describing helices involves listing the number of residues per turn and then denoting, by a subscript, the number of atoms (including the hydrogen atom) necessary to close the hydrogen bonding. By this scheme the α helix would be called the 3.6_{13}-helix. Another type of helix often found in protein structures is the 3_{10}-helix, which is less stable than the α helix and usually forms only a single turn. The helix radius, i.e. the distance of the Cα atoms from the helix axis, is about 2.3 Å for an α helix and smaller, about 1.9 Å, for a 3_{10}-helix [5]. The 3_{10}-helix is often found at the carboxy ends of α helices. It is found that α helices may be irregular and have bends in the middle as a result of losing one or two hydrogen bonds, and also tighten up or unravel in their last turn.

In the β sheet the polypeptide chains are nearly fully extended, and each is described as a β-strand which has a direction (from the amino end to the carboxy end) as shown in Figures 6(b) and 6(c) [60, 61]. Individual β-strands aggregate side by side, forming hydrogen bonds between the carbonyl group oxygen atom of one chain and the amide hydrogen atom of another chain [62–64]. The polypeptide chains in this side-by-side packing may be either parallel or antiparallel. When examined end on, the β-strand appears pleated, meaning that successive Cα atoms are slightly above or below the sheet plane. The β sheet is the most commonly encountered secondary structural feature in soluble proteins (with six amino acid residues in the average strand), and the α helix is the second most common [65]. Like α helices, β sheets can be irregular, and may be interrupted by β-bulges (in which two residues on one strand are opposed by one residue on the opposite strand) [66] or show irregularities at the end of the strand.

Antiparallel sheets have evenly spaced hydrogen bonds, while parallel strands have hydrogen bonds that are less evenly spaced. The side chains are opposite one another on neighboring strands and extend above and below the plane of the sheet, as shown in Figure 8(a). β-Sheets in globular proteins are generally not flat, but have a right-handed twist (when viewed along the sheet strands) [67] that varies from 0 to 30°, diagramed in Figure 8(b). As shown in Figure 8(c), which consists of views down the N–Cα and the C–Cα bonds of a peptide group, a left-handed twist would bring the main-chain carbonyl group near to the side chain on the Cα atom. This is energetically less favorable than a right-handed twist which places a main-chain amino hydrogen atom near the side chain, causing less steric hindrance.

To simplify reported crystal structures, α helices are often represented by cylinders, and β-strands by thick flat arrows, the arrow pointing to the carboxy

249

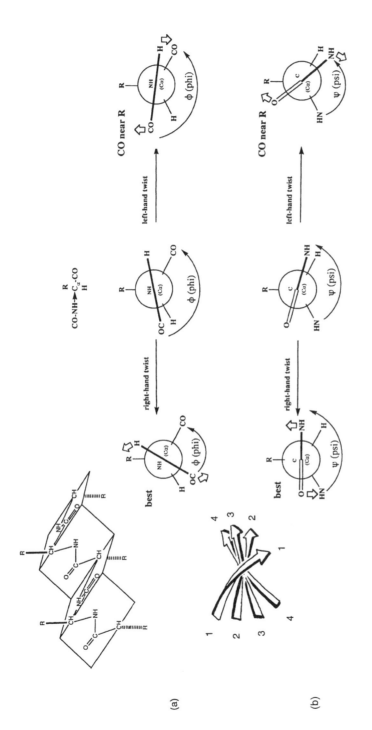

Figure 8 (a) The pleatedness of the β sheet, showing that the R side chains project alternately above and below the sheet. (b) The twist of the β sheet illustrated for four β-strands. (c) The reason for the twist in terms of the conformational angles φ and ψ. The right-handed twist relieves some steric overcrowding between the main-chain carbonyl group and the side chain

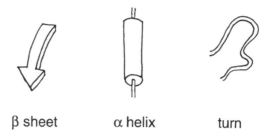

β sheet   α helix   turn

Figure 9   Stylizations used for the components of a protein structure

terminus (see Figure 9). Joining the α-helical and β sheet components of the protein to each other are loop regions which cause changes in the overall direction of the polypeptide chain [68]. Loops generally occur at and define the surface of the protein. They are the main sites of insertions and deletions of amino acids in like proteins from different species [69–72]. These reverse turns (loops) generally contain hydrophilic residues, particularly those such as glycine that are very flexible or like aspartic acid or asparagine that form favorable hydrogen bonds which stabilize the loop. Three types of reverse turns have been identified by Venkatachalam [69]; these are diagramed in Figure 10.

The result of the two types of folding of the polypeptide main chain — the α helix and the β sheet — is that the architect is provided with two different construction modules which can be joined together (by loops), given very stable entities that are often quite rigid. The loops serve as the main foci for flexibility and variability. The α helix, the β sheet and loops are the three components that, with slight variations (such as the 3_{10}-helix), will be used in the final conformation of the folded protein.

3. VARIABILITY IN THE BUILDING BLOCKS: THE 20 AMINO ACID SIDE CHAINS

All of the 20 amino acids that are available in the body for the building of a protein molecule (see Figure 11a) have different properties, and each will be considered in turn. These are the building blocks that can be considered to be laid on the framework made by the main polypeptide chain of the protein.

There are several ways to think about the amino acid residues in proteins: their length and size, their hydrophobicity or hydrophilicity, their charge and ionization state, their hydrogen-bonding potential and their possible metal-binding potentials [73–75]. It is possible to gain or lose charge, hydrophobicity or hydrophilicity in a controlled way, as is required in the active site of an enzyme. This is done by appropriate choices of amino acids with specific side-chain properties. Therefore amino acids must be chosen with care when a

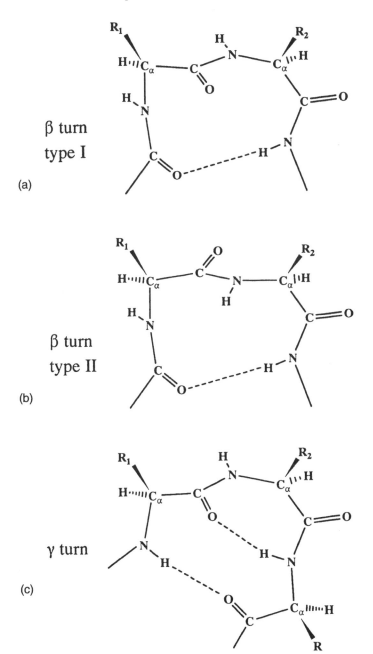

β turn
type I

(a)

β turn
type II

(b)

γ turn

(c)

Figure 10 Three types of turns that are found in proteins. Note that hydrogen bonds are formed if possible

cysteine Cys C   asparagine Asn N   glutamine Gln Q

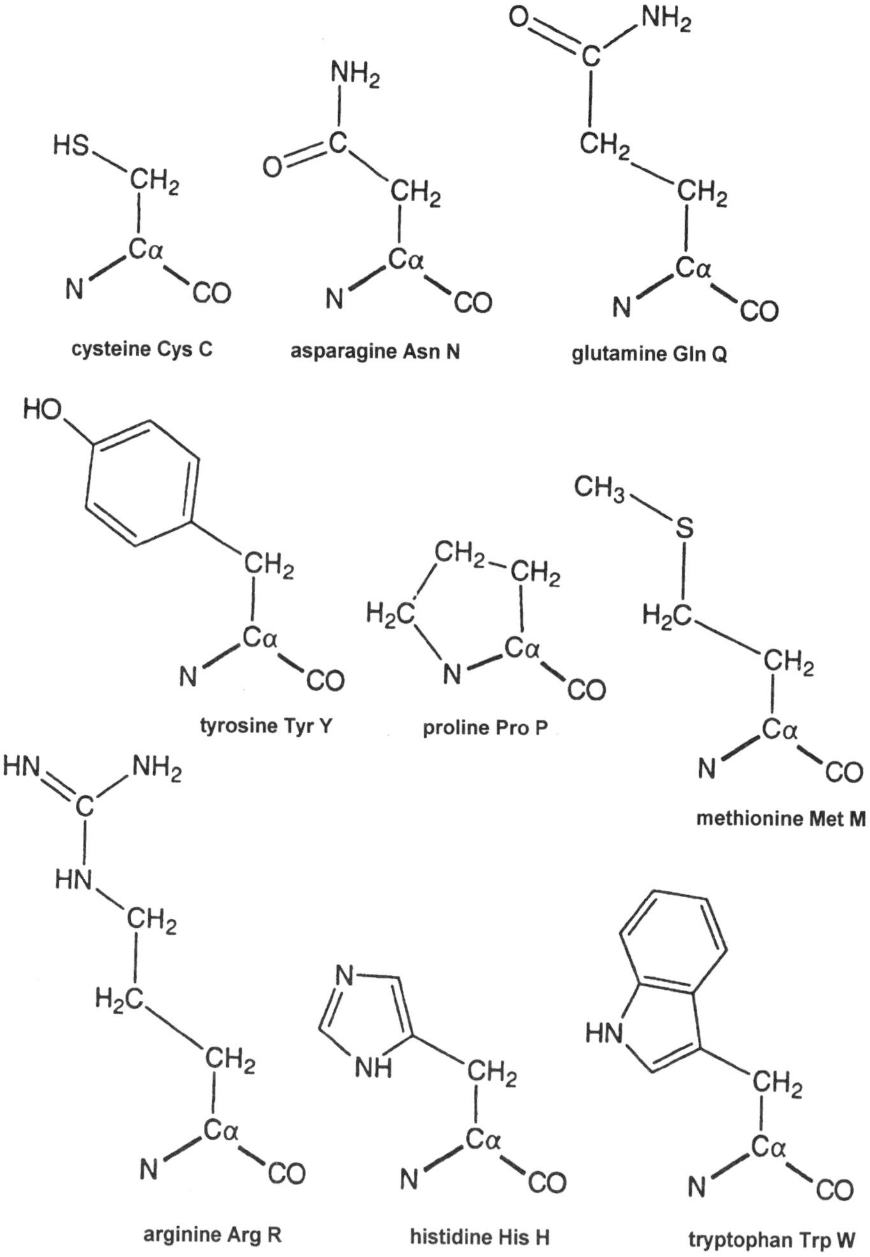

tyrosine Tyr Y   proline Pro P

methionine Met M

arginine Arg R   histidine His H   tryptophan Trp W

Figure 11   (a)

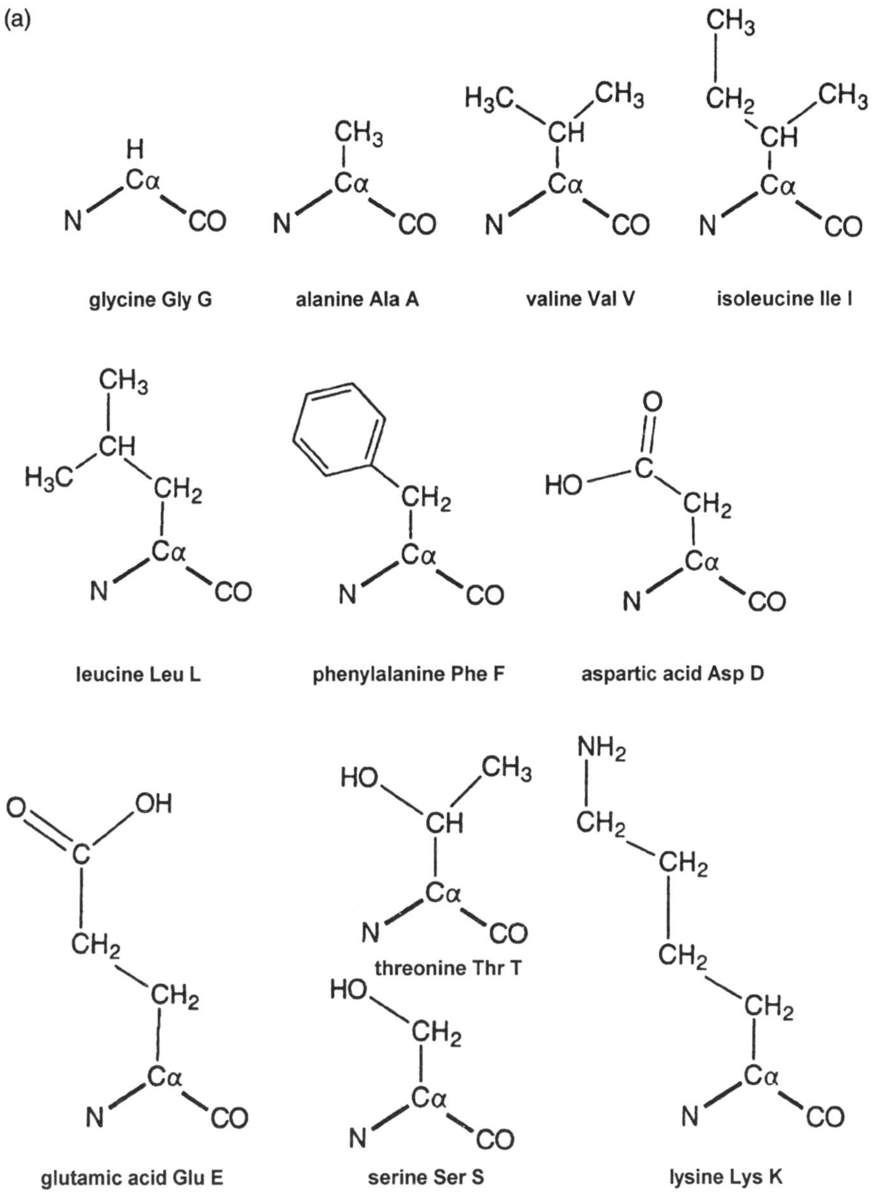

(a)

glycine Gly G alanine Ala A valine Val V isoleucine Ile I

leucine Leu L phenylalanine Phe F aspartic acid Asp D

glutamic acid Glu E threonine Thr T serine Ser S lysine Lys K

Figure 11 (a) (*continued*)

(b)

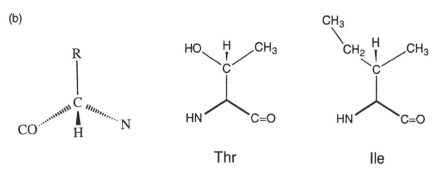

Thr   Ile

Figure 11 (*continued*) The 20 amino acids: (a) the formulae and the symbols (three-letter and one-letter codes) that are used; and (b) the absolute configurations of a general amino acid group and the side chains of threonine and isoleucine which contain asymmetric carbon atoms

protein with a required function is being made. Many of these amino acids can readily be replaced by others, but some play a vital role in the functioning of the protein and these amino acid residues generally cannot be substituted without loss of activity.

A detailed description of the way in which amino acid side chains pack one with another has been published [76] for all 20 amino acids. This compendium contains computer-drawn diagrams of the 400 (20 × 20) possible types of interactions for 62 proteins selected (for their precision) from the Protein Databank [16]; this implies 28 902 pairwise interactions. The results are tabulated in Table 1, in which the numbers of interactions of one side chain with another are listed from the data provided in the book [76]. The amino acids have been listed in an approximate order of hydrophobicity [77–80]; to do this each amino acid side chain is assigned a value equal to the free energy gained when it is moved from water to the interior of the protein. The convention is that the sign of the term is positive if the residue is apolar, since the transference of an apolar residue from water to the interior of a protein makes a net negative contribution to the free energy and the result has lower energy. The hydrophobicity of a protein side chain has several components because most amino acid residues contain some groups that are less polar than others (a methylene group versus a charged amino group in lysine, for example).

The most noticeable finding from Table 1 is that most of the interactions occur between hydrophobic side chains (isoleucine, valine, leucine and phenylalanine), especially the 616 interactions listed between pairs of leucine residues. These side chains congregate in the center of the protein molecule. By contrast, the hydrophilic side chains are dispersed throughout the outer core and extend into the solvent so that many of their interactions are with water molecules rather than with other protein side chains. Other peaks in Table 1 represent the large number of interactions between pairs of cysteine residues

Table 1 The numbers of interactions listed elsewhere [76] for each amino acid residue in the more precise crystal structure determinations of proteins. The last line gives the total numbers of interactions. The columns are arranged in approximate order of hydrophilicity. Note the large numbers of interactions between the most hydrophobic side chains, especially leucine, and also the numbers indicating 'salt bridges' between charged side chains

	Ile	Val	Leu	Phe	Cys	Met	Ala	Gly	Thr	Ser	Trp	Tyr	Pro	His	Glu	Gln	Asp	Asn	Lys	Arg
Ile	368	328	430	232	64	85	171	96	118	65	98	139	54	39	59	65	51	42	65	57
Val	328	388	466	240	74	75	181	125	145	90	110	144	91	54	64	77	42	65	83	58
Leu	430	466	616	335	84	123	214	143	145	98	134	178	94	63	72	90	57	74	87	86
Phe	232	240	335	214	58	82	134	71	94	66	86	115	55	52	46	45	57	47	53	57
Cys	64	74	84	58	234	29	28	42	31	28	25	57	35	20	9	23	19	27	15	21
Met	85	75	123	82	29	48	33	34	32	31	37	42	20	20	21	21	17	17	22	21
Ala	171	181	214	134	28	33	92	79	101	59	74	108	44	38	36	53	56	54	54	47
Gly	96	125	143	71	42	34	79	92	118	59	60	92	44	43	57	58	87	82	50	56
Thr	118	145	145	94	31	32	101	118	150	115	46	87	54	55	79	68	94	79	77	65
Ser	65	90	98	66	28	31	59	118	115	72	24	51	38	31	67	47	89	60	42	45
Trp	98	110	134	86	25	37	74	60	46	24	36	60	41	28	31	31	29	33	29	34
Tyr	139	144	178	115	57	42	108	92	87	51	60	94	78	51	46	60	68	82	78	57
Pro	54	91	94	55	35	20	44	44	54	38	41	78	38	22	28	45	29	37	35	37
His	39	54	63	52	20	20	38	43	55	31	28	51	38	72	51	17	53	38	24	29
Glu	59	64	72	46	9	21	36	37	79	67	31	46	51	51	42	33	40	61	118	91
Gln	65	77	90	45	23	21	53	58	68	47	31	60	28	17	33	34	31	54	37	53
Asp	51	42	57	57	19	17	56	87	94	89	29	68	45	53	40	31	31	76	119	102
Asn	42	65	74	47	27	17	54	82	79	60	33	82	29	38	61	54	50	80	74	46
Lys	65	83	87	53	15	22	54	50	77	42	29	78	37	24	118	37	76	74	18	21
Arg	57	58	86	57	21	21	47	56	65	45	34	57	37	29	91	53	102	46	21	40
Total	2626	2900	3589	2139	923	810	1656	1468	1753	1177	1046	1687	919	800	1051	942	1166	1128	1101	1023

and the pairwise interactions between side chains that are charged at neutral pH, i.e. lysine or arginine (positively charged) with aspartate or glutamate (negatively charged).

An important property of the amino acid side chains is their hydrogen-bonding capacity. Some 52% of side-chain atoms form hydrogen bonds to water, a larger fraction than the 43% for main-chain atoms. Bifurcated hydrogen bonds between a carbonyl group and two NH groups are generally only found in helices. Of the side chains, as shown in Figure 12, aspartate and glutamate are solely hydrogen bond acceptors; lysine, arginine and tryptophan are only hydrogen bond donors; but asparagine and glutamine are both donors and acceptors (although glutamine behaves more as a donor than as an acceptor). Serine, threonine and tyrosine can be both hydrogen bond donors and acceptors, but are more often found to act as donors. Cysteine and methionine only rarely form hydrogen bonds.

Carboxy groups have been shown by neutron diffraction studies of small molecules to form three or four hydrogen bonds per CO_2^- if the group is ionized, while the OH oxygen atom of a nonionized carboxy group (CO_2H) will form only one hydrogen bond [81]. This is very useful for assessing the ionization state of carboxy groups in proteins. For example, the ionization states of the active site groups Asp52 and Glu35 in the structure of hen egg-white lysozyme are shown, by this information, to be ionized and nonionized, respectively, as shown in Figure 13. This was confirmed by a subsequent neutron diffraction study of lysozyme [81]. Thus hydrogen-bonding patterns can give information on the ionization states of carboxyl groups.

Glycine is the simplest amino acid with two hydrogen atoms on the Cα atom, giving a methylene (CH_2) group. Therefore, because the methylene group does not contain a chiral carbon atom, glycine does not occur as D- and L-isomers. Since the side chain is only a hydrogen atom and hence extremely small, glycine in a polypeptide can take up conformations that are not possible for steric reasons for other amino acid residues. Therefore the presence of glycine increases the flexibility of the main chain. As a result, glycine is used profitably in loops and turns. It is also used, because of its small size, in the packing of α helices if the interhelical angle is near 90°, as will be described later. Alanine, with a methyl group on the Cα atom, is much less flexible than glycine and does not appear to have any remarkable characteristics. It occurs frequently in proteins and is small but not as small as glycine, so that it does not have the main-chain flexibility characteristics of glycine.

In proline the amide group is part of a five-membered ring, puckered with Cγ approximately 0.5 Å from the plane through the other ring atoms. Rotation about the C–N bond in a polypeptide at the site of a proline residue is highly restrained [82]. This means that the main chain is fairly rigid at this point. In addition, no amide NH group is available as a donor in hydrogen bonding. The proline ring takes up some of the space that would normally be occupied by a

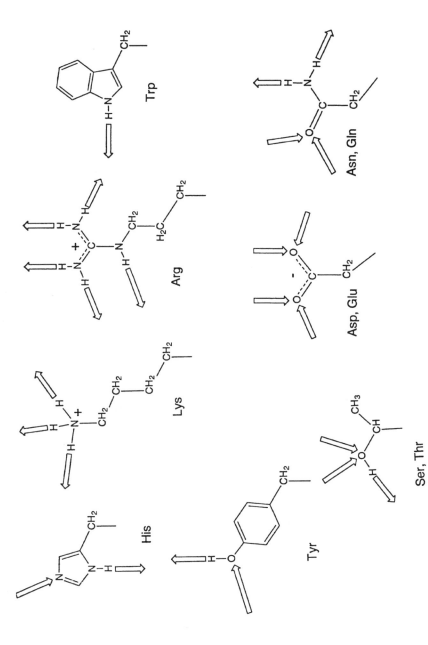

Figure 12 Hydrogen-bonding capabilities of various side chains. Open arrows indicate hydrogen bonds and their directionalities

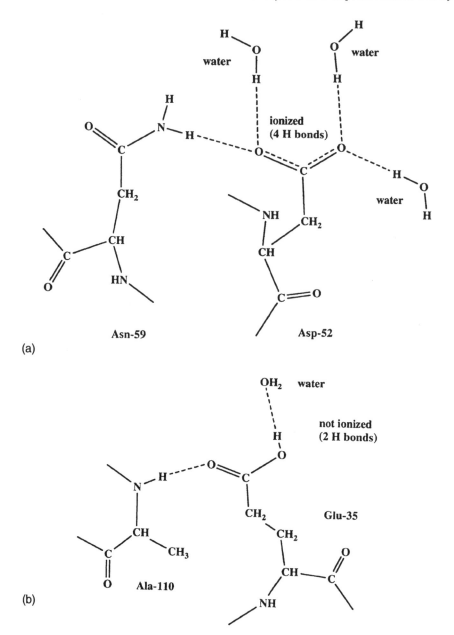

Figure 13 Deduction of the ionization state of carboxylic acids in proteins. The example of lysozyme is given here [81]. Two active site groups are Asp52 and Glu35, and it is inferred that the former (a) is ionized while the latter (b) is not

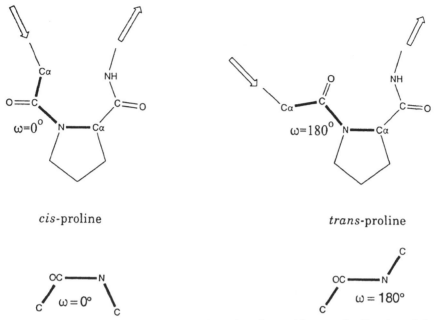

cis-proline   trans-proline

Figure 14 The effect of *cis–trans* isomerism of proline residues on the direction of the main chain of the polypeptide, indicated by open arrows. Proline is the only amino acid that so far has shown this isomerism in protein structures

neighboring chain, and therefore it is hard for proline to fit within an α helix, and it is considered to be a 'helix breaker'. When it is found in an α helix, there is usually a bend or kink of the helix at that point. One interesting property of proline is its ability to exist as a *cis*- as well as a *trans*-isomer and therefore control the general path of the main chain in a way that no other amino acid side chain can [83]. This *cis–trans* isomerism, diagramed in Figure 14, provides a slow step in protein folding. This is because the energy barrier between the *cis*-form and *trans*-form of proline is lower (13 kcal mol^{-1}) than for other amino acids (20 kcal mol^{-1}) [30]. *cis*-Proline (Figure 14) always forms a certain type of turn in the residue preceding it, and is itself in an extended conformation.

Serine and threonine are short-chain residues with hydroxy groups at their ends. These hydroxy groups are no more reactive than the hydroxy group in ethanol. Threonine has an asymmetric carbon atom (Cβ), but only one isomer is found naturally (see Figure 11b). The hydrogen-bonding schemes of serine and threonine are variable, but the hydroxy group often forms hydrogen bonds to main-chain carbonyl or amide NH groups, particularly carbonyl groups, thereby, for instance, disrupting α helix formation. For example, serine has a destabilizing effect on an α helix by forming a hydrogen bond to the main-chain

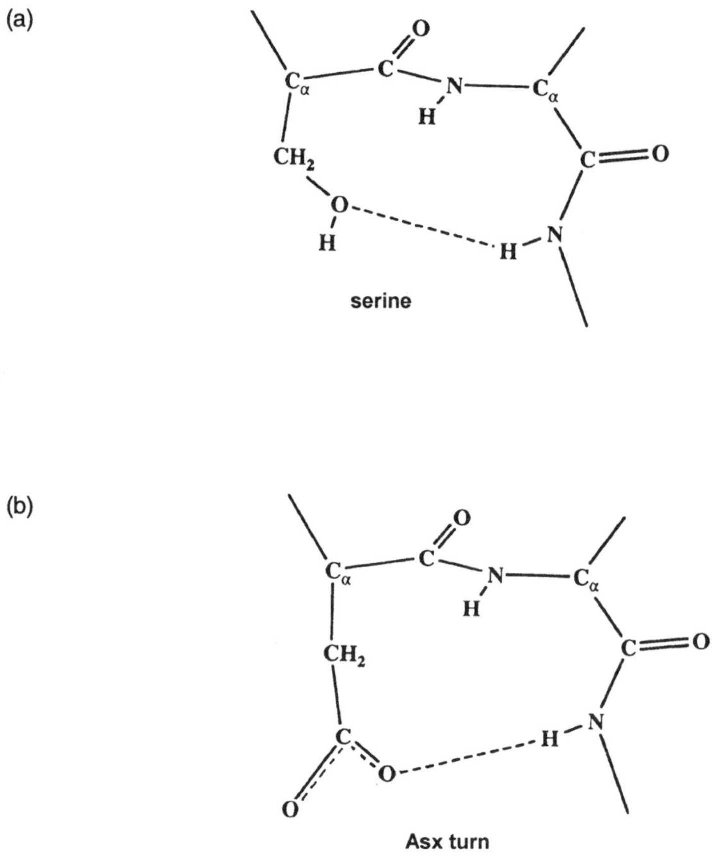

(a)

serine

(b)

Asx turn

Figure 15 Formation of turns by functional groups on side chains forming hydrogen bonds to main-chain groups for (a) serine and (b) aspartate (the 'Asx turn'). Note the similarity to the normal turns in polypeptide chains shown in Figure 10

hydrogen atom on the residue three amino acids earlier (see Figure 15a). Threonine and serine are important in the catalytic mechanisms of certain enzymes. The serine proteases are a good example. In addition, both serine and threonine can be phosphorylated and can be attachment sites for carbohydrates. They are both found in a number of transmembrane α helices.

Aspartic acid (pK_a 3.9) and glutamic acid (pK_a 4.1) are ionized under physiological conditions (approximately neutral) and provide negatively charged groups at pH 6 and higher, while lysine, histidine and arginine are positively charged near neutral pH [84]. Negatively charged aspartate and glutamate side chains are attracted by groups of the opposite charge, so that so-called 'salt bridges' or ion pairs result. In proteins these ion pairs are really

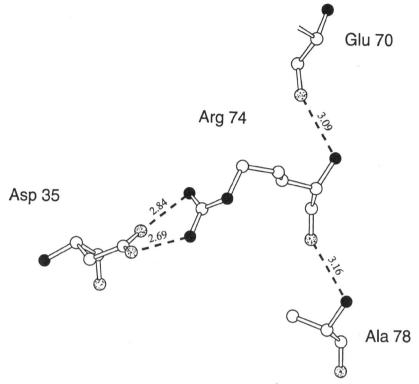

Glu 70

Arg 74

Asp 35

Ala 78

Figure 16 Formation of hydrogen bonds (lengths in Ångströms) between negatively charged carboxy groups and positively charged guanidinium groups in D-xylose isomerase. Some main-chain hydrogen bonding is also indicated

hydrogen bonded to each other, as shown in Figure 16. Ion pairs are formed by one third to one fifth of all arginine, histidine, aspartic acid and lysine residues [29]. Most of the ion pairs found in a protein are at the protein surface and they are poorly conserved, implying that they are generally not important to stability in folding pathways. Some ion pairs, however, serve to bind subunits together and even entire molecules to other molecules. The side chains of aspartate and glutamate differ in their lengths and therefore how they interact with neighboring groups and their effects on conformation and chemical reactivity. Aspartate has the shorter side chain of the two and one oxygen atom of the aspartate side chain can form a hydrogen bond to a main-chain NH group (see Figure 15b), while the other oxygen atom points into solution. Aspartate and glutamate are good sites for metal binding, and will tend to bind the metal ions in the plane of the carboxy group [85].

Lysine (pK_a 10.8) is ionized under physiological conditions. It is long and is the most flexible and mobile of all the side chains. It can form three N^+-H$\cdots$O hydrogen bonds in a tetrahedral manner. Arginine (pK_a 12.5) contains an ionized *gem*-amidino group and is planar as a result of resonance. Arginine and lysine are found both at the molecular surface and also in internal ion pairs. There are five hydrogen bond donors on an arginine side chain, and all five lie in a plane. When arginine binds carboxyl groups from aspartate, glutamate or a substrate, or other groups that it recognizes, such as a nucleic acid base, it maintains these groups also in this plane. Therefore arginine is a useful amino acid for tying different parts of a protein together [86], aligning substrates in a rigid manner (particularly those with carboxyl groups) and in protein–nucleic acid recognition [18, 87].

Asparagine and glutamine have hydrogen-bonding donors (NH_2 groups) and hydrogen-bonding acceptors (CO groups), and therefore are also useful for tying groups in different parts of the protein together. In addition, asparagine can also, by virtue of its chain length, bend back and form a hydrogen bond to a main-chain carbonyl group, giving a pseudoturn.

Phenylalanine, tyrosine and tryptophan have aromatic side chains and histidine can bind metal ions, notably copper and zinc, and can also serve in the transfer of hydrogen ions. Tryptophan appears to aid in aligning substrates in certain enzymes, such as the sugars that bind to D-xylose isomerase, by virtue of its aromatic heterocyclic ring system which provides a 'greasy surface'. It is the largest side-chain group and also the least common one. Histidine, with a pK_a of about 6.0, is one of the strongest bases that can exist at pH 7. Its nitrogen atom with a hydrogen attached is an electrophile and hydrogen bond donor, while if there is no attached hydrogen atom the nitrogen atom is a nucleophile and an acceptor for hydrogen bonding or hydrogen transfer. The nitrogen atom nearer to Cα (δ_1) has a pK_a value 0.6 units less than that of the one that is further away (ε_2). If both nitrogen atoms have a hydrogen atom on them the pK_a is 14.4 for both. When histidine is in position 2 of an α helix and is deprotonated, the α helix is destabilized. The helix dipole can affect the pK_a of a histidine residue (from 6.6 to 7.8 in carbon-monoxyhemoglobin A [88]). Tyrosine, with a pK_a of 11.1, has a phenolic hydroxyl group and is involved in electrophilic substitution reactions. It acts as a strong hydrogen bond donor rather than acceptor, generally interacting with CO rather than NH groups.

The aliphatic side chains of valine, isoleucine, leucine and methionine have no reactive groups, only methylene, methyl (Me) and SMe groups. They are important, however, because they do not interact well with water, preferring each other. In addition, if they are branched as are isoleucine and valine, they stiffen the main chain and decrease its flexibility [30]. Phenylalanine is nonpolar like benzene or toluene. It has a methylene group which lengthens the side chain and prevents steric hindrance with the main chain. These are hydrophobic residues found in the center of proteins and provide a core around which the functional parts of the protein are assembled. The conformational preferences of valine and isoleucine are different from those of leucine and methionine. Isoleucine has an extra asymmetric center at Cβ and only one isomer is found in nature (see Figure 11b). Methionine has a flexible side chain containing a thioether linkage. It is the only unbranched nonpolar side chain, and cannot be protonated.

An important function of cysteine is to form disulfide bonds, since these give additional stability, and sometimes also catalytic function [4, 30]. The thiol group in cysteine is the most reactive of the amino acids and has a pK_a of 8.3, loss of the proton giving the thiolate. Cysteine binds metal ions such as iron, zinc and copper. Disulfide bridges may be formed between different parts of one polypeptide chain or between different chains. The energy of a disulfide bridge relative to two SH groups depends on the redox potential of the environment. Inside the cell the environment is reducing, aided by the glutathione system [30], and SH groups are plentiful. As oxidative conditions prevail S–S bonds are formed, but to form such an S–S link the two SH groups need to be in the correct relative positions, and with the α-carbon atoms 4–7.5 Å apart. The C–S–S–C torsion angle needs to be near 90°; other torsion angles are unstable.

But, of course, the side chains play a role in the overall folding of the protein. The energies involved are small in general because, during protein folding, protein donor and acceptor groups replace hydrogen bonds to water by hydrogen bonds to other protein groups. Rules for the formation of an α-helical or β sheet structure by various combinations of amino acids in a main chain have been formulated by many [89–93], including Chou and Fasman. The rules that Chou and Fasman put forward are that four helix-forming residues adjacent in sequence can nucleate an α helix, while three out of five adjacent sheet-forming residues can nucleate a β sheet.

Main-chain hydrogen bonding will be affected by the surrounding environment of other peptide groups (both main chains and side chains), and these affect, restrain and control each other in complicated ways [94]. As a result, permissible conformational changes are complexly coupled, and are sensed by the neighboring portions of the protein. This interaction between main-chain folding and overall hydrogen bonding (and therefore overall energy) has allowed nature to engineer a molecule with a precise biological function. On the other hand, however, it makes structure prediction very difficult [95–98].

4. THE MODULAR UNITS: FOLDING MOTIFS AND THEIR PACKING

Now we can start considering how nature has folded the protein using the secondary structural units, such as α helices and β sheets, just described. This

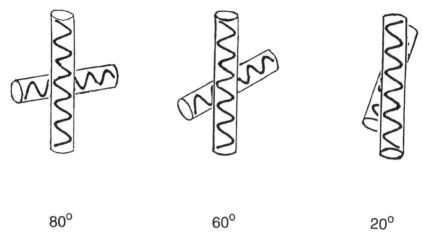

80°   60°   20°

Figure 17 Packing of α helices. Ridges and grooves on the surfaces of α helices lead to three main types of interaction, with angles of 80° (ridges formed by side chains i and $i+3$), 60° (ridges formed by side chains i and $i+4$) or 20° (ridges formed by side chains i and $i+5$) between the main axes of the helices, as shown

leads to a higher level of organization called 'supersecondary structure' [99–106]. How are the α helices and β sheets arranged within the more complex structure of each total protein? First we will consider interactions between α helices and α helices, between β sheets and β sheets, and those between β sheets and α helices, and then list a few of the overall structures that result. Only a few of them can be described here; for more details the reader is referred to the book by Brändén and Tooze [61].

Before any crystal structures of proteins had been determined, Crick suggested a model of how α helices would pack within a protein [107]. He based this on an idealized helical structure with knobs for the amino acid side chains and holes between the side chains. This concept was investigated further by Chothia who had an analogous model consisting of ridges and grooves [108, 109]. As shown in Figure 17, a ridge nearly parallel to the helix axis can be formed by amino acid residues four apart in the amino acid sequence, and other ridges can be formed when the residues are three or one apart. Two α helices pack together when the ridges of one helix fit neatly into the grooves of the other. Richmond and Richards extended these ideas using structural data on myoglobin as an example [110–112].

As a result of these analyses three general classifications of α helix–α helix interactions are found, as shown in Figure 17. These are described by the relative orientations of the helical axes (expressed as the angle between the two axes) and the nature of the amino acid side chains that are commonly found in that class. One class involves glycine, which has the smallest amino acid side chain, at the contact point on each helix; the two helices are nearly

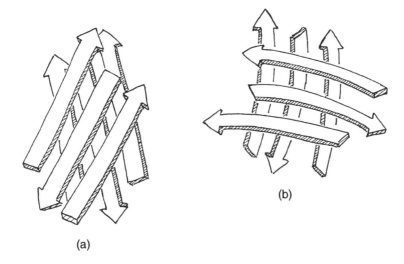

(a)

(b)

Figure 18 Packing of β sheets. The disposition of the side chains leads to the possibilities of alignment of the sheets either (a) nearly parallel (aligned) or (b) nearly perpendicular (orthogonal) to each other

perpendicular to each other, the angle between the main axes of the two α helices being approximately 80°. Alanine, valine, isoleucine and serine form a second group, and when they come into contact in two α helices the angle between the helical axes is 60°. In the third group, which contains leucine and threonine, the interhelical angle is small, approximately 20°, so that the helices are nearly parallel (Figure 17).

β Sheets pack together in proteins with individually variable extents of twist [67, 113, 114]. In order to pack well it is necessary for β sheets with different twists and residue compositions to pack so that the central part of each β sheet is closely packed as the two sheets make contact. When the twists of the β sheets are approximately 0°, two sheets are complementary to one another and can pack readily. When, however, the two β sheets have different degrees of twist, one β sheet needs to be rotated with respect to the other in order to optimize the overlap. If the main structure of a domain is β sheet, then it is likely that the domain will contain two such β sheets and that they will lie either almost parallel or almost perpendicular to each other (Figure 18) [115].

A third type of supersecondary structure results from the packing of α helices on β sheets [116]. As has been described, two parallel β-strands will form a sheet with a particular handedness. These β-structures necessarily have connections that can either be an α helix, a coil or another strand. For example, the arrangement of the two β-strands and the α helix (a βαβ-unit)

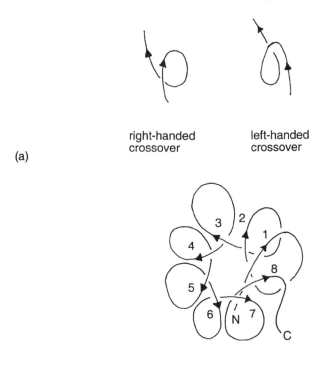

(a)

(b)   folding of (βα)$_8$ barrel

Figure 19 The handedness of the βαβ-unit crossover. (a) A right-handed crossover is commonly found, while a left-handed crossover is not. (b) In the folding of a (βα)$_8$ barrel this right-handed crossover is important, as shown

nearly always has a right-handed twist as the α helix passes over one β-strand to join up to the next β-strand. An examination of several protein structures has shown that no matter what the nature of the connection (helix, coil or strand), the topology is always right handed, as shown in Figure 19 [117]. The α helices follow the curvature of the β sheet and interact with β sheets, and the angle between the α helix and the β sheet is then approximately 20°, as shown in Figure 20(a). A stereoview of an α helix is shown in Figure 20(b).

The five basic protein families are α, β, α + β, α/β and 'other' [99–102, 118]. The α-proteins contain mainly α helices and no β-strands, while the β-proteins are those principally composed of β-strands with no α helices. The α/β-proteins have an alternation of α helices and β-strands along the main chain, while α + β

Figure 20 (*opposite*) An α helix and a β-strand in the (βα)^8 barrel of D-xylose isomerase: (a) a stereoview of an α helix; and (b) the interaction between an α helix and its contiguous β-strand

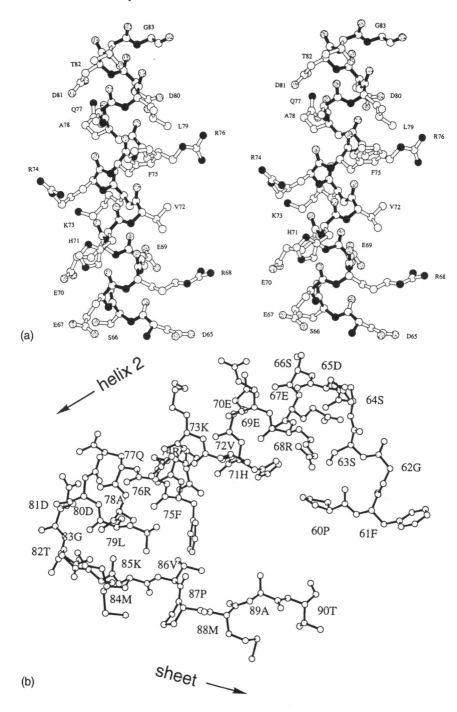

(a)

(b)

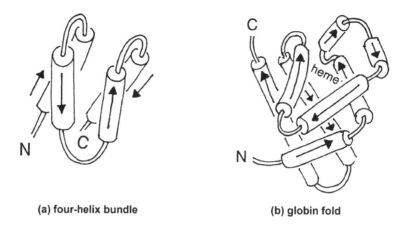

(a) four-helix bundle   **(b) globin fold**

Figure 21 Two helical motifs found in protein structures: (a) the four-helix bundle; and (b) the globin fold. Note how the α helices tend to line up antiparallel to each other

proteins are those in which the α helices and β-strands are segregated to different areas in the protein molecule. The remaining proteins fit into the fifth category.

Among the α-proteins the four-helix bundle and the globin fold, shown in Figure 21, are common motifs. They each consist of α helices that are lined up in an antiparallel manner, and if the protein is an enzyme its active site may be found between the α helices. The four-helix bundle contains four α helices in the sequence up–down–up–down. The globin fold, found in hemoglobin and myoglobin, contains eight α helices in a more complicated folding pattern, but with many of the characteristics of the four-helix bundle in that adjacent α helices are antiparallel (see Figure 21).

Antiparallel β sheets are, as was described earlier, twisted, and they can pack to form a barrel with a hydrophobic core. Three structures are commonly found for β-proteins; these are the up-and-down barrel (Figure 22a), the Greek key barrel (Figure 22b) and the jelly-roll barrel (Figure 22c). Another motif, shown in Figure 22(d), has been found in pectate lyase [119], even though it was thought too unstable to exist.

The folding of α/β-proteins falls into two major classes, illustrated in Figure 23; the closed $(\beta\alpha)_8$ barrel (the subject of this article) and the open twisted sheet which has α helices on both sides of it. They both involve parallel β sheets that are connected in most cases by an α helix (the βαβ-motif described earlier).

Each protein is built up of such units that are assembled to give the entire folding pattern. We can, therefore, from an examination of crystal structure determinations build up a large stockpile of motifs normally found in the folding patterns of proteins. Of course, these individual motifs interact with

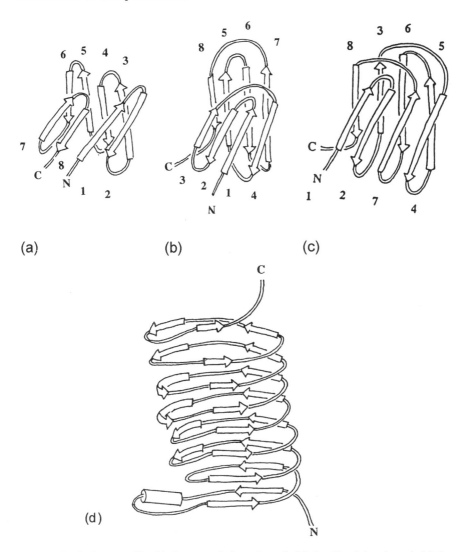

Figure 22 β Sheet motifs: (a) the up-and-down barrel; (b) the Greek key barrel; (c) the jelly-roll barrel; and (d) the structure in pectate lyase [119]

their surroundings in the protein and are affected by them both structurally and electronically.

So far the emphasis of such studies has been on soluble proteins because those are the ones that we know how to crystallize, but membrane-bound proteins are now beginning to be studied. It is becoming apparent, however, that there is not an infinite number of ways in which proteins can fold, but that

Table 2 The various reported crystal structures to date which show $(\beta\alpha)_8$ barrel structures

	Protein	References
Major axis near $\beta3$	Triosephosphate isomerase	126, 131–135
	α-Subunit of tryptophan synthase	136–138
	$N(5'$-phosphoribosyl) anthranilate isomerase–indole-3-glycerol phosphate synthase	139, 140
Require metal	D-Xylose isomerase	22, 23, 141–143
	Pyruvate kinase	144, 145
	Mandelate racemase	146
	Muconate-lactonizing enzyme (muconate cycloisomerase)	147
	Chloromuconate cycloisomerase	148
	Enolase	149–151
	Cyclodextrin glycosyltransferase	152
	Taka-amylase (α-amylase)	153, 154
	β-Amylase	155
Require FMN	Glycolate oxidase	156–158
	Ribulose-1,5-bisphosphate carboxylase/oxygenase	159, 160
	Flavocytochrome b_2	161, 162
	Trimethylamine dehydrogenase	163, 164
	2-Keto-3-deoxy-6-phosphogluconate aldolase (KDPG aldolase)	165
	Aldose reductase	166, 167
	Fructose bisphosphate aldolase	169
	Adenosine deaminase	168
	Narbonin	170

most belong to one of a limited number (possibly less than 100) of different groups [120–125]. It is found that if the amino acid sequences of two proteins do not differ by more than 40%, then the folding patterns will be similar. There are, however, folding patterns that are taken up by proteins with no apparent sequence homology.

5. INTERMEDIATE CONSTRUCTION: THE OVERALL FOLDING OF A BARREL DOMAIN

The protein-building principles that have been described so far are sufficient for the construction of a $(\beta\alpha)_8$ barrel, otherwise known as a TIM barrel (the acronym TIM denoting triosephosphate isomerase, the first protein found to contain this structure [126]). The $(\beta\alpha)_8$ barrel is a common structure [127–129], found in about 20 different proteins to date (see Table 2) [22, 23, 126–170]. It

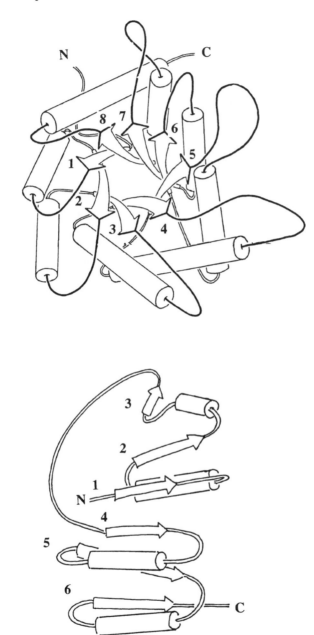

(a)

(b)

Figure 23 Some α/β-motifs: (a) the $(\beta\alpha)_8$ barrel; and (b) the open twisted sheet

was described by Chothia [128] as 'particularly simple and elegant'. The repeating motif is a β-strand connected by a loop to an amphipathic α helix which is connected to another loop. This β-strand–loop–α helix–loop motif is repeated eight times to give a barrel-like structure.

Imagine eight β-strands that form a β sheet. The β-strands have a right-handed twist, so that when eight of them form a β sheet the result is a closed bracelet-like structure, as shown in Figure 24. This makes the lining of the barrel in the middle of the protein molecule. Eight α helices that are interspersed in the amino acid sequence between the β-strands are arranged around this bracelet-like β sheet in an outer concentric circle. They lie nearly parallel to the β-strands, and span the barrel because they connect the strands of the parallel β sheet to each other. In this way they serve to rigidify the structure, like the staves of a barrel, as shown in Figure 25.

Since, as already described, βαβ-units generally have a right-handed connection [117], the direction of the β-strands around the $(\beta\alpha)_8$ barrels is found to be the same for the various protein molecules containing this motif. When the protein is viewed down the barrel from amino to carboxy ends, the β-strands are seen to be ordered in a clockwise fashion because of this right-handed connection between adjacent β-strands (see Figures 24 and 25). The eight β-strands are each tilted approximately 36° to the main axis of the barrel, and the axes of the α helices, which lie parallel to the β-strands, are also tilted in a similar manner. The side of the β sheet that packs against the α helices is fairly hydrophobic, while the side that points to the interior of the barrel is hydrophilic and exposed to solvent. Models of six types of $(\beta\alpha)_8$ barrels have been built [171], involving left- and right-handed tilts and crossovers (where the tilt is the orientational relationship of the β-strands to the main barrel axis and the crossover is the nature of the βαβ-connection). The right-tilted right-handed crossover model with a strong right-handed twist is found to be the lowest in energy and is the only one of the various possibilities that is found experimentally.

The general folding in D-xylose isomerase as a function of protein sequence is illustrated in Figure 26. This diagram shows the amino acid sequence and its folding into α helices, β sheets and turns [22]. The amino terminus of the protein is indicated in this figure by a large filled circle. The $(\beta\alpha)_8$ type of protein folding was first seen in the crystal structure of the glycolytic chicken-muscle enzyme triosephosphate isomerase (abbreviated to TIM) [126], and it has also been found for D-xylose isomerase [22]. All of the proteins listed in Table 2 are enzymes except, so far, for narbonin, a seed storage protein that has, to date, no identified enzymic function. Not all $(\beta\alpha)_8$ structures are identical, however, and the amino acid homology between them is very low, suggesting that this is a very stable folding pattern not dominated by a specific amino acid sequence. Some TIM barrels have an additional helix and a loop before helix 8 (H8), promoting binding of phosphate because of the macrodipole moment. In

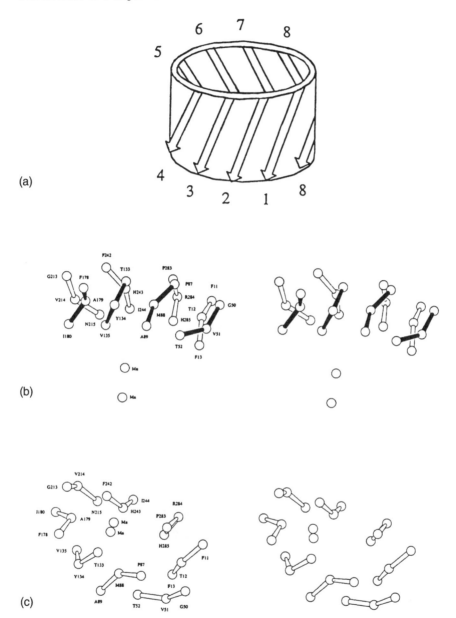

Figure 24 The β sheet interior of a (βα)₈ barrel. (a) Diagram of the strands and their inclination to the barrel axis. (b) Stereoview of the Cα positions of the amino acid residues that form the center of the (βα)₈ barrel, viewed perpendicular to the barrel axis. Amino acid residues nearer to the reader have filled connections. (c) Stereoview down the (βα)₈ barrel axis

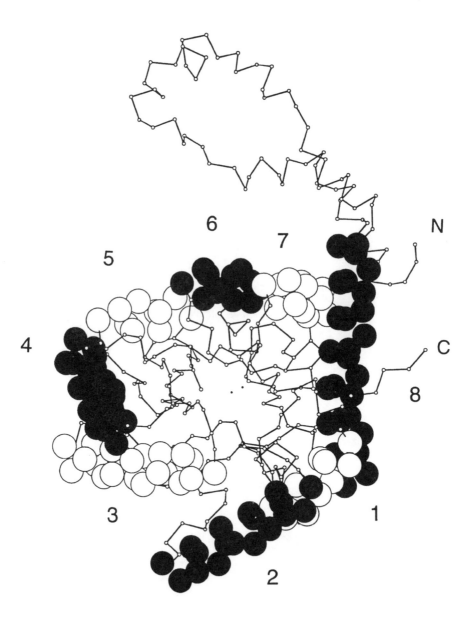

Figure 25 View down the β-barrel of D-xylose isomerase. The α helices have been illustrated with larger balls, odd-numbered helices being white, even-numbered helices black. Note that in this view down the axis from the amino to the carboxy ends of the β-strands the ordering of the helices is clockwise

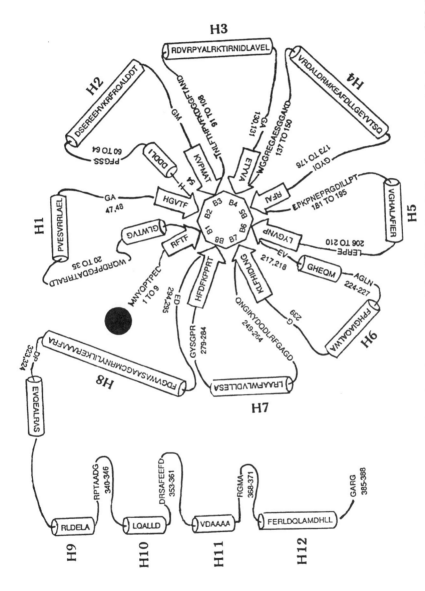

Figure 26 The amino acid sequence of D-xylose isomerase and the folding that is found in the crystal structure [22]. The amino terminus is marked by a large filled bullet. Note that there are additional small helices before helices 1, 2 and 6

muconate-lactonizing enzyme [147] one β-strand is missing, while in enolase one β-strand is inverted [151]. A protein structure distantly related to the $(\beta\alpha)_8$ barrel is that of cellobiohydrolase II [172].

In the $(\beta\alpha)_8$ structures of the proteins listed in Table 2 [129, 130, 173–178], two quantities are important for a description of the central β sheet [179]. These are the number of β-strands and the shear, illustrated in Figure 27(a). The shear is the difference along the sequence of residues when the barrel bracelet is completed (as if by superimposing the two copies of the first β-strand). All known $(\beta\alpha)_8$ barrels have eight β-strands, a shear of 8 and a tilt of the β-strands to the barrel axis of approximately 36°. A comparison of $(\beta\alpha)_8$ barrels in nine proteins showed that the mean radius of the barrel is fairly constant (6.5–7.5 Å), but the axial ratio can vary from 1.0 to 1.5. In TIM the smallest and largest diameters of the elliptical barrel are 11.5 Å and 16.5 Å (a ratio of 1.44 to 1.00). The best fit has been found to occur if the third β-strand of D-xylose isomerase is placed on the first β-strand of TIM [180]. The β sheet appears to provide a stable framework that defines active site residues that lie beyond it, as shown in Figure 27(b). If one follows the sheet to the center of the diagram, most of the active site residues lie in the center; they are underlined in this figure for clarity. The α helices that lie outside the β sheet are generally amphipathic, as shown in Figure 28. Helix 4 provides the best example of this.

The interior of the $(\beta\alpha)_8$ barrel is completely filled up by amino acid side chains, as shown in Figure 29. Solvent accessibility calculations show that water is excluded from the barrel center, although a large proportion of the rest of the barrel interior is exposed to solvent. When a space-filling model of the protein is sliced at 1 Å intervals, it is evident that the density of amino acid residues in the interior of the barrel is that normally expected for the interior of proteins.

Proceeding from the amino to the carboxyl end of the protein, the connecting loops can be described as $\beta\alpha$ or $\alpha\beta$. Some features of these loops are given in Table 3. The $\beta\alpha$-loops (at the carboxy ends of the β-strands) are generally much longer than the $\alpha\beta$-loops. These are the loops that have active site side chains in them, and, as shown in Table 3, they are the longer types of loops. These $\beta\alpha$-loops at the carboxy end of the barrel are commonly involved in subunit contacts, while $\alpha\beta$-loops are not [180]. The $(\beta\alpha)_8$ structure appears to be a very stable folding pattern [181–184]. It is found [185] that the β-strand contains a hydrophobic patch that fits into a hydrophobic pocket in the α helix.

The $(\beta\alpha)_8$ barrel is thus a large solid structure that provides a solid foundation for control of the conformation of side chains in the active site, which is generally found at the carboxy end of the β-strands and involves residues that are part of the β sheet or very near it. The side of the β sheet surface that packs against the α helices is mainly hydrophobic. The side that points towards the barrel interior is highly polar so that metal ions and substrates can bind. In several $(\beta\alpha)_8$ structures there are one or two divalent

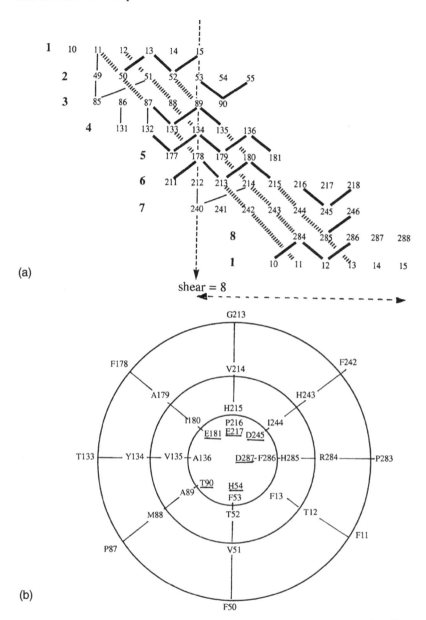

Figure 27 The β sheet in D-xylose isomerase showing the hydrogen bonding and the shear. (a) The sequence along the β-strands. The three amino acids in each strand that form the main portion of the center of the barrel are marked with hatched lines. (b) View down the barrel, showing how the active site of the enzyme lies at the carboxy end of the β-strands, presumably stabilized by the β sheet in the barrel itself. Active site groups are underlined

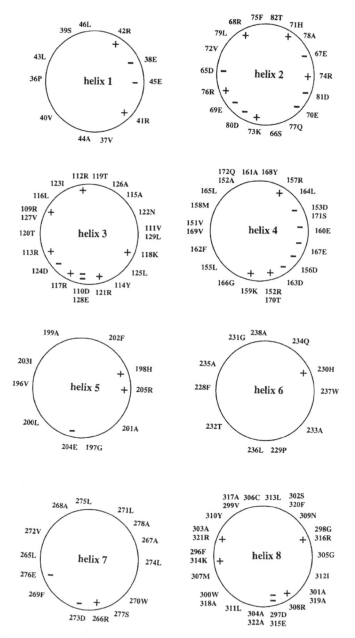

Figure 28 Helical wheels for helices 1 through 8 of the $(\beta\alpha)_8$ barrel. Positive signs are drawn near arginine, lysine and histidine and negative signs near aspartate and glutamate. Note the extent to which some helices are amphipathic with one side hydrophobic and the other hydrophilic

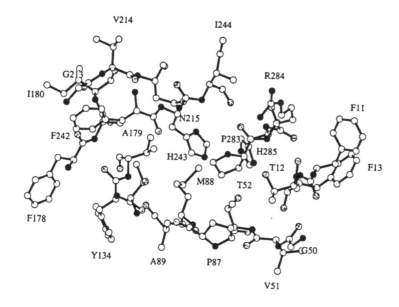

(a)

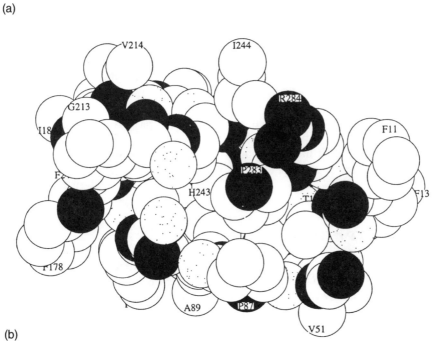

(b)

Figure 29 (a) View down the $(\beta\alpha)_8$ barrel. (b) Space-filling diagram showing that there is no space for another molecule to fit between the side chains in the $(\beta\alpha)_8$ barrel

Table 3 Numbers of amino acid residues in $\alpha\beta$- and $\beta\alpha$-loops in the $(\beta\alpha)_8$ structure of D-xylose isomerase

$\alpha\beta$	$\beta\alpha$	$\alpha\alpha$	α-Helix	Total
2	0	15	6	23
2	1	5	5	13
2	18	—	—	20
4	14	—	—	18
5	15	—	—	20
1	2	4	5	12
6	16	—	—	22
—	2	—	—	2

metal ions at the carboxy ends of the β-strands of the barrel. D-Xylose isomerase, mandelate racemase and muconate-lactonizing enzyme, for example, utilize magnesium or manganese in their active sites.

The high stability of the $(\beta\alpha)_8$ barrel is emphasized in the statement that 'you can do almost anything and still get an α/β-barrel' [186]. For example, a protein expected to contain a $(\beta\alpha)_{10}$ barrel has been prepared, but it really forms a $(\beta\alpha)_8$ barrel and the other two $\beta\alpha$-portions of the molecule form an additional dimer [187]. If the amino and carboxy ends of N-(5′-phosphoribosyl)anthranilate isomerase are moved to different loops between β-strands and α helices, there is not a large effect on enzymic activity [183]. Since this $(\beta\alpha)_8$ folding pattern is so common, its evolution has been the subject of much discussion, but no firm conclusions can yet be made [186]. It is found that $(\beta\alpha)_8$ barrel enzymes can be 'recruited for other purposes', such as the use of enolase in the τ-crystallin in the duck lens [186]. It has been noted that enolase and pyruvate kinase, both $(\beta\alpha)_8$ barrels, are consecutive enzymes in glycolysis [186, 188].

6. THE TOTAL ASSEMBLAGE: DOMAIN AND SUBUNIT INTERACTIONS

The entire enzyme can now be put together to give the functional product. The enzyme we are interested in here, D-xylose isomerase, is a tetramer. Each of its four subunits has two domains, i.e. regions with all the characteristics of an independent unit that behave like globular proteins but are covalently bound to another domain. The large amino terminal domain is the parallel eight-stranded $(\beta\alpha)_8$ barrel that has just been described. The carboxy terminal domain is an extended loop containing some α helices. This loop facilitates the aggregation of monomers to dimers and tetramers.

The four subunits may be considered as related by the coordinates

A	x	y	z
B	$-x$	$-y$	z
C	$-x$	y	$-z$
D	x	$-y$	$-z$

There are many interactions between subunits A and D related by the twofold axis parallel to the x-axis, fewer between A and B related by the twofold axis parallel to the z-axis and even fewer between A and C related by the twofold axis parallel to the y-axis. In several crystal structures of D-xylose isomerase the TIM-type barrel is formed by the first 320 amino acids and then a loop follows which helps to form the dimer. As a result, the protein can be described as a dimer of dimers. The monomer–monomer interactions and the complete tetramer are illustrated in Figure 30. The tetramers lie in the crystal at the origin of the unit cell where the symmetry of the space group, with three twofold axes of symmetry perpendicular to each other, generates the entire molecule, particularly the formation of dimers in a 'one-armed embrace'.

There is a reason why the overall size of a protein molecule is important for its correct and useful functioning. Molecules of different sizes go to different physiological locations because their size determines which barriers they can pass through. If a protein molecule needs to be large in order to be localized correctly in the body, it is better to build several smaller subunits and assemble them together than it is to make one big molecule. This is because it takes less DNA to code for a smaller molecule. Quality control is also easier because defective subunits can be discarded at less cost of protein material. In addition, if some evolutionary event is necessary for survival, it can work more readily on smaller units so that the organism can adapt better to its environment. Subunit aggregation to give an oligomer in this way reduces the surface-to-volume ratio so that the viscosity is lowered. This is a measure of the ability of protein molecules, with their surrounding water layer, to move past each other. In addition, forming a larger molecule, and thereby decreasing the number of particles, reduces the osmotic pressure. An alternative strategy for the control of proteins is to make larger precursors and cut off parts by a proteolytic mechanism to give a smaller but more active protein. This method is used particularly for digestive enzymes.

It is, of course, important to keep the subunits well cemented together. The subunits are cemented together through hydrophobic patches on their surfaces. They join with other such patches and the resulting interactions of subunits with other subunits give the total protein [189]. This, however, has to be done in a controlled way. Each subunit is folded into an apparently independent globular structure which then interacts with other similarly folded subunits. The resulting quaternary structure of a protein is often vital to its function.

What should one do if things go wrong during the assembly of the protein? The usual procedure is to call in an expert who knows how to control the

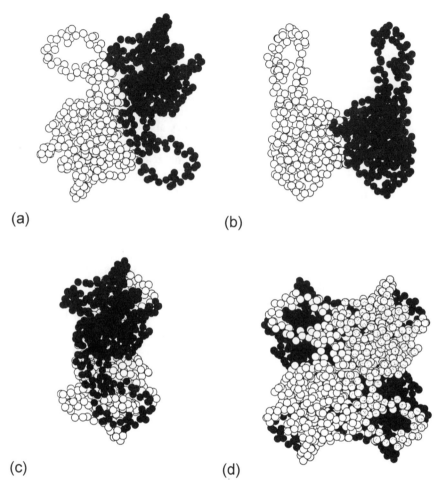

(a)   (b)

(c)   (d)

Figure 30 Dimer interactions and the total tetramer in D-xylose isomerase. The AB dimer (a), AC dimer (b), and AD dimer (c), showing the 'embracing dimer' formation (d). For clarity, in each dimer the original subunit is white and its symmetry-related subunit is black. In (d) the tetramer is shown with A white, B stippled, and C and D black. Only α-carbon atoms are shown

processes in the required manner. The question is how can the protein be effectively prevented from aggregating without blocking the forward pathway of folding to the native state? There are a number of proteins that directly influence the processes by which polypeptides attain their native structures within the cell. These include enzymes which catalyze defined steps during folding, such as disulfide isomerase and peptidylprolyl isomerase and molecular chaperones. A molecular chaperone is a protein that binds to and stabilizes the nonnative conformations of other proteins and then facilitates their correct folding by releasing them in a controlled manner [190, 191]. Chaperone molecules recognize, for example, hydrophobic surfaces in unfolded proteins, but they do not bind native proteins. These molecules provide an isolated compartment in which folding can proceed. They prevent the incorrect intermolecular association of unfolded polypeptide chains, i.e. they prevent aggregation. According to Gething and Sanibrook [192]:

> Chaperones are involved at all stages of cellular metabolism, during protein biosynthesis and maturation, in protection from environmental stress, in rearrangements of cellular macromolecules during functional cycles of assembly and disassembly, and finally in targeting proteins for degradation.

Some proteins, such as transmembrane and intracellular tyrosine phosphatases, important in signal transduction and needing to be regulated, have sequences outside the catalytic domain that serve as zip codes and direct the protein to the correct address [193].

The extent of dimer and tetramer interactions can be calculated from studies of the solvent accessibilities of the monomer, dimer and tetramer. Lee and Richards [194] showed how to measure the exposure of any atom (or amino acid residue) to solvent in the folded protein. Using protein coordinates they moved a sphere the size of a water molecule along the surface of the protein. If any atom can touch the probe it is considered accessible to solvent. The accessible surface area A for a protein with a molecular weight M can be roughly calculated [195] as $A = 5.3M^{0.76}$. Such measurements have been reported for the D-xylose isomerase from *Actinoplanes missouriensis* [143]. The total accessible surface of the protein is $48\,100\,\text{Å}^2$. The interface area between A and D is about $4700\,\text{Å}^2$, of which $1650\,\text{Å}^2$ is from the carboxy terminal loop. The carboxy terminal loop is also responsible for contacts between subunits A and C. The other subunit–subunit interface areas are of the order of $1600\,\text{Å}^2$. The contacts between subunits A and B involve pairs of barrels that face each other and touch through loops at the carboxy ends of the β-strands. The active site is, however, accessible to solvent. While there are several interactions between subunits, the packing between molecules in the crystalline state is described by very few interactions. They involve Arg340, Arg331 and Asp80, as shown in Figure 31(b). The most specific interaction is that involving hydrogen bonding of an arginine to a carboxyl group. This leaves food for thought for those who consider that the crystallization of a protein may vastly deform it, because the molecule-to-molecule interactions are so few in the crystal structure, and appear mainly to involve arginine residues. For example, the interactions between subunits A and D involve Arg368, Arg109, Arg112 and Asn107. The interaction between arginine and aspartate, for which an example is given in Figure 31(b), gives a closed planar hydrogen bond system. Backbone–backbone interactions within one subunit are often formed between a carbonyl group and the amino groups of two nearby amino acids. For example, the carbonyl group of Arg109 is hydrogen bonded to the NH groups of Arg112 and Arg 113, while the NH group of Arg112 is hydrogen bonded to the carbonyl groups of Asp108 and Asp109.

The building of supermolecules from subunits is well illustrated by the crystal structures of viruses. Viruses contain genetic information in the form of single- or double-stranded RNA or DNA. The nucleic acid is the agent by which the virus is infective. The virus must protect its nucleic acid from the environment, particularly nucleases, and therefore it constructs for itself a protective proteinaceous coat, called a capsid. Economy in the construction of the coat is necessary, since the nucleic acid must code for other proteins as well as the coat; the longer the nucleic acid that is needed to code for the coat, the larger the coat that must be constructed to cover it. To solve this dilemma the virus makes a coat composed of a large number of identical protein molecules [196] arranged in one of a limited number of possible efficient designs. This process of building a coat from smaller subunits has the additional advantage that it can be controlled biologically so that protein from defective mutants can be rejected. These protein structures and the viral nucleic acid together form a virus particle, called a virion. There must be significant protein–nucleic acid interactions, because it is essential that only the appropriate specific viral nucleic acid be incorporated in the particle. The driving energy for this assembly is provided by the formation of both protein–protein and protein–nucleic acid contacts, with the result that the stability of the viral particle is greater than that of its separate parts [197].

Caspar [197] showed that a particular virus, tomato bushy stunt virus, has a capsid structure with icosahedral symmetry. The icosahedron has a low surface-to-volume ratio, approximating that of a sphere. A regular three-dimensional icosahedron can be generated from an extended planar hexagonal net of equilateral triangles. On the surface of an icosahedron it is possible to put 60 equivalent objects in such a way that each is identically situated and related to the others by a rotational operation, yet many virus structures have a capsid with more than 60 units. To account for the additional subunits, it is necessary to relax the requirement that each subunit should have the same environment. Caspar and Klug [198] point out that 'molecular structures are not built to conform to exact mathematical concepts but, rather, to satisfy the condition that the system be in a minimum energy configuration'.

The role of water in the folding of the protein structure is highly significant [199–204]. The molecule essentially has an outer wrapping of solvent. Water,

---

Figure 31 (*opposite*) (a) Hydrogen bonds (lengths in Ångströms) formed by Arg31 bind together several portions of a subunit. (b) Packing between molecules in the crystalline state [22] also involves arginine residues, as shown. These few interactions are the only hydrogen bonds between different tetramer molecules in the crystalline state

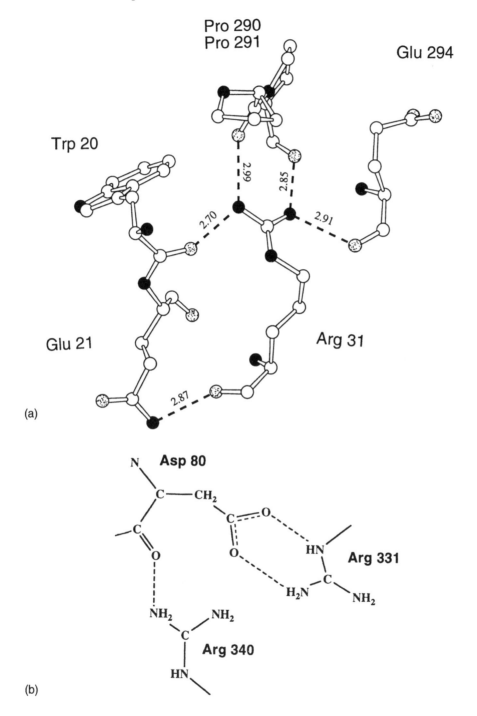

Pro 290
Pro 291

Glu 294

Trp 20

Glu 21

Arg 31

(a)

Asp 80

Arg 331

Arg 340

(b)

the solvent most generally encountered, acts both as a proton donor and as a proton acceptor. All protein–protein interactions are weak compared to protein–water and water–water interactions. Individual macromolecules capable of forming hydrogen bonds to each other do not tend to do so in aqueous solution because they can readily form stronger hydrogen bonds to water, and so do. In nonpolar solvents, however, they are more likely to form hydrogen bonds to each other. Proteins that mainly form strong intramolecular hydrogen bonds do not dissolve in water. Water molecules probably do not go into small nonpolar cavities in mutant enzymes [205].

7. THE NUTS AND BOLTS: KEEPING THE ASSEMBLAGE STABLE

The entire protein has been built, but its stability can be altered. This can be done by preparing mutant enzymes [206], checking their enzymic activities, determining their crystal structures and finding the structural effects of the mutations. This method was pioneered by Matthews. Protein stability is affected by pH, temperature, ionic strength and the concentration of any denaturant present. The total energy of a small protein is of the order of $107 \, \text{kcal mol}^{-1}$, but the stabilizing energies of proteins are only in the range 10–$20 \, \text{kcal mol}^{-1}$. Water, salts and other molecules strongly affect how proteins fold. The net free energy of stabilization of proteins is small and derives from a delicate balance between large stabilizing forces, principally due to hydrophobic interactions, and large destabilizing ones, principally due to chain entropy. The pH optimum of an enzyme can be adjusted by a judicious choice of mutants involving amino acids at or near the active site. This has been shown for subtilisin [207].

Most proteins are denatured above 50–60 °C. Some, especially thermophilic enzymes, are active at temperatures up to 80–90 °C. For the temperature-sensitive mutant of T4 lysozyme (Arg96His), except for the relacement of a partially exposed arginine by a histidine the three-dimensional structures of mutant and wild-type enzyme are almost the same. His96 lies parallel to the ring of Tyr88 and forms a hydrogen bond that destabilizes the helix from 82–90. The difference in thermal stability appears to be due to subtle changes. Deamidation of asparagine or glutamine can cause denaturation of the protein. For example, in homodimeric triosephosphate isomerase the deamidation of an asparagine residue near the interface between subunits caused complete thermal denaturation. On the other hand, when this asparagine was replaced by isoleucine, the thermal stability was increased [208]. Similarly, oxidation of some active site groups and hence loss of enzymic activity can be prevented by suitable replacements, such as methionine by a hydrophobic residue of a similar size [209]. Loops at the amino end of the barrel are more important for stability than loops at the carboxy end of the $(\beta\alpha)_8$ barrel [210]. In a study of the effects of the deletion of side chains between the enzyme and substrate, it was shown that an unpaired uncharged hydrogen bond donor or acceptor reduces the binding energy by only 0.5–1.5 kcal mol^{-1}. By contrast, an unpaired charged donor or acceptor reduces the binding energy by a further 3 kcal mol^{-1} [211].

Mutants can occur naturally as a result of random changes in DNA. It is also possible to modify the nucleic acid that codes for the protein by laboratory procedures [212]. The D-xylose from *Actinoplanes missouriensis* has been studied in this way. It is found that replacement of lysine residues (that form three hydrogen bonds to its ε-amino group) by arginine residues (that form five hydrogen bonds) gives increased stability. This is in line with the observation that enzymes from thermophilic organisms have higher arginine-to-lysine ratios than do enzymes from other organisms. Two examples [213, 214] are given in Figures 32 and 33. The one shown in Figure 32 involves binding between different subunits, while the one shown in Figure 33 involves interactions within a single subunit and these lie on the surface of the enzyme. The effects of the K253R mutant enzyme (in which lysine at position 253 has been replaced by arginine) illustrate the importance of intersubunit interactions in enzyme stability. Of the mutants prepared the most stable was the triple mutant K309R/K319R/K323R, which is more thermally stable than the wild-type enzyme. No loss of enzymic activity was found for changes of lysine to arginine at positions 253, 309, 319 and 323. In the examples in Figures 32 and 33 it can be seen that the additional hydrogen bonding that arginine can have over that possible for lysine plays an important role in increasing stability. In addition, arginine is longer but not more flexible, so that it can profitably replace water. In view of these results it is significant that the D-xylose isomerase from *T. thermophilus* has arginine residues at positions 253 and 309.

In another study [215] the metal-binding histidine at position 220 was replaced separately by serine, glutamic acid, asparagine and lysine. The last change gave an inactive enzyme and the other three only showed 0.3–0.5% activity. The H220S and H220N mutants are less stable than the wild-type enzyme and, because of the smaller size of the replacing group, bring water into position on the coordination sphere of the metal ion. This is shown in Figure 34. The H220E mutant enzyme is even less stable and binds the metal ion only poorly. Similar studies have been done for other $(\beta\alpha)_8$ structures [216].

Covalent cross-links can increase the thermal stability of a protein. One method that has been used in an attempt to stabilize a protein is to create new covalent disulfide bonds which result in cross-links between different parts of the polypeptide. The aim is to form these disulfide bonds without disrupting the overall protein conformation [217]. The method has worked well for several proteins, such as bacteriophage T4 lysozyme [218], and has increased their

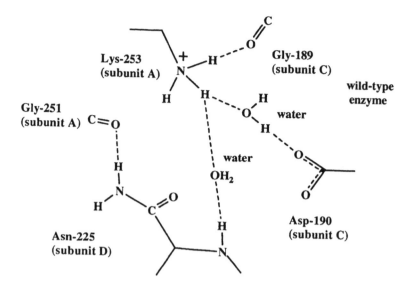

(a)

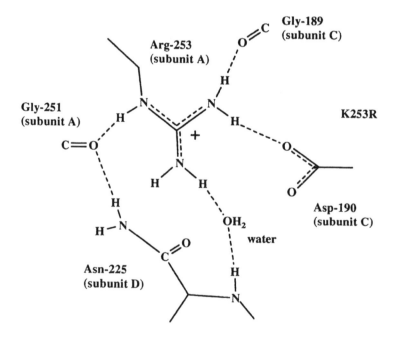

(b)

thermal stability. Computer graphics can be used to decide where to engineer a disulfide linkage. The burial of a disulfide group is favored over the burial of two cysteine residues by $0.5\,kcal\,mol^{-1}$, i.e. disulfide links are less polar from two cysteines. A disulfide bond has been engineered [219] in the $(\beta\alpha)_8$ barrel enzyme *N*-(5′-phosphoribosyl)anthranilate isomerase from *Saccharomyces cerevisiae*. This links the first and last α helices in the outer helical ring of the barrel. This cross-linked enzyme is about $1\,kcal\,mol^{-1}$ more stable than the wild-type enzyme. The metal in the active site has also been perturbed by putting lysine in that position [220].

Mutant enzymes have been produced that show a 10^7-fold switch in enzyme specificity [221]. As a result, a thermally stable lactate dehydrogenase from *Bacillus stearothermophilus* was converted to a malate dehydrogenase with twice the activity found in the malate dehydrogenase in the same organism. The mutations were [102–104]Gln–Lys–Pro to Met–Val–Ser and [236–237]Ala–Ala to Gly–Gly. This is an excellent example of the design of an enzyme.

Information from other methods than X-ray diffraction is providing us with a picture of how a protein folds. This appears to occur by a largely ordered (but not necessarily direct) pathway [11, 12]. The amino acid sequence dictates the native structure by specifying the series of steps comprising the folding pathway. Several proteins have been studied in this way. For example, barnase, an extracellular ribonuclease from *Bacillus amyloliquefaciens*, contains 116 amino acid residues in a single domain, but no cysteine or methionine residues. There are two α helices in the amino terminal region followed by a five-stranded antiparallel β sheet in the carboxy terminal region. There are three α helices and five antiparallel β sheets. Barnase can be completely unfolded by heat or chemical denaturants. Its folding mechanism has been analyzed in detail [222–226]. In early folding events an α helical region is rapidly formed in equilibrium with a more random conformation. The major α helix docks onto a β sheet to give an intermediate with a hydrophobic core. Leu63 in loop 2 packs against residues in the β sheet. Late events are the consolidation of the main hydrophobic core, the closing of loops and 'capping' of the amino ends of the α helices. Thus protein folding involves a collapse of hydrophobic regions into the interior of the molecule and formation of a stable secondary structure to provide a framework for subsequent folding. The final event is the formation of covalent interactions that stabilize the end product. There are a limited number of pathways to achieve the final folded protein involving distinct intermediates ('molten globules'). These have significant secondary structures and compact forms, but lack well-defined tertiary structures and expose more hydrophobic

Figure 32 (*opposite*) Structures of (a) the wild-type and (b) the K253R mutant D-xylose isomerase. The interactions involve three subunits

(a)

(b)

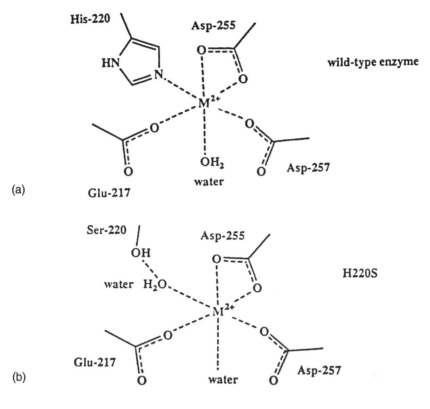

Figure 34 (a) and (b)

surface than do the fully folded molecules [227]. Adjustment to a more complicated three-dimensional structure is therefore required. The triad Asp8, Asp12 and Arg110 in barnase appears to contribute to the stability of both salt bridges [94]. The folded conformation of a protein is marginally the more stable (5–15 kcal mol^{-1}). Also, aromatic amino acids can affect the pK_a values of ionizable groups. Aromatic amino acids are found near histidine more often than expected and may have an effect on its pK_a value [228]. Another simple but interesting protein whose folding has been extensively studied, together with the folding of selected mutants, is ubiquitin [229].

Figure 33 (*opposite*) Structures of (a) the wild-type and (b) the K319R mutant D-xylose isomerase

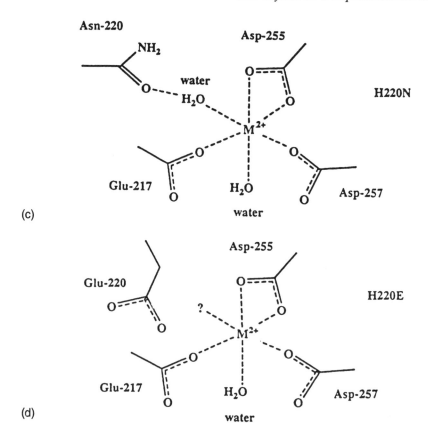

Figure 34 (*continued*)   Structures of wild-type and mutant enzymes involving His220: (a) the wild-type enzyme; (b) the H220S mutant enzyme; (c) the H220N mutant; and (d) the H220E mutant

8.   FINELY TUNING THE MECHANICAL PARTS: ASSURANCE OF PRECISE FUNCTIONING

The $(\beta\alpha)_8$ molecule has now been constructed, but what is its performance in carrying out the functions for which it was devised? How can it attract and bind substrates and convert them to products which are released to the surroundings? Each of these events must be favored energetically in the designed protein.

The binding of substrate, product or inhibitor involves, by Fischer's hypothesis [230], a complementarity of shape and electric charge. Lesk [5] lists

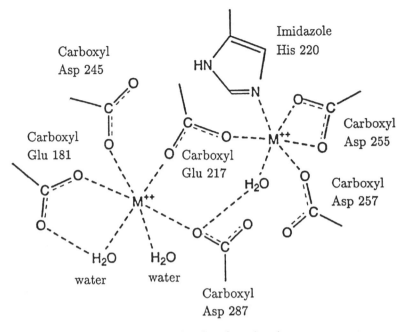

Figure 35 The active site of D-xylose isomerase

several uses of an understanding of protein structures since it is now possible, by molecular biology techniques to alter the protein. Such examples include the preparation of hybrid amino acid chains in order to test hypotheses such as those involving the Bohr effect for hemoglobin, the introduction of salt or sulfide bridges in order to make the protein more stable, and the modification of antibodies such as the changing of antigen-binding loops to human from rat species and the preparation of catalytic antibodies. Fersht and co-workers found that if a side chain is deleted leaving an unpaired hydrogen bond but no charge, the binding energy is decreased by 0.5–$1.5\,\mathrm{kcal\,mol^{-1}}$. If the environment is left charged, the binding is weakened by 3–$4\,\mathrm{kcal\,mol^{-1}}$ [211]. This understanding can also be used to design drugs that bind effectively with medicinally important protein receptors.

The overall crystal structure of D-xylose isomerase has been described in considerable detail. But the mechanism of action is still being debated. The active site (Figure 35) of the enzyme is at the base of the barrel where two divalent ions are bound, and across the diameter of the barrel lie the two amino acid residues His54 and Asp57 which hold each other firmly in place, as shown in Figure 36. The sugar substrate molecules bind across this site with one end bound to His54 and the other to one of two divalent metal ions. The metal ions

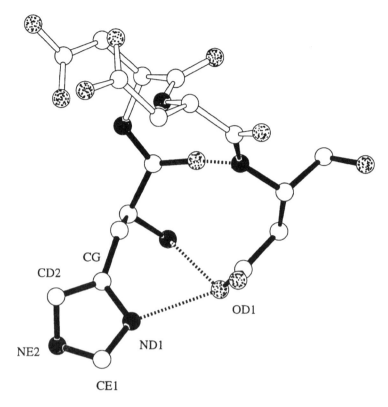

Figure 36   Interactions between His54 and Asp57 in the active site

are held in place by aspartate, glutamate and one histidine, and share the oxygen atom of one glutamate group (Glu217), as shown in Figures 37 and 38.

The mechanism of catalysis of proton transfer by this enzyme can be viewed as a proton transfer via an ene diol intermediate [231] or as a metal-assisted hydride transfer [232, 233], as shown in Figure 39. The enzyme moves the protons on O2 and C2 of D-xylose to O1 and C1 of D-xylulose. It is specific for the α-anomers of its substrates and appears to prefer pentoses to hexoses. The substrate binds in an open-chain form. Recent studies of the binding of substrates and their analogues suggest that binding can occur in either direction, with O1 binding either to or near His54 or to one of the metal ions. Our work [23] showed C1 binding near His54 and that was thought to be the proton-abstracting site (Figure 40). Later, a fluorinated derivative was synthesized that was first turned over by the enzyme to give a product that is an alkylating agent. The crystal structure of the alkylated enzyme [23] was determined and showed (Figure 16(c)) that His54 had been alkylated. But

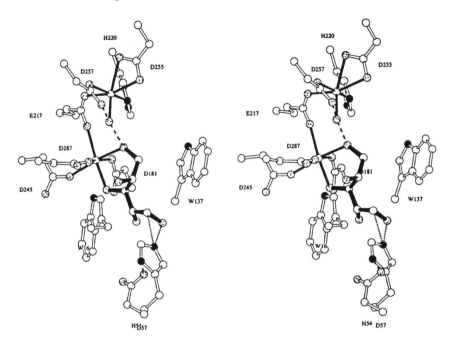

Figure 37 Stereoview of the active site of D-xylose isomerase with a bound substrate (heavy filled lines) and two metal sites. Note that two tryptophan side chains provide a greasy surface for the sugar to slide down on binding

evidence reported by others indicates that a mechanism in which substrates bind the other way round with O1 near a metal ion can give the product by a metal-assisted hydride shift. It appears that we will have to wait for more information before the mechanism of action of this enzyme can be deduced. Analogues studied at 1.6–1.9 Å resolution included vitamin C, which mimics a *cis*-ene diol complex [234]. This work indicated that Asp287 may play a significant role in the action of this enzyme.

The role of His54 is still being debated. Originally it had been suggested by us [23] that this was the catalytic base that abstracted a proton. Whitlow and co-workers, however, point out that His54 is not in an environment where any transferred proton can be protected from exchange with solvent [233]. They suggest that His54 may also act as an acid–base catalyst by removing a proton from Asp57 so that ND1 is protonated. His54 then shuttles a proton between the O1 hydroxy group of the substrate and the ring oxygen of α-D-xylose. When a ring oxygen binds to His54 it seems that the histidine must be protonated at NE2, suggesting that the role of His54 is more for ring opening than for catalysis.

The binding of substrates and their analogues has been studied in several different laboratories. Jenkins and coworkers [143, 235] studied the crystal

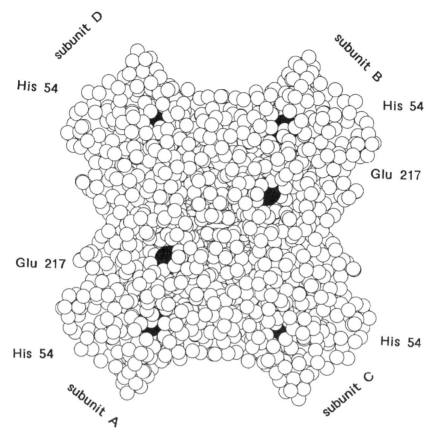

Figure 38 Locations of the active sites on the tetramer of D-xylose isomerase. Shown in black are His54 and Glu217 for each subunit, identified at the margins. There does not seem to be any interaction between the active sites of different subunits

structures of the enzyme from *Actinoplanes missouriensis* with bound xylose, xylitol and sorbitol, and found each bound in the open-chain form. These studies were at 2.2–2.6 Å resolution. The binding of the cyclic compounds 1,5-anhydroxylitol (no hydroxy group at position 2), α-D-pseudoxylose and β-D-pseudoglucose (ring oxygen replaced by methylene) was also studied. The enzyme was found to bind xylitol, sorbitol and xylose/xylulose to the metal through O2 and O4. The binding of 5-thioglucose is described as involving the metal binding to O3 and O4 [232].

The following mechanism has been proposed by Whitlow *et al.* [233]. A water molecule — initially hydrogen bonded to O1 and O2 of the substrate, a carboxyl oxygen atom of Asp257 and coordinated to the metal — transfers a proton to the carboxyl oxygen atom of Asp257 and, in so doing, is converted to

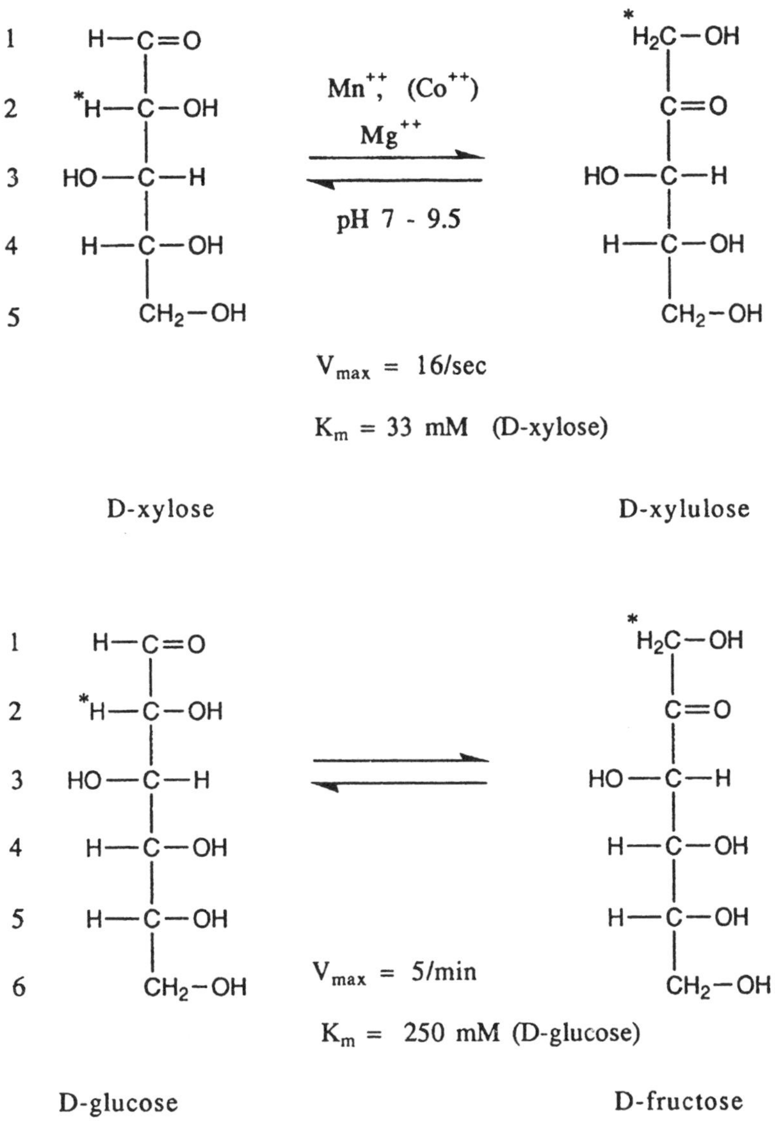

$$\text{D-xylose} \xrightarrow[\text{pH 7 - 9.5}]{\text{Mn}^{++}, \ (\text{Co}^{++}) \atop \text{Mg}^{++}} \text{D-xylulose}$$

$V_{max} = 16/\text{sec}$

$K_m = 33 \ \text{mM}$ (D-xylose)

D-xylose   D-xylulose

$V_{max} = 5/\text{min}$

$K_m = 250 \ \text{mM}$ (D-glucose)

D-glucose   D-fructose

Figure 39 (a)

a hydroxyl group. The O2 hydroxyl proton of D-xylose is then removed, and the negative charge that is formed on O2 and the hydroxyl group is stabilized by the metal. This metal ion moves 1.76 Å to coordinate directly to O1 and O2 of the sugar, displacing the water carrying the O2 proton. The C2 hydrogen atom of D-xylose then moves to the *pro-(R)*-position on C1 to give D-xylulose

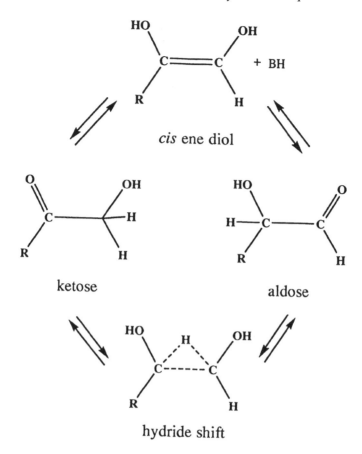

Figure 39 The reaction catalyzed by the enzyme D-xylose isomerase: (a) formulae of reactants and products; and (b) postulated mechanisms of action via a hydride shift or an ene diol intermediate. The former does not require an active site group but the latter requires an active site base to abstract a hydrogen ion and transfer it to the adjacent carbon atom

by a 1,2-hydride shift. The enzyme stabilizes the transition state by making C2 planar. The negative charge on O2 migrates to O1 during the hydride shift. Finally, D-xylulose is cyclized. The hydride shift mechanism was proposed because there did not seem to be a suitable base and because the proton transferred between C1 and C2 did not exchange with solvent. Trp16 excludes solvent. Collyer and co-workers [232] have reported the same mechanism for D-xylose isomerase. They suggest that Asp255 is repositioned when the metal ion is coordinated to O1 and O2 of the substrate. Whitlow *et al.* [233] do not see an alternative position for this metal ion. Thus the mechanism of action of this enzyme is still not clear, and is being investigated by biochemists as well as X-ray crystallographers.

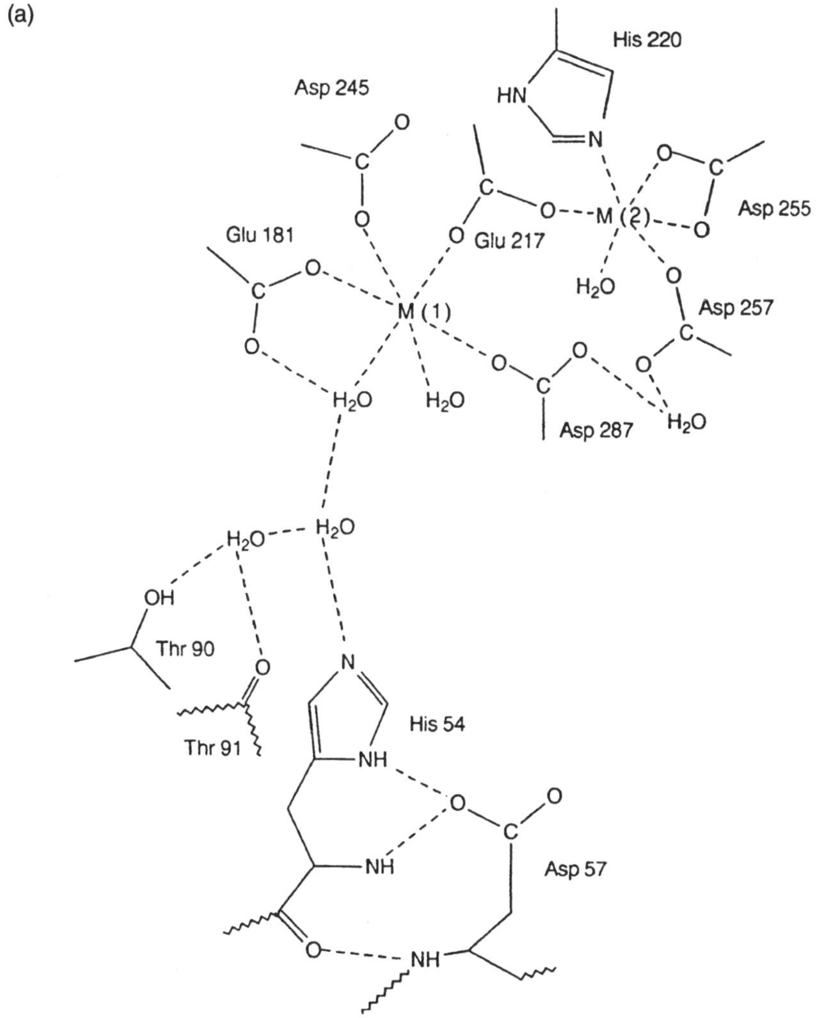

(a)

Figure 40 (a)

The crystallographic studies to 1.6 Å resolution of the same enzyme from different sources, in different space groups with different unit cell dimensions, give the same folding of the backbone and the same conformation of most of the conserved side chains. Thus we have found out how nature built this protein. We are still trying to find out how it effects the catalysis. Details of the actual enzymic mechanisms of action of proteins require study of the protein with and without bound ligand [236]. Lesk [5], however, warns us that even if

(b)

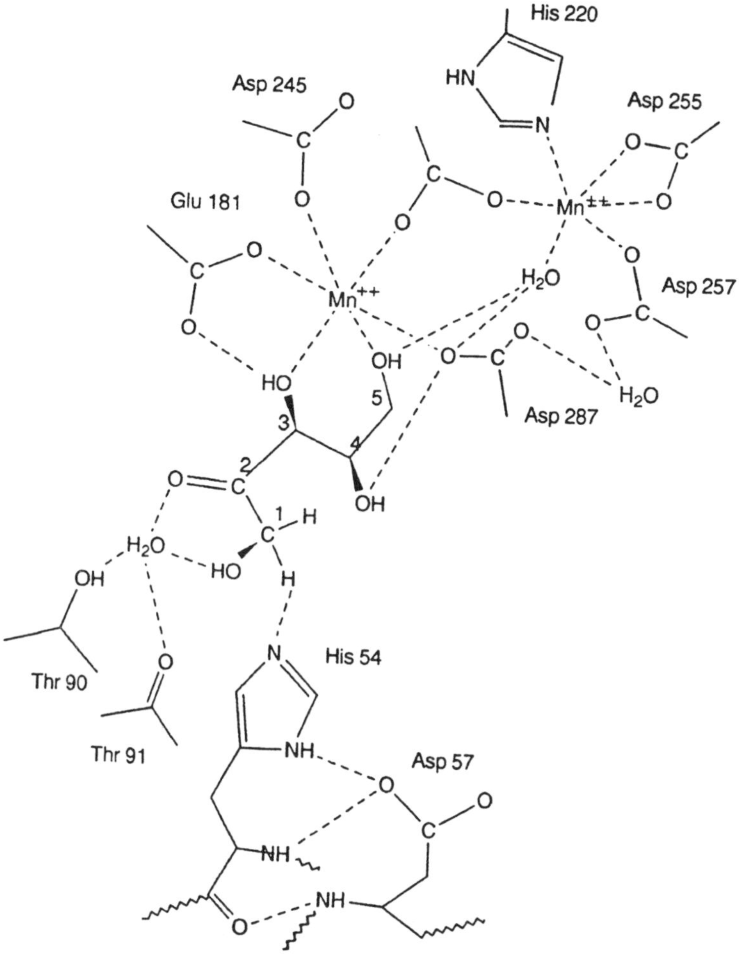

Figure 40 (*continued*) (b)

the two states are the same, we have no information on the transition between the states. This means that some time-resolved data are necessary, and the experimental hurdles for this requirement are presently high [237, 238].

9. OVERVIEW

The $(\beta\alpha)_8$ protein has been constructed as described. Now the economics and practicalities of this construction can be assessed, and this I will do by the

(c)

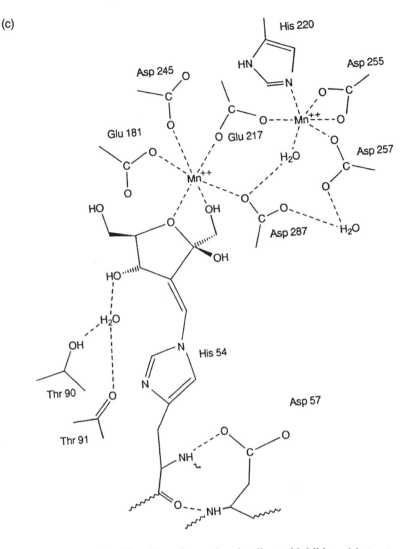

Figure 40 (*continued*) The effect of an active-site-directed inhibitor: (a) structure of the wild-type enzyme; (b) structure of the substrate-bound enzyme; and (c) structure of the alkylated enzyme [23]

analogy of building a house. A major consideration is the cost of raw materials, and it has been mentioned that the peptide unit is readily available — a cheap building block — and readily disposable. The product has to be unique because if several folding patterns were available for the same amino acid sequence, they would compete with each other and would be less controllable. The product must also be stable and durable and the $(\beta\alpha)_8$ structure fits these criteria, although further stabilization, such as by formation of disulfide bonds in appropriate places, is also possible. The amount of supervision required for the construction appears to vary with the protein being built, but assistance from, for example, chaperone molecules may aid the precision of the process, although they do not appear to be necessary for the specific example (D-xylose isomerase) described here. The time required for construction must not be too long. The product must be protected against environmental influences (fire, flood, windstorms and earthquakes in the case of a house; proteases, free radicals and certain inhibitors in the case of a protein), and this is done effectively in the $(\beta\alpha)_8$ structure. Nature decides how long the protein should exist in the cell.

The ultimate size of the protein determines its location, although some proteins have further 'zip code' information to assist in specifically localizing the protein. The protein supermolecule has to be inviting to a substrate and convert it efficiently and stereospecifically to the required product and not other useless products, while at the same time regenerating the protein in its original form for further work.

Nature designs thousands of enzymes in the manner described here for D-xylose isomerase. Account is taken of the overall thermal stability, of carefully controlled substrate binding and the elimination of product, of protection from other enzymes, of the correct solubility and of placement in the organism *in vivo*. All of these requirements, and many not listed here, are met. The palette is merely the 20 amino acids, but a wide range of electrostatic, hydrophilic or hydrophobic, space-filling and organizational considerations have been necessary for the final arrangement. The end product is unique and a marvel.

10. ACKNOWLEDGMENTS

I thank Dr H. L. Carrell for his helpful discussions. Most of the diagrams of molecules were drawn by the use of the computer program ICRVIEW [239]. This work was supported by grants CA-10925 (to J. P. G.) and CA-06927 (to the ICR) from the National Institutes of Health, and by an appropriation from the Commonwealth of Pennsylvania.

11. REFERENCES

1. F. Sanger and H. Tuppy *Biochem. J.* **49**, 463 (1951).
2. F. Sanger and H. Tuppy *Biochem. J.*, **49**, 481 (1951).
3. F. Sanger *Adv. Protein Chem.*, **7**, 1 (1952).

4. C. B. Anfinsen, E. Haber, M. Sela and F. H. White Jr *Proc. Natl. Acad. Sci. USA*, **47**, 1309 (1961).
5. A. M. Lesk *Protein Architecture. A Practical Approach*, Oxford University Press, Oxford, 1991.
6. D. L. Oxender and C. F. Fox (eds), *Protein Engineering*, Liss, New York, 1987.
7. T. E. Creighton, *Proteins*, Freeman, New York, 1984.
8. R. L. Baldwin and D. Eisenberg, in *Protein Engineering* (eds D. L. Oxender and C. F. Fox), Liss, New York, 1987, p. 127.
9. F. E. Cohen, R. M. Abarbanel, I. D. Kuntz and R. J. Fletterick, *Biochemistry*, **22**, 4894 (1983).
10. C. Levinthal, *J. Chem. Phys.*, **85**, 44 (1968).
11. H. Roder and G. Elöve, in *Mechanisms of Protein Folding* (ed. R. H. Pain), Oxford University Press, New York, 1994, p. 26.
12. M. S. Briggs and H. Roder, *Proc. Natl. Acad. Sci. USA*, **89**, 2017 (1992).
13. R. L. Baldwin, *Annu. Rev. Biochem.*, **44**, 453 (1975).
14. B. Gutte and R. B. Merrifield, *J. Am. Chem. Soc.*, **91**, 501 (1969).
15. T. E. Creighton and D. P. Goldenberg, *J. Mol. Biol.*, **179**, 497 (1984).
16. F. C. Bernstein, T. F. Koetzle, G. J. B. Williams, E. F. Meyer Jr, M. D. Brice, J. R. Rodgers, O. Kennard, T. Shimanouchi and M. Tasumi, *J. Mol. Biol.*, **112**, 535 (1977).
17. G. Párraga, S. J. Horvath, A. Eisen, W. E. Taylor, L. Hood, E. T. Young and R. E. Klevit, *Science*, **241**, 1489 (1988).
18. N. P. Pavletich and C. O. Pabo, *Science*, **252**, 809 (1991).
19. K. U. Linderström-Lang, *Proteins and Enzymes*, Stanford University Press, Stanford, CA, 1952, p. 58.
20. J. D. Bernal, *Faraday Discuss. Chem. Soc.*, **25**, 7 (1958).
21. J. M. A. Smith, R. F. D. Stansfield, G. C. Ford, J. L. White and P. M. Harrison, *J. Chem. Educ.*, **65**, 1083 (1988).
22. H. L. Carrell, B. H. Rubin, T. J. Hurley and J. P. Glusker, *J. Biol. Chem.*, **259**, 3230 (1984).
23. H. L. Carrell, J. P. Glusker, V. Burger, F. Manfre, D. Tritsch and J.-F. Biellmann, *Proc. Natl. Acad. Sci. USA*, **86**, 4440 (1989).
24. J. S. Richardson and D. C. Richardson, *Protein Engineering* (eds D. L. Oxender, and C. F. Fox), Liss, New York, 1987, p. 149.
25. G. D. Fasman, (ed.), *Prediction of Protein Structure and the Principles of Protein Conformation*, Plenum Press, New York, 1989.
26. F. M. Richards, *Sci. Am.* **264**, 54 (1991).
27. G. Nemethy and H. A. Scheraga, *Q. Rev. Biophys.*, **10**, 239 (1977).
28. M. G. Rossmann and P. Argos, *Annu. Rev. Biochem.*, **50**, 497 (1981).
29. D. Voet and J. G. Voet, *Biochemistry*, Wiley, New York, 1990.
30. G. E. Schulz and R. H. Schirmer, *Principles of Protein Structure*, Springer, New York, 1979.
31. L. Pauling and J. Sherman, *J. Chem. Phys.*, **1**, 606 (1933).
32. E. N. Baker and R. E. Hubbard, *Prog. Biophys. Mol. Biol.*, **44**, 97 (1984).
33. W. Kauzmann, *Adv. Protein Chem.*, **14**, 1 (1959).
34. C. Tanford, *The Hydrophobic Effect: Formation of Micelles and Biological Membranes*, Wiley, New York, 1973.
35. F. M. Richards, *Annu. Rev. Biophys. Bioeng.*, **6**, 151 (1977).
36. F. M. Richards, *J. Mol. Biol.*, **82**, 1 (1974).
37. B. W. Matthews, *J. Mol. Biol.*, **33**, 491 (1968).

38. B. Magdoff-Fairchild, P. H. Swerdlow and J. F. Bertles, *Nature (London)*, **239**, 217 (1972).
39. J. T. Finch, M. F. Perutz, J. F. Bertles and J. Dobler, *Proc. Natl. Acad. Sci. USA*, **70**, 718 (1973).
40. W. T. Astbury and H. J. Woods, *Nature (London)*, **126**, 913 (1930).
41. W. T. Astbury and W. A. Sisson, *Proc. R. Soc. London, Ser. A*, **150**, 333 (1935).
42. G. N. Ramachandran, and V. Sasisekharan, *J. Mol. Biol.*, **7**, 95 (1963).
43. G. N. Ramachandran and V. Sasisekharan, *Adv. Protein Chem.*, **23**, 283 (1968).
44. L. Pauling, R. B. Corey and H. R. Branson, *Proc. Natl. Acad. Sci. USA*, **37**, 205 (1951).
45. F. Hussain, *Trends Biochem. Sci.*, **1**, N37 (1976).
46. W. G. J. Hol, P. T. van Duijnen and H. J. C. Berendsen, *Nature (London)*, **273**, 443 (1978).
47. A. Wada, *Adv. Biophys.*, **9**, 1 (1976).
48. R. Fairman, K. R. Shoemaker, G. J. York, J. M. Stewart and R. L. Baldwin, *Proteins: Struct. Func. Genet.*, **5**, 1 (1989).
49. W. G. Hol, *Prog. Biophys. Mol. Biol.*, **45**, 149 (1985).
50. J. S. Richardson and D. C. Richardson, *Science*, **240**, 1648 (1988).
51. R. P. Sheridan R. M. Levy and F. R. Salemme, *Proc. Natl. Acad. Sci. USA*, **79**, 4545 (1982).
52. J. W. Pflugrath and F. A. Quiocho, *Nature (London)*, **314**, 257 (1985).
53. J. Sancho, L. Serrano and A. R. Fersht, *Biochemistry*, **31**, 2253 (1992).
54. P. Chakrabarti, *Protein Eng.*, **7**, 471 (1994).
55. R. Henderson and P. T. Unwin, *Nature (London)*, **257**, 28 (1975).
56. M. Schiffer and A. B. Edmundson, *Biophys. J.*, **7**, 121 (1967).
57. E. K. O'Shea, J. D. Klemm, P. S. Kim and T. Alber, *Science*, **254**, 539 (1991).
58. T. E. Ellenberger, C. J. Brandl, K. Struhl and S. C. Hanson, *Cell*, **71**, 1223 (1992).
59. A. R. Fersht and L. Serrano, *Curr. Opin. Struct. Biol.*, **3**, 75 (1993).
60. L. Pauling and R. B. Corey, *Proc. Natl. Acad. Sci. USA*, **37**, 251 (1951).
61. C. Brändén and J. Tooze, *Introduction to Protein Structures*, Garland, New York, 1991.
62. J. S. Richardson, *Nature (London)*, **268**, 495 (1977).
63. S. Lifson and C. Sander, *Nature (London)*, **282**, 109 (1979).
64. I. L. Karle, J. Karle, D. Mastropaolo, A. Camerman and N. Camerman, *Acta Crystallogr.*, Part B, **39**, 625 (1983).
65. T. J. Oldfield and R. E. Hubbard, *Proteins: Struct. Func. Genet.*, **18**, 324 (1994).
66. J. S. Richardson, G. D. Getzoff and D. Richardson, *Proc. Natl. Acad. Sci. USA*, **75**, 2574 (1978).
67. C. Chothia, *J. Mol. Biol.*, **75**, 295 (1973).
68. J. Leszczynski and G. Rose, *Science*, **234**, 849 (1986).
69. C. M. Venkatachalam, *Biopolymers*, **6**, 1425 (1968).
70. P. N. Lewis, F. A. Momany and H. A. Scheraga, *Biochim. Biophys. Acta*, **303**, 221 (1973).
71. J. A. Smith and L. G. Pease, *CRC Crit. Rev. Biochem.*, **8**, 315 (1980).
72. C. M. Wilmot and J. M. Thornton, *J. Mol. Biol.*, **203**, 221 (1988).
73. F. A. Momany, R. F. McGuire, A. W. Burgess and H. A. Scheraga, *J. Phys. Chem.*, **79**, 2361 (1975).
74. T. Ashida, Y. Tsunogae, I. Tanaka and T. Yamane, *Acta Crystallogr., Part B*, **43**, 212 (1987).
75. R. B. Corey and L. Pauling, *Proc. R. Soc. London, Ser. B*, **141**, 10 (1953).

76. J. R. Singh and J. M. Thornton, *Atlas of Protein Side-Chain Interactions*, Oxford University Press, Oxford, 1992.
77. Y. Nazaki and C. Tanford, *J. Biol. Chem.*, **246**, 2211 (1971).
78. J. Novotny, R. Bruccoleri and M. Karplus, *J. Mol. Biol.*, **177**, 787 (1984).
79. D. Eisenberg, R. M. Weiss and T. C. Tergwilliger, *Proc. Natl. Acad. Sci. USA*, **81**, 140 (1984).
80. J. Kyte and R. F. Doolittle, *J. Mol. Biol.*, **157**, 105 (1982).
81. M. Ramanadham, V. S. Jakkal and R. Chidambaram, *FEBS Lett.*, **323**, 203 (1993).
82. M. Levitt, *J. Mol. Biol.*, **145**, 251 (1981).
83. J. F. Brandts, H. R. Halvorson and M. Brennan, *Biochemistry*, **14**, 4953 (1975).
84. R. M. C. Dawson, D. C. Elliott, W. H. Elliott and K. M. Jones, *Data for Biochemical Research*, 2nd edn, Oxford University Press, Oxford, 1969, p. 1.
85. C. J. Carrell, H. L. Carrell, J. Erlebacher and J. P. Glusker, *J. Am. Chem. Soc.*, **110**, 8651 (1988).
86. C. L. Borders Jr, J. A. Broadwater, P. A. Bekeny, J. E. Salmon, A. S. Lee, A. M. Eldridge, and V. B. Pett, *Protein Sci.*, **3**, 541 (1994).
87. L. Shimoni and J. P. Glusker, *Protein Sci.*
88. M. F. Perutz, A. M. Gronenborn, G. M. Clore, J. H. Fogg and D. T.-B. Shih, *J. Mol. Biol.*, **183**, 491 (1985).
89. E. R. Blout, A. B. Lozé, S. M. Bloom and G. D. Fasman, *J. Am. Chem. Soc.*, **82**, 3787 (1960).
90. P. Y. Chou and G. D. Fasman, *Biochemistry*, **13**, 222 (1974).
91. P. Y. Chou and G. D. Fasman, *Annu. Rev. Biochem.*, **47**, 251 (1978).
92. M. Blaber, K. Zhang and B. W. Matthews, *Science*, **260**, 1637 (1986).
93. W. R. Taylor (ed.), *Patterns in Protein Sequence and Structure*, Springer, Berlin, 1992.
94. A. Horowitz and A. R. Fersht, *J. Mol. Biol.*, **224**, 733 (1992).
95. W. R. Taylor and C. A. Orengo, *J. Mol. Biol.*, **208**, 1 (1989).
96. T. E. Creighton, *Proc. Natl. Acad. Sci. USA*, **85**, 5082 (1988).
97. T. E. Creighton, *Prog. Biophys. Mol. Biol.*, **33**, 231 (1978).
98. C. A. Orengo, N. P. Brown and W. R. Taylor, *Proteins: Struct. Func. Genet.*, **14**, 139 (1992).
99. J. S. Richardson, *Adv. Protein Chem.*, **34**, 167 (1981).
100. J. S. Richardson, *Methods Enzymol.*, **115**, 349 (1985).
101. C. Chothia, *Nature (London)*, **254**, 304 (1975).
102. M. Levitt and C. Chothia, *Nature (London)*, **261**, 552 (1976).
103. J. S. Richardson and D. C. Richardson, in *Prediction of Protein Structure and the Principles of Protein Conformation* (ed. G. D. Fasman), Plenum Press, New York 1989, p. 1.
104. M. G. Rossmann and P. Argos, *J. Mol. Biol.*, **109**, 99 (1977).
105. S. T. Rao and M. G. Rossmann, *J. Mol. Biol.*, **76**, 241 (1971).
106. K. Nagano, *J. Mol. Biol.*, **109**, 235 (1977).
107. F. H. C. Crick, *Acta Crystallogr.*, **6**, 689 (1953).
108. C. Chothia, M. Levitt and D. Richardson, *J. Mol. Biol.*, **145**, 215 (1981).
109. C. Chothia, *Annu. Rev. Biochem.*, **53**, 537 (1984).
110. A. V. Efimov, *J. Mol. Biol.*, **134**, 23 (1979).
111. T. J. Richmond and F. M. Richards, *J. Mol. Biol.*, **119**, 537 (1978).
112. J. C. Kendrew, R. E. Dickerson, B. E. Strandberg, R. G. Hart, D. R. Davies, D. C. Phillips and V. C. Shore, *Nature (London)*, **185**, 422 (1960).
113. M. J. E. Sternberg, and J. M. Thornton, *J. Mol. Biol.*, **110**, 269 (1977).

114. M. J. E. Sternberg and J. M. Thornton, *J. Mol. Biol.*, **115**, 1 (1977).
115. S. Lifson and C. Sander, *J. Mol. Biol.*, **139**, 627 (1980).
116. C. Chothia, M. Levitt and D. Richardson, *Proc. Natl. Acad. Sci. USA*, **74**, 4130 (1977).
117. J. Richardson, *Proc. Natl. Acad. Sci. USA*, **73**, 2619 (1976).
118. M. Levitt and C. Chothia, *Nature (London)*, **261**, 552 (1976).
119. M. D. Yoder, N. T. Keen and F. Jurnak, *Science*, **260**, 1503 (1993).
120. L. Holm, C. Ouzounis, C. Sander, G. Tuparev and G. Vriend, *Protein Sci.*, **1**, 1691 (1992).
121. L. Holm and C. Sander, *Proteins: Struct. Func. Genet.*, **19**, 165 (1994).
122. A. G. Murzin, *Nature (London)*, **360**, 635 (1992).
123. E. M. Mitchell, P. J. Artymiuk, D. W. Rice and P. Willet, *J. Mol. Biol.*, **212**, 151 (1989).
124. G. Vriend and C. Sander, *Proteins: Struct. Func. Genet.*, **11**, 52 (1991).
125. M. B. Swindells, C. A. Orengo, D. T. Jones, L. H. Pearl and J. M. Thornton, *Nature (London)*, **362**, 299 (1993).
126. D. W. Banner, A. C. Bloomer, G. A. Petsko, D. C. Phillips, C. I. Pogson, I. A. Wilson, P. H. Corran, A. J. Furth, J. D. Milman, R. E. Offord, J. D. Priddle and S. G. Waley, *Nature (London)*, **255**, 609 (1975).
127. H. Muirhead, *Trends Biochem. Sci.*, **8**, 326 (1983).
128. C. Chothia, *Nature (London)*, **333**, 598 (1988).
129. T. Niermann and K. Kirschner, *Protein Eng.*, **4**, 359 (1991).
130. G. K. Farber and G. A. Petsko, *Trends Biochem. Sci.*, **15**, 228 (1990).
131. T. Alber, D. W. Banner, A. C. Bloomer, G. A. Petsko, D. C. Phillips, P. C. Rivers, and I. A. Wilson, *Philos. Trans. R. Soc. London, Ser. B*, **293**, 159 (1981).
132. S. C. Mande, V. Mainfroid, K. H. Kalk, K. Goraj, J. A. Martial and W. G. J. Hol, *Protein Sci.*, **3**, 810 (1994).
133. K. V. R. Kishan, J. P. Zeelen, M. E. M. Noble, T. V. Borchert and R. K. Wierenga, *Protein Sci.*, **3**, 779 (1994).
134. M. E. M. Noble, J. P. Zeelen, R. K. Wierenga, V. Mainfroid, K. Goraj, A. C. Gohimont and J. A. Martial, *Acta Crystallogr., Part D*, **49**, 403 (1993).
135. E. Lolis, T. Alber, R. C. Davenport, D. Rose, F. C. Hartman and G. A. Petsko, *Biochemistry*, **29**, 6609 (1990).
136. C. C. Hyde, S. A. Ahmed, E. A. Padlan, E. W. Miles and D. R. Davies, *J. Biol. Chem.*, **26**, 17 857 (1988).
137. M. R. Huble, C. R. Matthews, F. E. Cohen, I. D. Kuntz, A. Toumadje and W. C. Johnson Jr, *Proteins: Struct. Func. Genet.*, **2**, 210 (1987).
138. K. Yutani, K. Ogasahara and Y. Sugino, *J. Mol. Biol.*, **144**, 455 (1980).
139. J. P. Priestle, M. G. Grütter, J. L. White, M. G. Vincent, M. Kania, E. Wilson, T. S. Jardetzky, K. Kirschner and J. N. Jansonius, *Proc. Natl. Acad. Sci. USA*, **84**, 5690 (1987).
140. M. Wilmanns, J. P. Priestle, T. Niermann and J. N. Jansonius, *J. Mol. Biol.*, **223**, 477 (1992).
141. G. K. Farber, G. A. Petsko and D. Ringe, *Protein Eng.*, **1**, 459 (1987).
142. K. Henrick, D. M. Blow, H. L. Carrell and J. P. Glusker, *Protein Eng.*, **1**, 467 (1987).
143. J. Jenkins, J. Janin, F. Rey, M. Chiadmi, H van Tilbeurgh, I. Lasters, M. De Maeyer, D. Van Belle, S. J. Wodak, M. Lauwereys, P. Stanssens, N. T. Mrabet, J. Snauwaert, G. Matthyssens and A.-M. Lambeir, *Biochemistry*, **31**, 5449 (1992).
144. D. I. Stuart, M. Levine, H. Muirhead and D. K. Stammers, *J. Mol. Biol.*, **134**, 109 (1979).

145. H. Muirhead, D. A. Clayden, D. Barford, C. G. Lorimer, L. A. Fothergill-Gilmore, E. Schiltz and W. Schmitt, *EMBO J.*, **5**, 475 (1986).
146. D. J. Neidhart, P. L. Howell, G. A. Petsko, V. M. Powers, R. Li, G. L. Kenyon and J. A. Gerlt, *Biochemistry*, **30**, 9264 (1991).
147. A. Goldman, D. L. Ollis and T. A. Steitz, *J. Mol. Biol.*, **194**, 143 (1987).
148. H. Hoier, M. Schlömann, A. Hammer, J. P. Glusker, H. L. Carrell, A. Goldman, J. J. Stezowski and U. Heinemann, *Acta Crystallogr., Part D*, **50**, 75 (1994).
149. L. Lebioda, B. Stec and J. M. Brewer, *J. Biol. Chem.*, **264**, 3685 (1989).
150. B. Stec and L. Lebioda, *J. Mol. Biol.*, **211**, 235 (1990).
151. J. Sygusch, D. Beaudry and M. Allaire, *Proc. Natl. Acad. Sci. USA*, **84**, 7846 (1987).
152. C. Klein and G. E. Schulz, *J. Mol. Biol.*, **217**, 737 (1991).
153. Y. Matsuura, M. Kusunoki, W. Harada, N. Tanaka, Y. Iga, N. Yasuoka, H. Toda, K. Narita and M. Kakudo, *J. Biochem. (Tokyo)*, **87**, 1555 (1980).
154. R. L. Brady, A. M. Brzozowski, Z. S. Derewenda, E. J. Dodson and G. G. Dodson, *Acta Crystallogr., Part B*, **47**, 527 (1991).
155. B. Mikami, M. Sato, T. Shibata, M. Hirose, S. Aibara, Y. Katsube and Y. Morita, *J. Biochem. (Tokyo)*, **112**, 541 (1992).
156. Y. Lindqvist and C.-I. Brändén, *J. Mol. Biol.*, **143**, 201 (1980).
157. Y. Lindqvist and C.-I. Brändén, *Proc. Natl. Acad. Sci. USA*, **82**, 6855 (1985).
158. Y. Lindqvist, *J. Mol. Biol.*, **209**, 151 (1989).
159. G. Schneider, Y. Lindqvist, C.-I. Brändén and G. Lorimer, *EMBO J.*, **5**, 3409 (1986).
160. P. M. G. Curmi, D. Casico, R. M. Sweet, D. Eisenberg and H. Schreuder, *J. Biol. Chem.*, **267**, 16 980 (1992).
161. Z.-X. Xia, N. Shamala, P. H. Bethge, L. W. Lim, H. D. Bellamy, N. H. Xuong, F. Lederer and F. S. Mathews, *Proc. Natl. Acad. Sci. USA*, **84**, 2629 (1987).
162. Z.-X. Xia and F. S. Mathews, *J. Mol. Biol.*, **212**, 837 (1990).
163. L. W. Lim, N. Shamala, F. S. Mathews, D. J. Steenkamp,, R. Hamlin and N. H. Xuong, *J. Biol. Chem.*, **261**, 15 140 (1986).
164. H. D. Bellamy, L. W. Lim and F. S. Mathews, *J. Biol. Chem*, **264**, 11 887 (1989).
165. I. M. Mavridis, M. H. Hatada, A. Tulinsky and L. Lebioda, *J. Mol. Biol.*, **162**, 419 (1982).
166. D. K. Wilson, K. M. Bohren, K. H. Gabbay and F. A. Quiocho, *Science*, **257**, 81 (1992).
167. J. M. Rondeau, F. Tête-Favier, A. Podjarny, J. M. Reymann, P. Barth, J. F. Biellmann and D. Moras, *Nature (London)*, **355**, 469 (1992).
168. D. K. Wilson, F. B. Rudolph and F. A. Quiocho, *Science*, **252**, 1278 (1992).
169. G. Hester, O. Brenner-Holzach, F. A. Rossi, M. Struck-Donatz, K. H. Winterhalter, J. D. G. Smit and K. Piontek, *FEBS Lett.*, **292**, 237 (1991).
170. M. Hennig, B. Schlesier, Z. Dauter, S. Pfeffer, C. Betzel, W. E. Höhne and K. S. Wilson, *FEBS Lett.*, **306**, 80 (1992).
171. K.-C. Chou and L. Carlacci, *Proteins: Struct. Funct. Genet.*, **9**, 280 (1991).
172. J. Rouvinen, T. Bergfors, T. Teeri, J. K. C. Knowles and T. A. Jones, *Science*, **249**, 380 (1990).
173. L. Lebioda, M. H. Hatada, A. Tulinsky and I. Mavridis, *J. Mol. Biol.*, **162**, 445 (1982).
174. M. Wilmanns, C. C. Hyde, D. R. Davies, K. Kirschner and J. N. Jansonius, *Biochemistry*, **30**, 9161 (1991).
175. L. Lasters, S. J. Wodak, P. Alard and E. van Cutsem, *Proc. Natl. Acad. Sci. USA*, **85**, 3338 (1988).

176. A. M. Lesk, C.-I. Brändén and C. Chothia, *Proteins: Struct. Func. Genet.*, **5**, 139 (1989).
177. D. J. Neidhart, G. L. Kenyon, J. A. Gerlt and G. A. Petsko, *Nature (London)*, **347**, 692 (1990).
178. L. Lebioda and B. Stec, *Nature (London)*, **333**, 683 (1988).
179. A. D. McLachlan, *J. Mol. Biol.*, **128**, 49 (1979).
180. J.-P. Y. Scheerlinck, I. Lasters, M. Claessens, M. De Maeyer, F. Pio, P. Delhaise and S. J. Wodak, *Proteins: Struct. Func. Genet.*, **12**, 299 (1992).
181. A. Godzik, J. Skolnick and A. Kolinski, *Proc. Natl. Acad. Sci. USA*, **89**, 2629 (1992).
182. R. Rudolph, R. Siebendritt and T. Kiefhaber, *Protein Sci.*, **1**, 654 (1992).
183. K. Luger, U. Hommel, M. Herold, J. Hofsteenge and K. Kirschner, *Science*, **243**, 206 (1989).
184. J. Eder and K. Kirschner, *Biochemistry*, **31**, 3617 (1992).
185. P. A. Rice, A. Goldman and T. A. Steitz, *Proteins: Struct. Func. Genet.*, **8**, 334 (1990).
186. G. K. Farber, *Curr. Opin. Struct. Biol.*, **3**, 409 (1993).
187. K. Luger, H. Szadowski and K. Kirschner, *Protein Eng.*, **3**, 249 (1990).
188. N. H. Horowitz, *Proc. Natl. Acad., Sci. USA*, **31**, 153 (1945).
189. W. A. Lim and R. T. Sauer, *J. Mol. Biol.*, **219**, 359 (1991).
190. F.-U. Hartl, R. Hlodan and T. Langer, *Trends Biochem. Sci.*, **19**, 20 (1994).
191. J. Martin, T. Langer, R. Boteva, A. Schramel, A. L. Horwich and F. U. Hartl, *Nature (London)*, **352**, 36 (1991).
192. M.-J. Gething and J. Sanibrook, *Nature (London)*, **355**, 33 (1992).
193. L. J. Mauro and J. E. Dixon, *Trends Biochem. Sci.*, **19**, 151 (1994).
194. B. K. Lee and F. M. Richards, *J. Mol. Biol.*, **55**, 379 (1971).
195. S. Miller, A. M. Lesk and J. Janin, *Nature (London)*, **328**, 834 (1987).
196. F. H. C. Crick and J. D. Watson, *Nature (London)*, **177**, 473 (1956).
197. D. L. D. Caspar, *Trans. N. Y. Acad. Sci.*, **22**, 519 (1960).
198. D. L. D. Caspar and A. Klug, *Cold Spring Harbor Symp. Quant. Biol.*, **27**, 1 (1962).
199. F. Meyer, *Protein Sci.*, **1**, 1543 (1992).
200. R. Malin, P. Zielenkiewicz and W. Saenger, *J. Biol. Chem.*, **266**, 4848 (1991).
201. H. Savage and A. Wlodawer, *Methods Enzymol.*, **127**, 162 (1986).
202. G. Otting, E. Liepinah and K. Wüthrich, *Science*, **254**, 974 (1991).
203. R. Wolfenden, L. Anderson, P. M. Cullis and C. C. B. Southgate, *Biochemistry*, **20**, 849 (1981).
204. R. Parthasarathy, S. Chaturvedi and K. Go, *Proc. Natl. Acad. Sci. USA*, **87**, 871 (1990).
205. R. Wolfenden and A. Radzicka, *Science*, **265**, 936 (1994).
206. W. J. Rutter, S. J. Gardell, C. Roczniak, D. Hilvert, S. Sprang, R. J. Fletterick and C. S. Craik, in *Protein Engineering* (eds D. L. Oxender and C. F. Fox), Liss, New York, 1987, p. 257.
207. M. J. E. Sternberg, F. R. F. Hayes, A. J. Russell, P. G. Thomas and A. R. Fersht, *Nature (London)*, **330**, 86 (1987).
208. T. J. Ahern, J. I. Casal, G. A. Petsko and A. M. Klibanov, *Proc. Natl. Acad. Sci. USA*, **84**, 675 (1987).
209. D. A. Estell, T. P. Graycar and J. A. Wells, *J. Biol. Chem.*, **260**, 6518 (1985).
210. R. Urfer and K. Kirschner, *Protein Sci.*, **1**, 31 (1992).
211. A. R. Fersht, J.-P. Shi, J. Knill-Jones, D. M. Lowe, A. J. Wilkinson, D. M. Blow, P. Brick, P. Carter, M. M. Y. Way and G. Winter, *Nature (London)*, **314**, 235 (1985).

212. M. J. Zoller and M. Smith, *DNA*, **3**, 479 (1984).
213. A.-M. Lambeir, M. Lauwereys, P. Stanssens, N. T. Mrabet, J. Snauwaert, H. van Tilbeurgh, G. Matthyssens, I. Lasters, M. De Maeyer, S. J. Wodak, J. Jenkins, M. Chiadmi and J. Janin, *Biochemistry*, **31**, 5459 (1992).
214. H. van Tilbeurgh, J. Jenkins, M. Chiadmi, J. Janin, S. J. Wodak, N. T. Mrabet and A.-M. Lambeir, *Biochemistry*, **31**, 5467 (1992).
215. J. Cha, Y. Cho, R. D. Whitaker, H. L. Carrell, J. P. Glusker, P. A. Karplus and C. A. Batt, *J. Biol. Chem.*, **269**, 2687 (1994).
216. A. M. Beasty, M. R. Huble, J. T. Manz, T. Stackhouse, J. J. Onuffer and C. R. Matthews, *Biochemistry*, **25**, 2965 (1986).
217. A. J. Saunders, G. B. Young and G. J. Pielak, *Protein Sci.*, **2**, 1183 (1993).
218. L. J. Perry and R. Wetzel, *Science*, **226**, 555 (1984).
219. J. Eder and M. Wilmanns, *Biochemistry*, **31**, 4437 (1992).
220. K. N. Allen, A. Lavie, A. Glasfield, T. N. Tanada, D. P. Gerrity, S. C. Carlson, G. K. Farber, G. A. Petsko and D. Ringe, *Biochemistry*, **33**, 1488 (1994).
221. H. Wilks, D. J. Halsall, T. Atkinson, W. N. Chia, A. R. Clarke and J. J. Holbrook, *Biochemistry*, **29**, 8587 (1990).
222. A. R. Fersht, *FEBS Lett.*, **325**, 5 (1993).
223. S. Baudet and J. Janin, *J. Mol. Biol.*, **219**, 123 (1991).
224. J. Sancho and A. R. Fersht, *J. Mol. Biol.*, **224**, 741 (1992).
225. L. Serrano, J. T. Kellis Jr, P. Cann, A. Matouschek and A. R. Fersht, *J. Mol. Biol.*, **224**, 783 (1992).
226. J. Sancho, J. L. Neira and A. R. Fersht, *J. Mol. Biol.*, **224**, 749 (1992).
227. C. M. Dobson, P. A. Evans and S. E. Radford, *Trends Biochem. Sci.*, **19**, 31 (1994).
228. K. Loewenthal, J. Sancho and A. R. Fersht, *J. Mol. Biol.*, **224**, 759 (1992).
229. S. Khorasanizadeh, I. D. Peters, T. R. Butt and H. Roder, *Biochemistry*, **32**, 7054 (1993).
230. E. Fischer, *Ber. Dtsch. Chem. Ges.*, **27**, 2984 (1894).
231. I. A. Rose, *Philos. Trans. R. Soc. London, Ser. B*, **293**, 131 (1981).
232. C. A. Collyer, K. Henrick and D. M. Blow, *J. Mol. Biol.*, **211**, 211 (1990).
233. M. Whitlow, A. J. Howard, B. C. Finzel, T. L. Poulos, E. Winborne and G. L. Gilliland, *Proteins: Struct. Func. Genet.*, **9**, 153 (1991).
234. H. L. Carrell, H. Hoier and J. P. Glusker, *Acta Crystallogr., Part D*, **50**, 113 (1994).
235. F. Rey, J. Jenkins, J. Janin, I. Lasters, P. Alard, M. Claessens, G. Matthyssens and S. J. Wodak, *Proteins: Struct. Func. Genet.*, **4**, 165 (1988).
236. A. J. Wilkinson, A. R. Fersht, D. M. Blow, P. Carter and G. Winter, *Nature (London)*, **307**, 187 (1984).
237. H. Frauenfelder, G. A. Petsko and D. Tsernoglou, *Nature (London)*, **280**, 558 (1979).
238. G. K. Farber, P. Machin, S. C. Almo, G. A. Petsko and J. Hajdu, *Proc. Natl. Acad. Sci. USA*, **85**, 112 (1988).
239. J. Erlebacher and H. L. Carrell, *ICRVIEW*, Program for UNIX systems for the Silicon Graphics System, The Institute for Cancer Research, Philadelphia, PA, 1992.

Index